The Handbook of Geoscience

Frank R. Spellman
Melissa L. Stoudt

THE SCARECROW PRESS, INC.
Lanham • Toronto • Plymouth, UK
2013

Published by Scarecrow Press, Inc.
A wholly owned subsidiary of The Rowman & Littlefield Publishing Group, Inc.
4501 Forbes Boulevard, Suite 200, Lanham, Maryland 20706
www.rowman.com

10 Thornbury Road, Plymouth PL6 7PP, United Kingdom

British Library Cataloguing in Publication Information Available

Library of Congress Cataloging-in-Publication Data
Spellman, Frank R.
  The handbook of geoscience / Frank R. Spellman, Melissa L. Stoudt.
    p. cm.
  Includes bibliographical references and index.
  ISBN 978-0-8108-8614-8 (cloth : alk. paper) — ISBN 978-0-8108-8615-5 (ebook)
  1. Earth sciences--Handbooks, manuals, etc. I. Stoudt, Melissa L. II. Title.
  QE5.S65 2013
  550—dc23

                                                                        2012029485

∞™ The paper used in this publication meets the minimum requirements of American
National Standard for Information Sciences—Permanence of Paper for Printed Library
Materials, ANSI/NISO Z39.48-1992.

Printed in the United States of America

For

Suzanne Wilson

Contents

PART III: EARTH'S GEOSPHERE (LITHOSPHERE)

PART V: PEDOSPHERE

PART VI: BIOSPHERE

PART VII: GEODESY

Preface

Acts of creation are ordinarily reserved for gods and poets, but humbler folk may circumvent this restriction if they know how. To plant a pine, for example, one need be neither god nor poet; one need only own a good shovel. By virtue of this curious loophole in the rules, any clodhopper may say: Let there be a tree—and there will be one. If his back be strong and his shovel sharp, there may eventually be ten thousand. And in the seventh year he may lean upon his shovel, and look upon his trees, and find them good. God passed on his handiwork as early as the seventh day, but I notice He has since been rather noncommittal about its merits. I gather either that He spoke too soon, or that trees stand more looking upon than do fig leaves and firmaments.

—Aldo Leopold, *A Sand County Almanac, and Sketches Here and There,* 1948.

We read this passage from Leopold often. It not only grounds us to the majesty of acts of creation but also to the reality of things as they are and to the way things could be—or, in our opinion, to the way they should be. Along with Leopold's reverberating prose about the "Nature" of things on Earth, we are also reminded of Steinbeck's statement apropos to the theme of this text, stated in his *The Log from the Sea of Cortez:* "None of it is important or all of it is."

Geoscience (also known as the *geosciences, Earth science,* or *the Earth sciences*) is a huge discipline to study. It covers Earth's major spheres:

- Atmosphere (atmospheric chemistry, meteorology, paleoclimatology)
- Biosphere (biogeography, paleontology)
- Hydrosphere (hydrology, limnology, oceanography)

- Lithosphere (geosphere, geology)
- Pedosphere (soil science)
- Systems (environmental science, geography)
- Miscellaneous (geodesy and surveying, cartography)

Generally, geoscience is studied or approached holistically or delegated to reductionist theory. In *The Handbook of Geoscience,* we present a holistic treatment of the major scientific fields that make up this all-embracing discipline, including geology, physical geography, biology, ecology, geodesy, soil science, glaciology, oceanography and hydrology, and atmospheric sciences. We also discuss many of the subdisciplines contained within these major scientific fields.

Written in an engaging, highly readable style, *The Handbook* is ideal for students, administrators, legal professionals, nonscience professionals, and general readers with little or no science background. *The Handbook* is a user-friendly overview of our physical, biological, and ecological environment that offers up-to-date coverage of the major scientific fields that in combination form the structure of geoscience. Students who are enrolled in a geoscience course or one of its many subdisciplines will find this book to be an invaluable resource and reference to supplement classroom instruction and provide greater insight into many of the topics usually discussed. The emphasis is on readability, with clear, example-driven explanations refined by more than 35 years of experience of instruction and student feedback.

Frank R. Spellman
Norfolk, Virginia

Melissa L. Stoudt
New York, New York

To the Reader

In reading this text, you are going to spend some time in following an essence as it wafts through the air, a drop of water on its travels, a ped as it settles in its complex intricacy to support roots that support all life on Earth, and the mysteries of life as they leave their footprint on Earth's surface and subsurface areas. When you breathe in a liter of air, you bring into your lungs a chemical mixture critical to life. When you dip a finger in a basin of water and lift it up again, you bring with it a small, glistening drop out of the water below and hold it before you. When you plant a Royal Empress tree in rich soil, you watch the organic mixture provide sustenance to a rapidly growing member of nature at its best.

Do you have any idea where this waft of air, drop of water, or soil particle has been?

What changes has each undergone during all the long ages they have surrounded or lain on and under the face of the Earth?

Part I

THE BASICS

CHAPTER 1

Introduction

> Bathed in such beauty, watching the expression ever varying on the faces of the mountains, watching the stars, which here have a glory that the lowlander never dreams of, watching the circling seasons, listening to the songs of the waters and winds and birds, would be endless pleasure. And what glorious cloud-lands I would see, storms and calms, a new heaven and a new earth every day.
>
> —John Muir, 1912

Just the Facts, Ma'am

You may wonder why we begin this massive tome with the classic TV show *Dragnet's* catchphrase, attributed to (but never actually uttered by) Sergeant Joe Friday, the deadpan, fast-talking cop's cop. Simply, we feel it is appropriate (absolutely essential) at this early stage in the text to present bulk data and other parameters (facts) about the Earth—the topic of focus in this study of geoscience.

EARTH: BULK PARAMETERS (NASA 2010)

Mass (10^{24} kg): 5.9736
Volume (10^{10} km^3): 108.321
Equatorial radius (km): 6,378.1
Polar radius (km): 6,356.8
Volumetric mean radius (km): 6,371.0
Core radius (km): 3,485

Ellipticity (Flattening): 0.00335
Mean density (kg/m^3): 5515
Surface gravity (m/s^2): 9.798
Surface acceleration (m/s^2): 9.780
Escape velocity (km/s): 11.186
Gm ($\times$ 10^6 km3/s^2): 0.3986
Bond albedo: 0.306
Visual geometric albedo: 0.367
Visual magnitude: V(1,0) −3.86
Solar irradiance (W/m^2): 1367.6
Black-body temperature (K): 254.3
Topographic range (km): 20
Moment of inertia (I/MR^2): 0.3308
J$_2$ ($\times$ 10^{-4}): 1082.63
Number of natural satellites: 1
Planetary ring system: No

EARTH: ORBITAL PARAMETERS

Semimajor axis (10^6 km): 149.60
Sidereal orbit period (days): 365.256
Tropical orbit period (days): 365.242
Perihelion (10^6 km): 147.09
Aphelion (10^6 km): 152.10
Mean orbital velocity (km/s): 29.78
Max. orbital velocity (km/s): 30.29
Min. orbital velocity (km/s): 29.29
Orbit inclination (deg): 0.000
Orbit eccentricity: 0.0167
Sidereal rotation period (hrs): 23.9345
Length of day (hrs): 24.0000
Obliquity to orbit (deg): 23.44

EARTH: MEAN ORBITAL ELEMENTS (J2000)

Semimajor axis (AU): 1.00000011
Orbital eccentricity: 0.01671022
Orbital inclination (deg): 0.00005
Longitude of ascending node (deg): −11.26064

Longitude of perihelion (deg): 102.94719
Mean Longitude (deg): 100.46435

NORTH POLE OF ROTATION

Right Ascension: 0.00–0.641T
Declination: 90.00–0.557T
Reference Date: 12:00 UT 1 Jan 2000 (JD 2451545.0)
T = Julian centuries from reference date

TERRESTRIAL MAGNETOSPHERE

Dipole field strength: 0.3076 gauss-Re^3
Latitude/Longitude of dipole: N78.6 degrees N/70.1 degrees W
Dipole offset (planet center to dipole center) distance: 0.0725 Re
Latitude/Longitude of offset vectors: 18.3 degrees N/147.8 degrees E

✓ **Note:** Re denotes Earth radii, 6,378 km

TERRESTRIAL ATMOSPHERE

Surface pressure: 1014 mb
Surface density: 1.217 kg/m^3
Total mass of atmosphere: 5.1×10^{18} kg
Total mass of hydrosphere: 1.4×10^{21} kg
Average temperature: 283 K (15 C)
Diurnal temperature range: 283 K to 293 K (10 to 20 C)
Wind speeds: 0 to 100 m/s
Mean molecular weight: 28.97 g/mole
Atmospheric composition (by volume, dry air)
Major
- 79.08% Nitrogen (N_2)
- 20.95% Oxygen (O_2)

Minor (ppm)
- Argon (Ar)
- Carbon dioxide
- Neon
- Helium

- Methane
- Krypton
- Hydrogen

Measurements and Units

> "If you cannot measure it, then it is not science."
>
> —Lord Kelvin, 1883

Lord Kelvin wastes few words in making one of the critical points regarding science: We must be able to measure what we observe. That is, science involves the gathering of information such as facts, statistics, measurements, evidence; this is generally obtained through observation using our senses.

In regard to Earth facts or parameters listed previously, one might jump to the conclusion that such measurements are no big deal—especially in the present era with our sophisticated measuring devices, computers, GPS, GIS, and instrument-laden circling satellites. Thus the impression might be that the listed measurements are rather recent, certainly not based on ancient measurements. Well, this thinking would be incorrect. Measurement is nothing new. Humans have been making measurements long before they learned to read or write. Early humans learned to measure out of necessity. When they found a weapon or tool of the right weight and length, they copied them by performing comparison sight-and-feel measurements. To measure distance, they used their appendages: fingers, hands, arms, legs, and so on. Weight measurement was based on what a person could lift or carry; each person learned their comfortable limit. Later, great thinkers were able to measure the diameter and radius of Earth; this occurred long before the inventions of Galileo and the theories of Newton, Kepler, and Einstein, and many others.

Early units of measurement were based on 10 because humans had 10 fingers and learned to count in this manner and relate units to each other. Even though each person is different, some of the earliest units of measurement were based on familiar body part measurements. A few of the well-known of the early units of measurement were:

inch—the width of the thumb
digit—the width of the middle finger (~ 3/4 inch)
span—the distance covered by the spread hand (~ 9 inches)
foot—the length of the foot
cubit—distance from the elbow to the tip of the middle finger (~ 18 inches)
fathom—distance spanned by the outstretched arms (~ 72 inches)

These early units of measurement are interesting, but the question arises: What do these simplistic ancient units have to do with the profound and exact Earth facts and parameters presented earlier? Knowledge evolves and is a function of experience, experiment, thought, luck, time, and genius. One such genius was Eratosthenes of Cyrene (c. 276 BCE–c. 195 BCE); he was a Greek mathematician, poet, athlete, geographer, astronomer, and musician. Eratosthenes is not only credited with inventing a system of latitude and longitude, he is also credited with inventing the discipline of geography. More than 2,000 years ago he was the first person to calculate the circumference of the Earth and the tilt of the Earth's axis (with remarkable accuracy). Well before Columbus and other ocean navigators proved that the Earth is round, Eratosthenes determined this fact on his own.

Another question presents itself: How was Eratosthenes able to measure many of Earth's parameters with such a high degree of accuracy before the invention of the modern measuring instruments, computers, and satellites that are so common today, but not even imagined in his day?

The most important instrument Eratosthenes possessed was his mind, which he trained through observation, measurement, and inference to make conclusions. In regard to his measurement of the size of Earth, he heard an account of an unusual water well in a distant city. On the summer solstice (the longest day of the year), at exactly noon, the sun would shine directly down the well. Eratosthenes, at exactly the same time, looked down the well in his own city. He noticed that the sun was *not* shining directly down the well in his town, at the same time that the sun *was* shining straight down the well in the distant town.

Eratosthenes correctly inferred that the explanation for this was that the surface of the Earth was curved and that the distant well was on the location of the curve directly under the sun, whereas his hometown well was off to the side a bit. Although he had little information available to him, Eratosthenes developed an equation to calculate the entire size of the Earth.

$$C = \frac{360°}{A} \times D \qquad (1.1)$$

where:

A = is the angle between the two cities as measured from the core
D = is the distance between the two cities
C = is the circumference of Earth

Although Eratosthenes's method was well founded, the accuracy of his calculation was limited, causing his final result to be off a little. That is, his calculation would have been more accurate if the two cities had been aligned exactly north and south of each other.

UNITS OF MEASUREMENT

Like much of life itself, science is all about measurement. Simply, we do not go through a day without measuring something. Whether counting the money in our pockets or in our bank accounts or in our 401Ks, counting the number of shopping days left until Christmas, weighing ourselves to determine how many pounds we have lost (or gained), using a measuring device to add ingredients to our food or coffee, or determining where to buy the cheapest gallon of gasoline, we are always measuring. We constantly take a measure of our world. In geoscience (and all the other sciences), measurement is fundamental—measurement is what geoscience is all about; measurement is why geoscience exists. From the atomic scale to clusters of stars and galaxies, geoscientists must measure to probe the natural world.

In this section, we begin where we should: at the beginning—with a presentation of fundamental math principles and operations. We can't even approach the fringe of comprehending geoscience unless we understand the basics of measurement, unit conversions, and basic math operations.

The units most commonly used by geoscientists are based on the complicated English system of weights and measures. However, bench work is usually based on the metric system or the international system of units (SI) because of the convenient relationship between milliliters (ml), cubic centimeters (cm^3), and grams (gm).

The SI (short for *système international d'unités*) is a modernized version of the metric system established by International agreement. The metric system of measurement was developed during the French Revolution and was first promoted in the United States in 1866. In 1902, proposed congressional legislation requiring the U.S. government to use the metric system exclusively was defeated by a single vote.

The most common measurement systems in geoscience are the centimeter-gram-second (CGS) and meter-kilogram-second (MKS) systems.

Did You Know?

The meter (m) was originally measured to be one ten-millionth of the distance from the North Pole to the equator along the meridian running near Dunkirk, Paris, and Barcelona. It was redefined in 1971 as the length of the path traveled by light in a vacuum during the time internal of 1/299,792,458 second (General Conference on Weights and Measures, 1987).

While we use both systems in this text, *SI* provides a logical and interconnected framework for all measurements in engineering, science, industry, and commerce. The metric system is much simpler to use than the existing English system because all its units of measurement are divisible by 10. That is, instead of having a large number of units of different sizes, such as inches, feet, years, fathoms, furlongs, and miles in the English system, it is easier to use prefixes that multiply base units by multiples of 10 for larger measurements and decimal transition for smaller measurements.

Did You Know?

The kilogram (kg) is the mass of a particular cylinder of platinum-iridium alloy, called the *International Prototype Kilogram,* kept at the International Bureau of Weights and Measures in Sèrves, France. The kilogram, the only unit still defined by an artifact, was derived from the mass of a cubic decimeter of water (*Comptes Rendus de la 1ᵉ* [1889], 1890, 34).

Before listing the various conversion factors commonly used in geoscience, it is important to describe the prefixes commonly used in the SI system. As mentioned, these prefixes are based on the power 10. For example, a kilo means 1,000 grams, and a centimeter means one-hundredth of 1 meter. The 20 SI prefixes used to form decimal multiples and submultiples of SI units are given in Table 1.1.

Note that the kilogram is the only SI unit with a prefix as part of its name and symbol. Because multiple prefixes may not be used, in the case of the kilogram, the prefix names of Table 1.1 are used with the unit name "gram" and the prefix symbols are used with the unit symbol "g." With this exception, any SI prefix may be used with any SI unit, including the degree Celsius and its symbol (° C).

Example 1.1

10^{-6} kg = 1 mg (one milligram), but not 10^{-6} kg = 1 mkg (one microkilogram)

Example 1.2

Consider the height of the Washington Monument. We may write h_w = 169,000 mm = 16,900 cm = 169 m = 0.169 km using the millimeter (SI prefix *milli,* symbol *m*), centimeter (SI prefix *centi,* symbol *c*), or kilometer (SI prefix *kilo,* symbol *k*).

Table 1.1. SI Prefixes

Factor	Name	Symbol
10^{24}	yotta	Y
10^{21}	zetta	Z
10^{18}	exa	E
10^{15}	peta	P
10^{12}	tera	T
10^{9}	giga	G
10^{6}	mega	M
10^{3}	kilo	k
10^{2}	hecto	h
10^{1}	dekad	a
10^{-1}	deci	d
10^{-2}	centi	c
10^{-3}	milli	m
10^{-6}	micro	μ
10^{-9}	nano	n
10^{-12}	pico	p
10^{-15}	femto	f
10^{-18}	atto	a
10^{-21}	zepto	z
10^{-24}	yocto	y

Did You Know?

The Kelvin (K) is 1/273.16 of the temperature interval between absolute zero and the triple point of water (the temperature at which ice, liquid water, and water vapor are in equilibrium). The Celsius scale is derived from the Kelvin scale. An interval of 1 K is equal to 1° C (National Institute of Standards and Technology).

CONVERSION FACTORS

Conversion factors refer to the conversion of units between different units of measurement for the same quantity. It is important to note that the process of making a conversion cannot produce a more precise result than the original figure. Appropriate rounding of the results is normally performed after conversion. Conversion factors are given in alphabetical order in Table 1.2 and in unit category listing order in Table 1.3.

Table 1.2. Alphabetical Listing of Conversion Factors

Factors	Metric (SI) or English Conversions
1 atm (atmosphere) =	1.013 bars
	10.133 newtons/cm² (newtons/square centimeter)
	33.90 ft of H₂O (feet of water)
	101.325 kp (kilopascals)
	1,013.25 mg (millibars)
	13.70 psia (pounds/square inch—absolute)
	760 torr
	760 mm Hg (millimeters of mercury
1 bar =	0.987 atm (atmospheres)
	1 × 10⁶ dynes/cm² (dynes/square centimeter)
	33.45 ft of H₂O (feet of water)
	1 × 10⁵ pascals [nt/m²] (newtons/square meter)
	750.06 torr
	750.06 mm Hg (millimeters of mercury)
1 Bq (becquerel) =	1 radioactive disintegration/second
	2.7 × 10⁻¹¹ Ci (curie)
	2.7 × 10⁻⁸ mCi (millicurie)
1 BTU (British Thermal Unit) =	252 cal (calories)
	1,055.06 j (joules)
	10.41 liter-atmosphere
	0.293 watt-hours
1 cal (calories) =	3.97 × 10⁻³ BTUs (British Thermal Units)
	4.18 j (joules)
	0.0413 liter-atmospheres
	1.163 × 10⁻³ watt-hours
1 cm (centimeters) =	0.0328 ft (feet)
	0.394 in (inches)
	10,000 microns (micrometers)
	100,000,000 Å =10⁸ Å (Ångstroms)
1 cc (cubic centimeter) =	3.53 × 10⁻⁵ ft³ (cubic feet)
	0.061 in³ (cubic inches)
	2.64 × 10⁻⁴ gal (gallons)
	52.18 ℓ (liters)
	52.18 ml (milliliters)
1 ft³ (cubic foot) =	28.317 cc (cubic centimeters)
	1,728 in³ (cubic inches)
	0.0283 m³ (cubic meters)
	7.48 gal (gallons)
	28.32 ℓ (liters)
	29.92 qts (quarts)

(continued)

Table 1.2.   (continued)

Factors	Metric (SI) or English Conversions
1 in³	16.39 cc (cubic centimeters)
	16.39 ml (milliliters)
	5.79×10^{-4} ft³ (cubic feet)
	1.64×10^{-5} m³ (cubic meters)
	4.33×10^{-3} gal (gallons)
	0.0164 ℓ (liters)
	0.55 fl oz (fluid ounces)
1 m³ (cubic meter) =	1,000,000 cc = 10^6 cc (cubic centimeters)
	33.32 ft³ (cubic feet)
	61,023 in³ (cubic inches)
	264.17 gal (gallons)
	1,000 ℓ (liters)
1 yd³ (cubic yard) =	201.97 gal (gallons)
	764.55 ℓ (liters)
1 Ci (curie) =	3.7×10^{10} radioactive disintegrations/second
	3.7×10^{10} Bq (becquerel)
	1,000 mCi (millicurie)
1 day =	24 hr (hours)
	1,440 min (minutes)
	86,400 sec (seconds)
	0.143 weeks
	2.738×10^{-3} yr (years)
1° C (expressed as an interval) =	1.8° F = [9/5] °F (degrees Fahrenheit)
	1.8 °R (degrees Rankine)
	1.0 K (degrees Kelvin)
° C (degree Celsius) =	[(5/9)(° F – 32°)]
1° F (expressed as an interval) =	0.556° C = [5/9]° C (degrees Celsius)
	1.0° R (degrees Rankine)
	0.556 K (degrees Kelvin)
° F (degree Fahrenheit) =	[(9/5)(° C) + 32°
1 dyne =	1×10^{-5} nt (newton)
1 ev (electron volt) =	1.602×10^{-12} ergs
	1.602×10^{-19} j (joules)
1 erg =	1 dyne-centimeters
	1×10^{-7} j (joules)
	2.78×10^{-11} watt-hours
1 fps (feet/second) =	1.097 kmph (kilometers/hour)
	0.305 mps (meters/second)
	0.01136 mph (miles/hour)
1 ft (foot) =	30.48 cm (centimeters)
	12 in (inches)
	0.3048 m (meters)
	1.65×10^{-4} nt (nautical miles)
	1.89×10^{-4} mi (statute miles)

Factors	Metric (SI) or English Conversions
1 gal (gallon) =	3,785 cc (cubic centimeters)
	0.134 ft^3 (cubic feet)
	231 in^3 (cubic inches)
	3.785 ℓ (liters)
1 g (gram)	0.001 kg (kilogram)
	1,000 mg (milligrams)
	1,000,000 ng = 10^6 ng (nanograms)
	2.205 × 10^{-3} lb (pounds)
1 g/cc (grams/cubic cent.) =	62.43 lb/ft^3 (pounds/cubic foot)
	0.0361 lb/in^3 (pounds/cubic inch)
	8.345 lb/gal (pounds/gallon)
1 Gy (gray) =	1 j/kg (joules/kilogram)
	100 rad
	1 Sv (sievert)—(unless modified through division by an appropriate factor, such as Q or N)
1 hp (horsepower) =	745.7 j/sec (joules/sec)
1 hr (hour) =	0.0417 days
	60 min (minutes)
	3,600 sec (seconds)
	5.95 × 10^{-3} weeks
	1.14 × 10^{-4} yr (years)
1 in (inch) =	2.54 cm (centimeters)
	1,000 mils
1 inch of water =	1.86 mm Hg (millimeters of mercury)
	249.09 pascals
	0.0361 psi (lb/in^2)
1 j (joule) =	9.48 × 10^{-4} BTUs (British Thermal Units)
	0.239 cal (calories)
	10,000,000 ergs = 1 × 10^7 ergs
	9.87 × 10^{-3} liter-atmospheres
	1.0 nt-m (newton-meters)
1 kcal (kilocalories) =	3.97 BTUs (British Thermal Units)
	1,000 cal (calories)
	4,186.8 j (joules)
1 kg (kilogram) =	1,000 g (grams)
	2,205 lb (pounds)
1 km (kilometer) =	3,280 ft (feet)
	0.54 nt (nautical miles)
	0.6214 mi (statute miles)
1 kw (kilowatt) =	56.87 BTU/min (British Thermal Units)
	1.341 hp (horsepower)
	1,000 j/sec (kilocalories)
1 kw-hr (kilowatt-hour) =	3,412.14 BTU (British Thermal Units)
	3.6 × 10^6 j (joules)
	859.8 kcal (kilocalories)

(continued)

Table 1.2. *(continued)*

Factors	Metric (SI) or English Conversions
1 ℓ (liter) =	1,000 cc (cubic centimeters)
	1 dm^3 (cubic decimeters)
	0.0353 ft^3 (cubic feet)
	61.02 in^3 (cubic inches)
	0.264 gal (gallons)
	1,000 ml (milliliters)
	1.057 qts (quarts)
1 m (meter) =	1 × 10^{10} Å (Ångstroms)
	100 cm (centimeters)
	3.28 ft (feet)
	39.37 in (inches)
	1 × 10^{-3} km (kilometers)
	1,000 mm (millimeters)
	1,000,000 μ = 1 × 10^6 μ (micrometers)
	1 × 10^9 nm (nanometers)
1 mps (meters/second) =	196.9 fpm (feet/minute)
	3.6 kmph (kilometers/hour)
	2.237 mph (miles/hour)
1 mph (mile/hour) =	88 fpm (feet/minute)
	1.61 kmph (kilometers/hour)
	0.447 mps (meters/second)
1 kt (nautical mile) =	6,076.1 ft (feet)
	1.852 km (kilometers)
	1.15 mi (statute miles)
	2,025.4 yd (yards)
1 mi (statute mile) =	5,280 ft (feet)
	1.609 km (kilometers)
	1,609.3 m (meters)
	0.869 nt (nautical miles)
	1,760 yd (yards)
1 mi Ci (millicurie) =	0.001 Ci (curie)
	3.7 × 10^{10} radioactive disintegrations/ second
	3.7 × 10^{10} Bq (becquerel)
1 mm Hg (mm of mercury) =	1.316 × 10^{-3} atm (atmosphere)
	0.535 in H$_2$O (inches of water)
	1.33 mb (millibars)
	133.32 pascals
	1 torr
	0.0193 psia (pounds/square inch – absolute
1 min (minute) =	6.94 × 10^{-4} days
	0.0167 hr (hours)
	60 sec (seconds)
	9.92 × 10^{-5} weeks
	1.90 × 10^{-6} yr (years)

Factors	Metric (SI) or English Conversions
1 nt (newton) =	1×10^5 dynes
1 nt-m (newton-meter) =	1.00 j (joules)
	2.78×10^{-4} watt-hours
1 ppm (parts/million-volume) =	1.00 ml/m^3 (milliliters/cubic meter)
1 ppm [wt] (parts/million-weight) =	1.00 mg/kg (milligrams/kilograms)
1 pascal =	9.87×10^{-6} atm (atmospheres)
	4.015×10^{-3} in H$_2$O (inches of water)
	0.01 mb (millibars)
	7.5×10^{-3} mm Hg (milliliters of mercury)
1 lb (pound) =	453.59 g (grams)
	16 oz (ounces)
l lb/ft^3 (pounds/cubic foot) =	16.02 g/l (grams/liter)
1 lb/ft^3 (pounds/cubic inch) =	27.68 g/cc (grams/cubic centimeter)
	1,728 lb/ft^3 (pounds/cubic feet)
1 psi (pounds/square inch)=	0.068 atm (atmospheres)
	27.67 in H$_2$O (inches or water)
	68.85 mb (millibars)
	51.71 mm Hg (millimeters of mercury)
	6,894.76 pascals
1 qt (quart) =	946.4 cc (cubic centimeters)
	57.75 in^3 (cubic inches)
	0.946 ℓ (liters)
1 rad =	100 ergs/g (ergs/gram)
	0.01 Gy (gray)
	1 rem (unless modified through division by an
	appropriate factor, such as Q or N)
1 rem	1 rad (unless modified through division by an
	appropriate factor, such as Q or N)
1 Sv (sievert) =	1 Gy (gray) (unless modified through division by an
	appropriate factor, such as Q and/or N)
1 cm^2 (square centimeter) =	1.076 $\times 10^{-3}$ft^2 (square feet)
	0.155 in^2 (square inches)
	1×10^{-4} m^2 (square meters)
1 ft^2 (square foot) =	2.296×10^{-5} acres
	9.296 cm^2 (square centimeters)
	144 in^2 (square inches)
	0.0929 m^2 (square meters)
1 m^2 (square meter) =	10.76 ft^2 (square feet)
	1,550 in^2 (square inches)
1 mi^2 (square mile) =	640 acres
	2.79×10^7 ft^2 (square feet)
	2.59×10^6 m^2 (square meters)

(continued)

Table 1.2. *(continued)*

Factors	Metric (SI) or English Conversions
1 torr =	1.33 mb (millibars)
1 watt =	3.41 BTI/hr (British Thermal Units/hour)
	1.341 × 10⁻³ hp (horsepower)
	52.18 j/sec (joules/second)
1 watt-hour =	3.412 BTUs (British Thermal Unit)
	859.8 cal (calories)
	3,600 j (joules)
	35.53 liter-atmosphere
1 week =	7 days
	168 hr (hours)
	10,080 min (minutes)
	6.048 × 10⁵ sec (seconds)
	0.0192 yr (years)
1 yr (year) =	365.25 days
	8,766 hr (hours)
	5.26 × 10⁵ min (minutes)
	3.16 × 10⁷ sec (seconds)
	52.18 weeks

Table 1.3.   Conversion Factors by Unit Category

Units of Length	
1 cm (centimeter) =	0.0328 ft (feet)
	0.394 in (inches)
	10,000 microns (micrometers)
	100,000,000 Å = 10⁸ Å (Ångstroms)
1 ft (foot) =	30.48 cm (centimeters)
	12 in (inches)
	0.3048 m (meters)
	1.65 × 10⁻⁴ nt (nautical miles)
	1.89 × 10⁻⁴ mi (statute miles)
1 in (inch) =	2.54 cm (centimeters)
	1,000 mils
1 km (kilometer) =	3,280.8 ft (feet)
	0.54 nt (nautical miles)
	0.6214 mi (statute miles)
1 m (meter) =	1 × 10¹⁰ Å (Ångstroms)
	100 cm (centimeters)
	3.28 ft (feet)
	39.37 in (inches)
	1 × 10⁻³ km (kilometers)
	1,000 mm (millimeters)
	1,000,000 μ = 1 × 10⁶ μ (micrometers)
	1 × 10⁹ nm (nanometers)

Units of Length

1 kt (nautical mile) =	6,076.1 ft (feet)
	1.852 km (kilometers)
	1.15 km (statute miles)
	2.025.4 yd (yards)
1 mi (statute mile) =	5,280 ft (feet)
	1.609 km (kilometers)
	1.690.3 m (meters)
	0.869 nt (nautical miles)
	1,760 yd (yards)

Units of Area

1 cm^2 (square centimeter) =	1.076×10^{-3} ft^2 (square feet)
	0.155 in^2 (square inches)
	1×10^{-4} m^2 (square meters)
1 ft^2 (square foot) =	2.296×10^{-5} acres
	929.03 cm^2 (square centimeters)
	144 in^2 (square inches)
	0.0929 m^2 (square meters)
1 m^2 (square meter) =	10.76 ft^2 (square feet)
	1,550 in^2 (square inches)
1 mi^2 (square mile) =	640 acres
	2.79×10^7 ft^2 (square feet)
	2.59×10^6 m^2 (square meters)

Units of Volume

1 cc (cubic centimeter) =	3.53×10^{-5} ft^3 (cubic feet)
	0.061 in^3 (cubic inches)
	2.64×10^{-4} gal (gallons)
	0.001 ℓ (liters)
	1.00 ml (milliliters)
1 ft^3 (cubic foot) =	28,317 cc (cubic centimeters)
	1,728 in^3 (cubic inches)
	0.0283 m^3 (cubic meters)
	7.48 gal (gallons)
	28.32 ℓ (liters)
	29.92 qt (quarts)
1 in^3 (cubic inch) =	16.39 cc (cubic centimeters)
	16.39 ml (milliliters)
	5.79×10^{-4} ft^3 (cubic feet)
	1.64×10^{-5} m^3 (cubic meters)
	4.33×10^{-3} gal (gallons)
	0.0164 ℓ (liters)
	0.55 fl oz (fluid ounces)

(continued)

Table 1.3. (*continued*)

Units of Volume	
1 m³ (cubic meter) =	1,000,000 cc = 10^6 cc (cubic centimeters)
	35.31 ft³ (cubic feet)
	61,023 in³ (cubic inches)
	264.17 gal (gallons)
	1,000 ℓ (liters)
1 yd³ (cubic yards) =	201.97 gal (gallons)
	764.55 ℓ (liters)
1 gal (gallon) =	3,785 cc (cubic centimeters)
	0.134 ft³ (cubic feet)
	231 in³ (cubic inches)
	3.785 ℓ (liters)
1 ℓ (liter) =	1,000 cc (cubic centimeters)
	1 dm³ (cubic decimeters)
	0.0353 ft³ (cubic feet)
	61.02 in³ (cubic inches)
	0.264 gal (gallons)
	1,000 ml (milliliters)
	1.057 qt (quarts)
1 qt (quart) =	946.4 cc (cubic centimeters)
	57.75 in³ (cubic inches)
	0.946 ℓ (liters)

Units of Mass	
1 g (grams) =	0.001 kg (kilograms)
	1,000 mg (milligrams)
	1,000,000 mg = 10^6 ng (nanograms)
	2.205×10^{-3} lb (pounds)
1 kg (kilogram) =	1,000 g (grams)
	2.205 lb (pounds)
1 lb (pound) =	453.59 g (grams)
	16 oz (ounces)

Units of Time	
1 day =	24 hr (hours)
	1440 min (minutes)
	86,400 sec (seconds)
	0.143 weeks
	2.738×10^{-3} yr (years)
1 hr (hours) =	0.0417 days
	60 min (minutes)
	3,600 sec (seconds)
	5.95×10^{-3} yr (years)

Units of Time	
1 hr (hour) =	0.0417 days
	60 min (minutes)
	3,600 sec (seconds)
	5.95×10^{-3} weeks
	1.14×10^{-4} yr (years)
1 min (minutes) =	6.94×10^{-4} days
	0.0167 hr (hours)
	60 sec (seconds)
	9.92×10^{-5} weeks
	1.90×10^{-6} yr (years)
1 week =	7 days
	168 hr (hours)
	10,080 min (minutes)
	6.048×10^5 sec (seconds)
	0.0192 yr (years)
1 yr (year) =	365.25 days
	8,766 hr (hours)
	5.26×10^5 min (minutes)
	3.16×10^7 sec (seconds)
	52.18 weeks

Units of the Measure of Temperature	
°C (degrees Celsius) =	$[(5/9)(°F - 32°)]$
1° C (expressed as an interval) =	$1.8° F = [9/5]° F$ (degrees Fahrenheit)
	1.8° R (degrees Rankine)
	1.0 K (degrees Kelvin)
°F (degree Fahrenheit) =	$[(9/5)(°C) + 32°]$
1° F (expressed as an interval) =	$0.556° C = [5/9]° C$ (degrees Celsius)
	1.0° R (degrees Rankine)
	0.556 K (degrees Kelvin)

Example 1.3

Problem:
Find degrees in Celsius of water at 72.

Solution:
°C = (F − 32) × 5/9 = (72 − 32) × 5/9 = 22.2

Units of Force	
1 dyne =	1×10^{-5} nt (newtons)
1 nt (newton) =	1×10^5 dynes

Units of Work or Energy

1 BTU (British Thermal Unit) =	252 cal (calories)
	1,055.06 j (joules)
	10.41 liter-atmospheres
	0.293 watt-hours
1 cal (calories) =	3.97×10^{-3} BTUs (British Thermal Units)
	4.18 j (joules)
	0.0413 liter-atmospheres
	1.163×10^{-3} watt-hours
1 ev (electron volt) =	1.602×10^{-12} ergs
	1.602×10^{-19} j (joules)
0 erg =	1 dyne-centimeter
	1×10^{-7} j (joules)
	2.78×10^{-11} watt-hours
1 j (joule) =	9.48×10^{-4} BTUs (British Thermal Units)
	0.239 cal (calories)
	10,000,000 ergs = 1×10^{7} ergs
	9.87×10^{-3} liter-atmospheres
	1.00 nt-m (newton-meters)
1 kcal (kilocalorie) =	3.97 BTUs (British Thermal Units)
	1,000 cal (calories)
	4,186.8 j (joules)
1 kw-hr (kilowatt-hour) =	3,412.14 BTU (British Thermal Units)
	3.6×10^{6} j (joules)
	859.8 kcal (kilocalories)
1 nt-m (newton-meter) =	1.00 j (joules)
	2.78×10^{-4} watt-hours
1 watt-hour =	3.412 BTUs (British Thermal Units)
	859.8 cal (calories)
	3,600 j (joules)
	35.53 liter-atmospheres

Units of Power

1 hp (horsepower) =	745.7 j/sec (joules/sec)
1 kw (kilowatt) =	56.87 BTU/min (British Thermal Units/minute)
	1.341 hp (horsepower)
	1,000 j/sec (joules/sec)
1 watt =	3.41 BTU/hr (British Thermal Units/hour)
	1.341×10^{-3} hp (horsepower)
	1.00 j/sec (joules/second)

Units of Pressure	
1 atm (atmosphere) =	1.013 bars
	10.133 newtons/cm² (newtons/square centimeters)
	33.90 ft. of H_2O (feet of water)
	101.325 kp (kilopascals)
	14.70 psia (pounds/square inch – absolute)
	760 torr
	760 mm Hg (millimeters of mercury)
1 bar =	0.987 atm (atmospheres)
	1×10^6 dynes/cm² (dynes/square centimeter)
	33.45 ft of H_2O (feet of water)
	1×10^5 pascals [nt/m²] (newtons/square meter)
	750.06 torr
	750.06 mm Hg (millimeters of mercury)
1 inch of water =	1.86 mm Hg (millimeters of mercury)
	249.09 pascals
	0.0361 psi (lb/in²)
1 mm Hg (millimeter of merc.) =	1.316×10^{-3} atm (atmospheres)
	0.535 in H_2O (inches of water)
	1.33 mb (millibars)
	133.32 pascals
	1 torr
	0.0193 psia (pounds/square inch—absolute)
1 pascal =	9.87×10^{-6} atm (atmospheres)
	4.015×10^{-3} in H_2O (inches of water)
	0.01 mb (millibars)
	7.5×10^{-3} mm Hg (millimeters of mercury)
1 psi (pounds/square inch) =	0.068 atm (atmospheres)
	27.67 in H_2O (inches of water)
	68.85 mb (millibars)
	51.71 mm Hg (millimeters of mercury)
	6,894.76 pascals
1 torr =	1.33 mb (millibars)

Units of Velocity or Speed	
1 fps (feet/second) =	1.097 kmph (kilometers/hour)
	0.305 mps (meters/second)
	0.01136 mph (miles/hours)
1 mps (meters/second) =	196.9 fpm (feet/minute)
	3.6 kmph (kilometers/hour)
	2.237 mph (miles/hour)

Units of Velocity or Speed

1 mph (mile/hour) =	88 fpm (feet/minute)
	1.61 kmph (kilometers/hour)
	0.447 mps (meters/second)

Units of Density

1 g/cc (grams/cubic cent.) =	62.43 lb/ft^3 (pounds/cubic foot)
	0.0361 lb/in^3 (pounds/cubic inch)
	8.345 lb/gal (pounds/gallon)
1 lb/ft^3 (pounds/cubic foot) =	16.02 g/ℓ (grams/liter)
1 lb/in^2 (pounds/cubic inch) =	27.68 g/cc (grams/cubic centimeter)
	1.728 lb/ft^3 (pounds/cubic foot)

Units of Concentration

| 1 ppm (parts/million-volume) = | 1.00 ml/m^3 (milliliters/cubic meter) |
| 1 ppm (wt) = | 1.00 mg/kg (milligrams/kilograms) |

Radiation & Dose Related Units

1 Bq (becquerel) =	1 radioactive disintegration/second
	2.7×10^{-11} Ci (curie)
	2.7×10^{-8} (millicurie)
1 Ci (curie) =	3.7×10^{10} radioactive disintegration/second
	3.7×10^{10} Bq (becquerel)
	1,000 mCi (millicurie)
1 Gy (gray) =	1 j/kg (joule/kilogram)
	100 rad
	1 Sv (sievert)—(unless modified through division by an appropriate factor, such as Q or N)
1 mCi (millicurie) =	0.001 Ci (curie)
	3.7×10^{10} radioactive disintegrations/second
	3.7×10^{10} Bq (becquerel)
1 rad =	100 ergs/g (ergs/g)
	0.01 Gy (gray)
	1 rem (unless modified through division by an appropriate factor, such as Q or N)
1 rem =	1 rad (unless modified through division by an appropriate factor, such as Q or N)
1 Sv (sievert) =	1 Gy (gray) (unless modified through division by an appropriate factor, such as Q or N)

The Scientific Method

We certainly cannot have a discussion about geology (or any other science) without a brief discussion of the scientific method of investigation. This is the case even though many scientists believe that there is no single universally accepted "scientific method" (Figure 1.1).

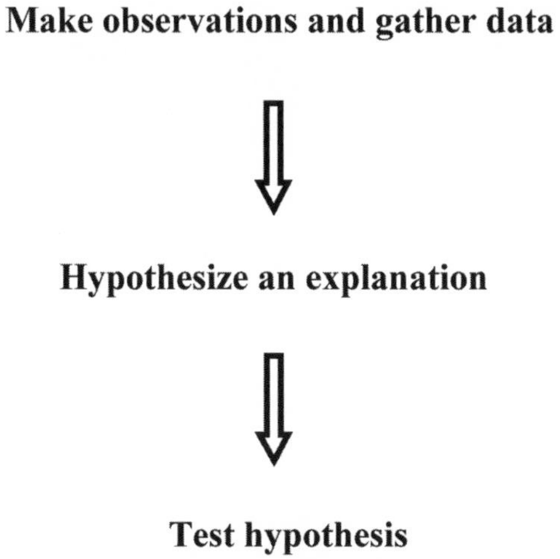

Make observations and gather data

Hypothesize an explanation

Test hypothesis

Figure 1.1. The Scientific method simplified. This process cycles; it is re-evaluated on an on-going basis.

The step-by-step procedure used in the scientific method follows:

1. Observe and define the problem.
2. Hypothesize a theory (make reasonable explanations based on observations).
3. Test (experiment) the theory. (Can it be reproduced?)
4. Can others reproduce the same results?
5. If steps 3 and 4 are unsuccessful, modify or reject the theory (go back to step 1).
6. If it is consistently reproducible and most of the scientific community agrees with the theory, it becomes a natural law.

Most scientific thought is based on two forms of logical reason: inductive and deductive reasoning. The goal of science is to establish general principles for the study of specific cases. The classic example is Newton's apple. He examined how specific objects fall, from which he inferred the general theory of gravitation.

- **Inductive reasoning** is the inference of general principles based on specific cases; the observer moves from particular observations to general principles (discovering the rules). In other words, no matter how much evidence exists for a conclusion, the conclusion could still conceivably be false. Such that,

> *Suppose someone eats 4 oranges out of a box of 100 and finds each of the 4 oranges to be tasty. From this, the person concludes that all the oranges are tasty.*

The field of geology has many principles based on inductive reasoning.

- **Deductive reasoning** is the analysis of a specific case based on specific cases; the observer moves from general laws to specific predictions (making predictions based on the rules). In other words, if the premises are true, then the conclusion has to be true. Such that,

> *If all Maytag washing machines built in 1969 have switches installed that shut down the machines any time the lid is opened (lifted), and Mabel's Maytag washer was built in 1969, then Mabel's washing machine has a shutdown switch that is activated when the lid cover is opened (or at least originally had a shutdown switch if it wasn't removed or jerry-rigged).*

*   *   *

The Power of Science is its Power of Prediction.

*   *   *

References and Recommended Reading

Comptes Rendus de la 1ᵉ. (1889, 1890). Paris, France, French Academy of Science.
General Conference on Weights and Measures. (1987). *Metre Convention*. Paris, France.
National Aeronautics and Space Administration (NASA). (2010). Earth fact sheet. Retrieved from at http://nssdc.gsfc.nasa.gov/planetary/factsheet/earthfact.html
National Institute of Standards and Technology (NIST) (2012). *Weights and Measures*. Gaithersburg, MD.

CHAPTER 2

Basic Math Operations

Most calculations required by geoscientists (as with many others) start with the basics, such as addition, subtraction, multiplication, division, and sequence of operations. Although many of the operations are fundamental tools within each geoscientist's toolbox, using these tools on a consistent basis is important to remaining sharp in their use. Geoscientists should master basic math definitions and the formation of problems; daily practice requires calculation of percentage, average, simple ratio, geometric dimensions, force, pressure, and head, as well as the use of dimensional analysis and advanced math operations.

Basic Math Terminology and Definitions

The following basic definitions will aid in understanding the material in this chapter.

- *Integer,* or an *integral number:* a whole number. Thus 1, 2, 3, 4, 5, 6, 7, 8, 9, 10, 11, and 12 are the first 12 positive integers.
- *Factor* or *divisor* of a whole number: any other whole number that exactly divides it. Thus, 2 and 5 are factors of 10.
- *Prime number:* in math, a number that has no factors except itself and 1. Examples of prime numbers are 1, 3, 5, 7, and 11.
- *Composite number:* a number that has factors other than itself and 1. Examples of composite numbers are 4, 6, 8, 9 and 12.
- *Common factor* or *common divisor* of two or more numbers: a factor that exactly divides each of them. If this factor is the largest factor possible, it is called the *greatest common divisor.* Thus, 3 is a common divisor of 9 and 27, but 9 is the greatest common divisor of 9 and 27.

25

- *Multiple* of a given number: a number that is exactly divisible by the given number. If a number is exactly divisible by two or more other numbers, it is a common multiple of them. The least (smallest) such number is called the *lowest common multiple.* Thus 36 and 72 are common multiples of 12, 9, and 4; however, 36 is the lowest common multiple.
- *Even number:* is a number exactly divisible by 2. Thus, 2, 4, 6, 8, 10, and 12 are even integers.
- *Odd number:* an integer that is not exactly divisible by 2. Thus, 1, 3, 5, 7, 9, and 11 are odd integers.
- *Product:* the result of multiplying two or more numbers together. Thus, 25 is the product of 5 × 5. Also, 4 and 5 are factors of 20.
- *Quotient:* the result of dividing one number by another. For example, 5 is the quotient of 20 divided by 4.
- *Dividend:* a number to be divided; a *divisor* is a number that divides. For example, in 100 ÷ 20 = 5, 100 is the dividend, 20 is the divisor, and 5 is the quotient.
- *Area:* the area of an object, measured in square units—the amount of surface an object contains or the amount of material required to cover the surface.
- *Base:* a term used to identify the bottom leg of a triangle, measured in linear units.
- *Circumference:* the distance around an object, measured in linear units. When determined for other than circles, it may be called the *perimeter* of the figure, object, or landscape.
- *Cubic units:* measurements used to express volume, cubic feet, cubic meters, and so on.
- *Depth:* the vertical distance from the bottom of the tank to the top. This is normally measured in terms of liquid depth and given in terms of sidewall depth (SWD), measured in linear units.
- *Diameter:* the distance from one edge of a circle to the opposite edge passing through the center, measured in linear units.
- *Height:* the vertical distance from the base or bottom of a unit to the top or surface.
- *Linear units:* measurements used to express distances: feet, inches, meters, yards, and so on.
- *Pi (π):* a number in the calculations involving circles, spheres, or cones: $\pi = 3.14$.
- *Radius:* the distance from the center of a circle to the edge, measured in linear units.
- *Sphere:* a container shaped like a ball.
- *Square units:* measurements used to express area, square feet, square meters, acres, and so forth.

- *Volume:* the capacity of the unit (how much it will hold) measured in cubic units (cubic feet, cubic meters) or in liquid volume units (gallons, liters, million gallons).
- *Width:* the distance from one side of the tank to the other, measured in linear units.

KEY WORDS FOR MATH OPERATIONS

- *of:* means to multiply
- *and:* means to add
- *per:* means to divide
- *less than:* means to subtract

Sequence of Operations

Mathematical operations such as addition, subtraction, multiplication, and division are usually performed in a certain order, or sequence. Typically, multiplication and division operations are done prior to addition and subtraction operations. In addition, mathematical operations are also generally performed from left to right using this hierarchy. The use of parentheses is also common to set apart operations that should be performed in a particular sequence.

✓ **Note:** We assume that the reader has a fundamental knowledge of basic arithmetic and math operations. Thus, the purpose of the following sections is to provide a brief review only of the mathematical concepts and applications frequently employed by geoscientists.

Sequence of Operations

RULES

Rule 1: In a series of additions, the terms may be placed in any order and grouped in any way. Thus,

$$4 + 3 = 7 \text{ and } 3 + 4 = 7; (4 + 3) + (6 + 4) = 17,$$
$$(6 + 3) + (4 + 4) = 17, \text{ and } [6 + (3 + 4) + 4 = 17.$$

Rule 2: In a series of subtractions, changing the order or the grouping of the terms may change the result. Thus,

$$100 - 30 = 70, \text{ but } 30 - 100 = -70; (100 - 30) - 10 = 60,$$
$$\text{but } 100 - (30 - 10) = 80.$$

Rule 3: When no grouping is given, the subtractions are performed in the order written, from left to right. Thus,

$$100 - 30 - 15 - 4 = 51; \text{ or by steps,}$$
$$100 - 30 = 70, 70 - 15 = 55, 55 - 4 = 51.$$

Rule 4: In a series of multiplications, the factors may be placed in any order and in any grouping. Thus,

$$[(2 \times 3) \times 5] \times 6 = 180 \text{ and } 5 \times [2 \times (6 \times 3)] = 180.$$

Rule 5: In a series of divisions, changing the order or the grouping may change the result. Thus,

$$100 \div 10 = 10, \text{ but } 10 \div 100 = 0.1; (100 \div 10) \div 2 = 5,$$
$$\text{but } 100 \div (10 + 2) = 20.$$

Again, if no grouping is indicated, the divisions are performed in the order written, from left to right. Thus, $100 \div 10 \div 2$ is understood to mean $(100 \div 10) \div 2$.

Rule 6: In a series of mixed mathematical operations, the convention is as follows: whenever no grouping is given, multiplications and divisions are to be performed in the order written, then additions and subtractions in the order written.

EXAMPLES

In a series of additions, the terms may be placed in any order and grouped in any way.

Examples:

$$3 + 6 = 10 \text{ and } 6 + 4 = 10$$
$$(4 + 5) + (3 + 7) = 19, (3 + 5) + (4 + 7) = 19, \text{ and } [7 + (5 + 4)] + 3 = 19$$

In a series of subtractions, changing the order or the grouping of the terms may change the result.

Examples:

$$100 - 20 = 80, \text{ but } 20 - 100 = -80$$
$$(100 - 30) - 20 = 50, \text{ but } 100 - (30 - 20) = 90$$

When no grouping is given, the subtractions are performed in the order written—from left to right.

Examples:

$$100 - 30 - 20 - 3 = 47$$

or by steps,

$$100 - 30 = 70, 70 - 20 = 50, 50 - 3 = 47$$

In a series of multiplications, the factors may be placed in any order and in any grouping.

Example:

$$[(3 \times 3) \times 5] \times 6 = 270 \text{ and } 5 \times [3 \times (6 \times 3)] = 270$$

In a series of divisions, changing the order or the grouping may change the result.

Example:

$$100 \div 10 = 10, \text{ but } 10 \div 100 = 0.1$$
$$(100 \div 10) \div 2 = 5, \text{ but } 100 \div (10 \div 2) = 20$$

If no grouping is indicated, the divisions are performed in the order written—from left to right.

Example:

$$100 \div 5 \div 2 \text{ is understood to mean } (100 \div 5) \div 2$$

In a series of mixed mathematical operations, the rule of thumb is whenever no grouping is given, multiplications and divisions are to be performed in the order written, and then additions and subtractions in the order written.

Percent

The word *percent* means "by the hundred." Percentage is often designated by the symbol %. Thus, 15% means 15 percent or 15/100 or 0.15. These equivalents may be written in the reverse order: 0.15 = 15/100 = 15%. When working with percent, the following key points are important:

- Percents are another way of expressing a part of a whole.
- As mentioned, the percent means "by the hundred," so a percentage is the number out of 100. To determine percent, divide the quantity we wish to express as a percent by the total quantity then multiply by 100:

$$Percent\ (\%) = \frac{Part}{Whole} \qquad (2.1)$$

For example, 22 percent (22%) means 22 out of 100, or 22/100. Dividing 22 by 100, results in the decimal 0.22:

$$22\% = \frac{22}{100} = 0.22$$

- When using percentages in calculations, the percentage must be converted to an equivalent decimal number; this is accomplished by dividing the percentage by 100.
- In a chemical dosing example, calcium hypochlorite contains 65% available chlorine. What is the decimal equivalent of 65%? Because 65% means 65 per hundred, divide 65 by 100: 65/100, which is 0.65.
- Decimals and fractions can be converted to percentages. The fraction is first converted to a decimal, and then the decimal is multiplied by 100 to get the percentage.
- For example, if a 50-foot high water tank has 26 feet of water in it, how full is the tank in terms of the percentage of its capacity?

$$\frac{26\ ft}{50\ ft} = 0.52\ (decimal\ equivalent)$$

$$0.52 \times 100 = 52$$

The tank is 52% full.

Example 2.1

Problem:

The plant operator removes 6,500 gal. of a chemical mixture from the storage tank. The chemical mixture contains 325 gallons of solids. What is the percentage of solids in the chemical mixture?

Solution:

$$Percent = \frac{325 \text{ gal}}{6,500 \text{ gal}} \times 100 = 5\%$$

Example 2.2

Problem:

Convert 65% to decimal percent.

Solution:

$$Decimal\ percent = \frac{Percent}{100}$$

$$= \frac{65}{100} = 0.65$$

Example 2.3

Problem:

A solution contains 5.8% solids. What is the concentration of solids in decimal percent?

Solution:

$$Decimal\ percent = \frac{5.8\%}{100} = 0.058$$

✓ **Key Point:** Unless otherwise noted, all calculations in the text using percent values require the percent be converted to a decimal before use.

✓ **Key Point:** To determine what quantity a percent equals, first convert the percent to a decimal then multiply by the total quantity.

$$Quantity = Total \times Decimal\ Percent \qquad (2.2)$$

Example 2.4

Problem:

A chemical mixture drawn from the settling tank is 5% solids. If 2,800 gallons of solids are withdrawn, how many gallons of solids are removed?

Solution:

$$Gallons = \frac{5\%}{100} \times 2,800 \text{ gallons} = 140 \text{ gal}$$

Example 2.5

Problem:

Convert 0.55 to percent.

Solution:

$$0.55 = \frac{55}{100} = 0.55 = 55\%$$

In converting 0.55 to 55%, we simply moved the decimal point two places to the right.

Example 2.6

Problem:

Convert 7/22 to a percent.

Solution:

$$\frac{7}{22} = 0.318 \times 100 = 31.8\%$$

Significant Digits

When rounding numbers, remember the following key points:

• Numbers are rounded to reduce the number of digits to the right of the decimal point. This is done for convenience, not for accuracy.

• *Rule:* A number is rounded off by dropping one or more numbers from the right and adding zeroes if necessary to place the decimal point. If the last figure dropped is 5 or more, increase the last retained figure by 1. If the last digit dropped is less than 5, do not increase the last retained figure. If the digit 5 is dropped, round off the preceding digit to the nearest *even* number.

Example 2.7

Problem:
    Round off 10,546 to 4, 3, 2, and 1 significant figures.

Solution:
    10,546 = 10,550 to four significant figures
    10,546 = 10,500 to three significant figures
    10,546 = 11,000 to two significant figures
    10,547 = 10,000 to one significant figure

Significant figures are those digits in a final calculation that have physical meaning. In determining significant figures, remember the following key points:

• The concept of significant figures is related to rounding.
• It can be used to determine where to round off.

✓ **Key Point**: No answer can be more accurate than the least accurate piece of data used to calculate the answer.

• *Rule*: significant figures are those numbers that are known to be reliable. The position of the decimal point does not determine the number of significant figures.

Example 2.8

Problem:
    How many significant figures are in a measurement of 1.35 in?

Solution:
    Three significant figures: 1, 3, and 5.

Example 2.9

Problem:
    How many significant figures are in a measurement of 0.000135?

Solution:

Again, three significant figures: 1, 3, and 5. The three zeros are used only to place the decimal point.

Example 2.10

Problem:

How many significant figures are in a measurement of 103,500?

Solution:

Four significant figures: 1, 0, 3, and 5. The remaining two zeros are used to place the decimal point.

Powers and Exponents

In working with powers and exponents, important key points include:

- Powers are used to identify area (as in square feet) and volume (as in cubic feet).
- Powers can also be used to indicate that a number should be squared, cubed, and so on. This later designation is the number of times a number must be multiplied times itself. For example, when several numbers are multiplied together, as $4 \times 5 \times 6 = 120$, the numbers 4, 5, and 6 are the *factors;* 120 is the *product.*
- If all the factors are alike, as $4 \times 4 \times 4 \times 4 = 256$, the product is called a *power.* Thus, 256 is a power of 4, and 4 is the *base* of the power. A *power* is a *product* obtained by using a base a certain number of times as a factor.
- Instead of writing $4 \times 4 \times 4 \times 4$, it is more convenient to use an *exponent* to indicate that the factor 4 is used as a factor four times. This exponent, a small number placed above and to the right of the base number, indicates how many times the base is to be used as a factor. Using this system of notation, the multiplication $4 \times 4 \times 4 \times 4$ is written as 4^4. The ^{4} is the *exponent,* showing that 4 is to be used as a factor four times.
- These same considerations apply to letters (a, b, x, y, etc.) as well. For example:

$$z^2 = (z)\,(z) \text{ or } z^4 = (z)\,(z)\,(z)\,(z)$$

- When a number or letter does not have an exponent, it is considered to have an exponent of one.

The powers of 1:

$1^0 = 1$

$1^1 = 1$

$1^2 = 1$

$1^3 = 1$

$1^4 = 1$

The powers of 10:

$10^0 = 1$

$10^1 = 10$

$10^2 = 100$

$10^3 = 1000$

$10^4 = 10,000$

Example 2.11

Problem:

How is the term 2^3 written in expanded form?

Solution:

The power (exponent) of 3 means that the base number (2) is multiplied by itself three times.

$$2^3 = (2)(2)(2)$$

Example 2.12

Problem:

How is the term $(3/8)^2$ written in expanded form?

✓ **Key Point:** When parentheses are used, the exponent refers to the entire term within the parentheses. Thus, in this example, $(3/8)^2$ means

Solution:

$$(3/8)^2 = (3/8)\,(3/8)$$

✓ **Key Point:** When a negative exponent is used with a number or term, a number can be re-expressed using a positive exponent.

$$6^{-3} = 1/6^3$$

Another example is

$$11^{-5} = 1/11^5$$

Example 2.13

Problem:
    How is the term 8^{-3} written in expanded form?

$$8^{-3} = \frac{1}{8^3} = \frac{1}{(8)(8)(8)}$$

✓ **Key Point:** Any number or letter such as 3^0 or X^0 does not equal 3×1 or X^1, but simply 1.

Averages (Arithmetic Mean)

Whether we speak of *harmonic mean, geometric mean,* or *arithmetic mean,* each is designed to find the "center" or the "middle" of the set of numbers. They capture the intuitive notion of a central tendency that may be present in the data. In statistical analysis, an average of data is a number that indicates the middle of the distribution of data values.

An *average* is a way of representing several different measurements as a single number. Although averages can be useful by telling "about" how much or how many, they can also be misleading, as we demonstrate next.

Example 2.14

Problem:
    When working with averages, the mean (again, what we usually refer to as an *average*) is the total of values of a set of observations divided by the number of observations. We simply add up all of the individual measurements and divide by the total number of measurements we took. For example, the operator of a waterworks or wastewater treatment plant takes a chlorine residual measurement every day, and part of his or her operating log is shown in Table 2.1. Find the mean.

Solution:
    Add up the seven chlorine residual readings: 0.9, 1.0, 0.9, 1.3, 1.1, 1.4, 1.2 = 7.8. Next, divide by the number of measurements, in this case seven: 7.8 , 7 = 1.11. The mean chlorine residual for the week was 1.11 mg/l.

Table 2.1.   Daily Chlorine Residual Results

Day	Chlorine Residual (mg/l)
Monday	0.9
Tuesday	1.0
Wednesday	0.9
Thursday	1.3
Friday	1.1
Saturday	1.4
Sunday	1.2

Example 2.15

Problem:
    A water system has four wells with the following capacities: 115 gallons per minute (gpm), 100 gpm, 125 gpm, and 90 gpm. What is the mean?

Solution:

$$\frac{115 \text{ gpm} + 100 \text{ gpm} + 125 \text{ gpm} + 90 \text{ gpm}}{4} = \frac{430}{4} = 107.5 \text{ gpm}$$

Example 2.16

Problem:
    A water system has four storage tanks. Three of them have a capacity of 100,000 gallons each, while the fourth has a capacity of 1 million gallons. What is the mean capacity of the storage tanks?

Solution:
    The mean capacity of the storage tanks is:

$$\frac{100,000 + 100,000 + 100,000 + 1,000,000}{4} = 325,000 \text{ gal}$$

    Notice that no tank in Example 2.16 has a capacity anywhere close to the mean.

Solving for the Unknown

Many calculations in geoscience involve the use of formulae and equations. To make these calculations, you must first know the values for all but one of the terms of the equation to be used. The obvious question is "What is an equation?" Simply, an equation is a mathematical statement that tells us that what is on one side of an equal sign (=) is equal to what is on the other side.

✓ **Key Point**: What we do to one side of the equation, we must do to the other side. This is the case, of course, because the two sides, by definition, are always equal.

An *equation* is a statement that two expressions or quantities are equal in value. The statement of equality $6x + 4 = 19$ is an equation; that is, it is algebraic shorthand for, "The sum of 6 times a number plus 4 is equal to 19." It can be seen that the equation $6x + 4 = 19$ is much easier to work with than the equivalent sentence.

When thinking about equations, it is helpful to consider an equation as being similar to a balance. The equal sign tells you that two quantities are "in balance" (i.e., they are equal). The solution to the problem $6x + 4 = 19$ may be summarized in three steps.

Step (1) $6x + 4 = 19$
Step (2) $6x = 15$
Step (3) $x = 2.5$

✓ **Note:** Step 1 expresses the whole equation. In step 2, 4 has been subtracted from both members of the equation. In step 3, both members have been divided by 6.

✓ **Key Point:** An equation is, therefore, kept in balance (both sides of the equal sign are kept equal) by subtracting the same number from both members (sides), adding the same number to both, or dividing or multiplying by the same number.

The expression $6x + 4 = 19$ is called a *conditional equation,* because it is true only when x has a certain value. The number to be found in a conditional equation is called the *unknown number* the *unknown quantity* or, more briefly, the *unknown*.

✓ **Key Point:** Solving an equation is finding the value or values of the unknown that make the equation true.

Let's look a look at another equation:

$$W = F \times D \qquad\qquad (2.3)$$

where

W = work
F = force
D = distance

$Work$ = $Force$ (lb) $\times$ $Distance$ (ft or in)
= ft-lb or in-lb

Suppose we have this equation:

$$60 = (x)\,(2)$$

How can we determine the value of x? By using the following axioms, the solution to the unknown is quite simple.

✓ **Key Point:** It is important to point out that the following discussion includes only what the axioms are and how they work.

AXIOMS

1. If equal numbers are added to equal numbers, the sums are equal.
2. If equal numbers are subtracted from equal numbers, the remainders are equal.
3. If equal numbers are multiplied by equal numbers, the products are equal.
4. If equal numbers are divided by equal numbers (except zero), the quotients are equal.
5. Numbers that are equal to the same number or to equal numbers are equal to each other.
6. Like powers of equal numbers are equal.
7. Like roots of equal numbers are equal.
8. The whole of anything equals the sum of all its parts.

✓ **Note:** Axioms 2 and 4 were used to solve the equation $6x + 4 = 19$.

✓ **Key Point:** As mentioned, solving an equation is determining the value or values of the unknown number or numbers in the equation.

Example 2.17

Problem:
Find the value of x if $x - 8 = 2$.

Solution:
Here it can be seen by inspection that $x = 10$, but inspection does not help in solving more complicated equations. However, if we notice that to determine that $x = 10$, 8 is added to each member of the given equation, we have acquired a method or procedure that can be applied to similar but more complex problems.
Given equation:

$$x - 8 = 2$$

Add 8 to each member (axiom 1),

$$x = 2 + 8$$

Collecting the terms (that is, adding 2 and 8).

$$x = 10$$

Example 2.18

Problem:
Solve for x, if $4x - 4 = 8$ (each side is in simplest terms)

Solution:

$4x = 8 + 4$ [the term (-4) is moved to the right of the equal sign as $(+4)$]
$4x = 12$

$$\frac{4x}{4} = \frac{12}{4} \text{(divide both sides)}$$

$x = 3$ (x is alone on the left and is equal to the value on the right)

Example 2.19

Problem:
   Solve for x, if $x + 10 = 15$

Solution:
   Subtract 10 from each member (axiom 2),

$$x = 15 - 10$$

   Collect the terms,

$$x = 5$$

Example 2.20

Problem:
   Solve for x, if $5x + 5 - 7 = 3x + 6$

Solution:
   Collect the terms (+5) and (-7):

$$5x - 2 = 3x + 6$$

   Add 2 to both members (axiom 2)

$$5x = 3x + 8$$

   Subtract $3x$ from both members (axiom 2)

$$2x = 8$$

   Divide both members by 2 (axiom 4):

$$X = 4$$

   After obtaining a solution to an equation, we should always check it. This is an easy process. All we need to do is substitute the solution for the unknown quantity in the given equation. If the two members of the equation are then identical, the number substituted is the correct answer.

Example 2.21

Problem:
   Solve and check $4x + 5 - 7 = 2x + 6$

Solution:

$$4x + 5 - 7 = 2x + 6$$
$$4x - 2 = 2x + 6$$
$$4x = 2x + 8$$
$$2x = 8$$
$$x = 4$$

Substituting the answer $x = 4$ in the original equation,

$$4x + 5 - 7 = 2x + 6$$
$$4(4) + 5 - 7 = 2(4) + 6$$
$$16 + 5 - 7 = 8 + 6$$
$$14 = 14$$

Because the statement $14 = 14$ is true, the answer $x = 4$ must be correct.

The equations discussed to this point were expressed in *algebraic* language. It is important to learn how to set up an equation by translating a sentence into an equation (into algebraic language) and then solving this equation.

In setting up an equation properly, the following suggestions and examples should help.

1. Always read the statement of the problem carefully.
2. Select the unknown number and represent it by some letter. If more than one unknown quantity exists in the problem, try to represent those numbers in terms of the same letter—that is, in terms of one quantity.
3. Develop the equation, using the letter or letters selected, and then solve.

Example 2.22

Problem:
   Given: One number is eight more than another. The larger number is two less than three times the smaller. What are the two numbers?

Solution: Let n represent the small number.

Then $n + 8$ must represent the larger number.

$$n + 8 = 3n - 2$$
$$n = 5 \text{ (small number)}$$
$$n + 8 = 13 \text{ (large number)}$$

Example 2.23

Problem:
Given: If five times the sum of a number and 6 is increased by 3, the result is two less than 10 times the number. Find the number.

Solution: Let n represent the number.

$$5(n + 6) + 3 = 10n - 2$$
$$n = 5$$

Example 2.24

Problem:
If $2x + 5 = 10$, solve for x.

Solution:

$$2x + 5 = 10$$
$$2x = 5$$
$$0.x = 5/2$$
$$x = 2\frac{1}{2}$$

Example 2.25

Problem:
If $0.5x - 1 = -6$, find x.

Solution:

$$0.5x - 1 = -6$$
$$0.5x = -5$$
$$x = -10$$

Ratio

A *ratio* is the established relationship between two numbers. For example, if someone says, "I'll give you four to one the Redskins over the Cowboys in the Super Bowl," what does that person mean? Four to one, or 4:1, is a ratio. If someone gives you 4 to 1, it's his or her $4 to your $1.

As another more pertinent example, if an average of 3 cu ft of screenings are removed from each million gallons of wastewater treated, the ratio of screenings removed (cu ft) to treated wastewater (MG) is 3:1. Ratios are normally written using a colon (such as 2:1), or written as a faction (such as 2/1).

A *proportion* is a statement that two ratios are equal. For example, 1 is to 2 as 3 is to 6, so 1:2 = 3:6. In this case, 1 has the same relation to 2 that 3 has to 6. And what exactly is that relation? What do you think? You're right: 1 is half the size of 2, and 3 is half the size of 6. Or, alternately, 2 is twice the size of 1, and 6 is twice the size of 3.

When working with ratio and proportion, the following key points are important to remember.

1. One place where fractions are used in calculations is when **ratios** and **proportions** are used, such as calculating solutions.
2. A ratio is usually stated in the form *A* is to *B* as *C* is to *D*, and we can write it as two fractions that are equal to each other:

$$\frac{A}{B} = \frac{C}{D}$$

3. Cross-multiplying solves ratio problems; that is, we multiply the left numerator (*A*) by the right denominator (*D*) and say that is equal to the left denominator (*B*) times the right numerator (*C*): </nl>

$$A \times D = B \times C$$
$$AD = BC$$

4. If one of the four items is unknown, dividing the two known items that are multiplied together by the known item that is multiplied by the unknown solves the ratio. For example, if 2 pounds of alum are needed to treat 500 gallons of water, how many pounds of alum will we need to treat 10,000 gallons? We can state this as a ratio: 2 pounds of alum is to 500 gallons of water as "pounds of alum is to 10,000 gallons."

This is set up in this manner:

$$\frac{1 \text{ lb alum}}{500 \text{ gal water}} = \frac{x \text{ lb alum}}{10,000 \text{ gal water}}$$

Cross-multiplying:

$$(500)\,(x) = (1) \times (10,000)$$

Transposing:

$$x = \frac{1 \times 10,000}{500}$$

$$x = \quad 20 \text{ lb alum}$$

For calculating proportion, for example, five gallons of fuel costs $5.40. How much does 15 gallons cost?

$$\frac{5 \text{ gal}}{\$5.40} = \frac{15 \text{ gal}}{\$y}$$

$$5 \times y = 15 \times 5.40 = 81$$

$$y = \frac{81}{5} = \$16.20$$

Example 2.26

Problem:
If a pump will fill a tank in 20 hours at 4 gpm, how long will it take a 10-gpm pump to fill the same tank?

Solution:
First, analyze the problem. Here, the unknown is some number of hours. But should the answer be larger or smaller than 20 hours? If a 4-gpm pump can fill the tank in 20 hours, a larger pump (10 gpm) should be able to complete the filling in less than 20 hours. Therefore, the answer should be less than 20 hours.

Now set up the proportion:

$$\frac{x\text{ h}}{20\text{ h}} = \frac{4\text{ gpm}}{10\text{ gpm}}$$

$$x = \frac{(4)(20)}{10}$$

$$x = 8\text{ h}$$

Example 2.27

Problem:

Solve for x in the following proportion problem.

Solution:

$$\frac{36}{180} = \frac{x}{4450}$$

$$\frac{(4450)(36)}{180} = x$$

$$= 890$$

Example 2.28

Problem:

Solve for the unknown value x in the problem given below.

Solution:

$$\frac{3.4}{2} = \frac{6}{x}$$

$$(3.4)\,(x) = (2)\,(6)$$

$$x = \frac{(2)\,(6)}{3.4}$$

$$x = 3.53$$

Dimensional Analysis

Dimensional analysis is a problem-solving method that uses the fact that one, without changing its value, can multiply any number or expression. It is a useful technique used to check if a problem is set up correctly. In using dimensional analysis to check a math setup, we work with the dimensions (units of measure) only—not with numbers.

An example of dimensional analysis common to everyday life is the unit pricing found in many hardware stores. A shopper can purchase a 1-lb box of nails for $0.98 in one store, whereas a warehouse store sells a 5-lb bag of the same nails for $3.50. The shopper will analyze this problem almost without thinking about it. The solution calls for reducing the problem to the price per pound. The pound is selected as the unit common to both stores. Knowing the unit price, which is expressed in dollars per pound ($/lb), is implicit in the solution to this problem.

To use the dimensional analysis method, we must know how to perform three basic operations:

✓ **Note:** Unit factors may be made from any two terms that describe the same or equivalent "amounts" of what we are interested in. For example, we know that 1 inch = 2.54 centimeters.

1. *Basic Operation:* To complete a division of units, always ensure that all units are written in the same format; it is best to express a horizontal fraction (such as gal/ft²) as a vertical fraction. Horizontal to vertical:

$$gal/ft^3 \text{ to } \frac{gal}{ft^3}$$

$$psi \text{ to } \frac{lb}{in^2}$$

2. *Basic Operation:* We must know how to divide by a fraction. For example,

$$\frac{\dfrac{lb}{d}}{\dfrac{min}{d}} \text{ becomes } \frac{lb}{d} \times \frac{d}{min}$$

In the preceding example, notice that the terms in the denominator were inverted before the fractions were multiplied. This is a standard rule that must be followed when dividing fractions. An example is,

$$\frac{\dfrac{mm^2}{mm^2}}{m^2} \text{ becomes } mm^2 \times \frac{m^2}{mm^2}$$

3. *Basic Operation*: We must know how to cancel or divide terms in the numerator and denominator of a fraction. After fractions have been rewritten in the vertical form and division by the fraction has been re-expressed as multiplication, as shown previously, then the terms can be canceled (or divided) out.

✓ **Key Point:** For every term that is canceled in the numerator of a fraction, a similar term must be canceled in the denominator and vice versa, as shown in the following example:

$$\frac{Kg}{d} \times \frac{d}{min} = \frac{kg}{min}$$

$$mm^2 \times \frac{m^2}{mm^2} = m^2$$

$$\frac{gal}{min} \times \frac{ft^3}{gal} = \frac{ft^3}{min}$$

How are units that include exponents calculated?

When written with exponents, such as ft^3, a unit can be left as is or put in expanded form, (ft)(ft)(ft), depending on other units in the calculation. The point is that it is important to ensure that square and cubic terms are expressed uniformly, as sq ft, cu ft, or as ft^2. For dimensional analysis, the latter system is preferred. For example, if we wish to convert 1,400 ft^3 to gallons, and we will use 7.48 gal/ft^3 in the conversions, the question becomes whether to multiply or divide by 7.48.

In this instance, it is possible to use dimensional analysis to answer this question; that is, are we to multiply or divide by 7.48? In order to determine if the math setup is correct, only the dimensions are used.

First, try dividing the dimensions:

$$\frac{ft^3}{gal/ft^3} = \frac{ft^3}{\dfrac{gal}{ft^3}}$$

Then, the numerator and denominator are multiplied to get

$$= \frac{ft^6}{gal}$$

Thus, by dimensional analysis we determine that if we divide the two dimensions (ft^3 and gal/ft^3), the units of the answer are ft^6/gal, not gal. It is clear that division is not the right way to go in making this conversion.

What would have happened if we had multiplied the dimensions instead of dividing?

$$(ft^3)\,(gal/ft^3) = (ft^3) \left(\frac{gal}{ft^3} \right)$$

Then multiply the numerator and denominator to obtain

$$= \frac{(ft^3)\,(gal)}{ft^3}$$

and cancel common terms to obtain

$$= \frac{(ft^3)\,(gal)}{ft^3}$$

Obviously, by multiplying the two dimensions (ft^3 and gal/ft^3), the answer will be in gallons, which is what we want. Thus, because the math setup is correct, we would then multiply the numbers to obtain the number of gallons.

$$(1{,}400\ \text{ft}^3)\,(7.48\ \text{gal/ft}^3) = 10{,}472\ \text{gal}$$

Now let's try another problem with exponents. We wish to obtain an answer in square feet. If we are given the two terms 70 ft³/s and 4.5 ft/s, is the following math setup correct?

$$(70 \text{ ft}^3/\text{sec}) \ (4.5 \text{ ft}/\text{sec})$$

First, only the dimensions are used to determine if the math setup is correct. Multiplying the two dimensions' yields:

$$(\text{ft}^3/s) \ (\text{ft}/s) = \left(\frac{\text{ft}^3}{s} \right) \left(\frac{\text{ft}^3}{s} \right)$$

Then the terms in the numerators and denominators of the fraction are multiplied:

$$= \frac{(\text{ft}^3) \ (\text{ft})}{(s) \ (s)}$$

$$= \frac{\text{ft}^4}{s^2}$$

Obviously, the math setup is incorrect because the dimensions of the answer are not square feet. Therefore, if we multiply the numbers as shown previously, the answer will be wrong.

Let's try division of the two dimensions instead.

$$\text{ft}^3/\text{sec} = \frac{\dfrac{\text{ft}^3}{\text{sec}}}{\dfrac{\text{ft}}{\text{sec}}}$$

Invert the denominator and multiply to get

$$= \left(\frac{\text{ft}^3}{(\text{sec})} \right) \left(\frac{\text{sec}}{(\text{ft})} \right)$$

$$= \frac{(\text{ft})(\text{ft})(\text{ft})(\text{sec})}{(\text{sec})(\text{ft})}$$

$$= ft^2$$

Because the dimensions of the answer are square feet, this math setup is correct. Therefore, by dividing the numbers as was done with units, the answer will also be correct.

$$\frac{70 \text{ ft}^3/\text{s}}{4.5 \text{ ft}/\text{s}} = 15.56 \text{ ft}^2$$

Mass

The terms *weight* and *mass* are often confused. The weight of an object depends on its mass and the gravitational force pulling on it. This is why our weight on the Moon would be less: It exerts less gravitational pull on our bodies—weight is a measure of how heavy an object is (our bodies or any other object), measured in units of force. However, mass (a fundamental concept in science) on the Moon does not change or vary. This is the case because mass is a measure of the number of atoms in an object. Simply, unless we add to or subtract atoms from our bodies (by eating too much or dieting, for example), it does not matter what we are doing or where we are; our mass remains the same.

The fundamental (SI) unit of mass is the kilogram (kg). Many other units of mass are also employed, such as:

- gram: 1 gm = 0.001 kg
- ton: 1 ton = 1,000 kg
- eV/c^2

SPECIFIC GRAVITY AND DENSITY

Specific gravity is the ratio of the *density* (ratio of a body's mass [m] to its volume [V]; in layman's terms, density is a measurement of how much mass is packed into a given volume: Density = Mass/Volume) of a substance to that of a standard material under standard conditions of temperature and pressure. The specific gravity of water is 1.0 (one). Any substance with a density greater than that of water will have a specific gravity greater than 1.0, and any substance with a density less than that of water sill have specific gravity less than 1.0. Specific gravity can be used to calculate the weight of a gallon of liquid chemical.

Chemical, lb/gal = *Water,* lb/gal × *Specific Gravity* (chemical)   (2.4)

Example 2.29

Problem:
If we say that the density of gasoline is 43-lb/cu ft, what is the specific gravity of gasoline?

Solution:
The specific gravity of gasoline is the comparison—or ratio—of the density of gasoline to that of water:

$$Specific\ Gravity = \frac{Density\ of\ Gasoline}{Density\ of\ Water}$$

$$= \frac{43\ lb/cu\ ft\ (density\ of\ gasoline)}{62.4\ lb/cu\ ft\ (density\ of\ water)} = 0.69$$

✓ **Key Point:** Because gasoline's specific gravity is less than 1.0 (lower than water's specific gravity), it will float in water. If gasoline's specific gravity were greater than water's specific gravity it would sink.

✓ **Key Point:** The density of water is 1,000 g per 1,000 cubic centimeters (cm^3) or, more simply, 1,000/1,000 = 1 gram per cubic centimeter (gm/cm^3) at a temperature of 4° C.

Flow

Flow is expressed in many different terms (English system of measurements). The most commonly used flow terms are as follows:

- gpm—gallons per minute
- cfs—cubic feet per second
- gpd—gallons per day
- MGD—million gallons per day

In converting flow rates, the most common flow conversions are 1 cfs = 448 gpm and 1 gpm =1440 gpd.

To convert gallons per day to MGD, divide the gpd by 1,000,000. For instance, convert 150,000 gallons to MGD.

$$\frac{150{,}000 \text{ gpd}}{1{,}000{,}000} = 0.150 \text{ MGD}$$

In some instances, flow is given in MGD but is needed in gpm. To make the conversion (MGD to gpm), two steps are required.

Step 1: Convert the gpd by multiplying by 1,000,000.
Step 2: Convert to gpm by dividing by the number of minutes in a day (1,440 min/day).

Example 2.30

Problem:
Convert 0.135 MGD to gpm.

Solution:
First convert the flow in MGD to gpd.

$$0.135 \text{ MGD} \times 1{,}000{,}000 = 135{,}000 \text{ gpd}$$

Now convert to gpm by dividing by the number of minutes in a day (24 hr per day × 60 min per hour) = 1,440 min/day.

$$\frac{135{,}000 \text{ gpd}}{1{,}440 \text{ min/day}} = 93.8 \text{ or } 94 \text{ gpm}$$

In determining flow through a pipeline, channel, or stream, we use the following equation:

$$Q = VA \qquad\qquad (2.5)$$

where

Q = cubic feet per second (cfs)
V = velocity in feet per second (ft/sec)
A = area in square feet (ft²)

Example 2.31

Problem:
Find the flow in cfs in an 8-inch line, if the velocity is 3 feet per second.

Solution:

- Step 1: Determine the cross-sectional area of the line in square feet. Start by converting the diameter of the pipe to inches.
- Step 2: The diameter is 8 inches; therefore, the radius is 4 inches and 4 inches is 4/12 of a foot or 0.33 feet.
- Step 3: Find the area in square feet.

$$A = \pi r^2$$
$$A = \pi (0.33 \text{ ft})^2$$
$$A = \pi \times 0.109 \text{ ft}^2$$
$$A = 0.342 \text{ ft}^2$$

- Step 4: $Q = VA$

$$Q = 3 \text{ ft/sec} \times 0.342 \text{ ft}^2$$
$$Q = 1.03 \text{ cfs}$$

Example 2.32

Problem:
Find the flow in gpm when the total flow for the day is 75,000 gpd.

Solution:

$$\frac{75,000 \text{ gpd}}{1,440 \text{ min/day}} = 52 \text{ gpm}$$

Example 2.33

Problem:
Find the flow in gpm when the flow is 0.45 cfs.

Solution:

$$0.45 \, \frac{\text{cfs}}{1} \times \frac{448 \text{ gpm}}{1 \text{ cfs}} = 202 \text{ gpm}$$

Horsepower and Work

Horsepower is a common expression for power. One horsepower is equal to 33,000 foot pounds of work per minute. This value is determined, for example, for selecting a pump or combination of pumps to ensure an adequate pumping capacity. Pumping capacity depends on the flow rate desired and the feet of head against which the pump must pump (aka *effective height*).

Calculations of horsepower are made in conjunction with many industrial operations. The basic concept from which the horsepower calculation is derived is the concept of work. *Work* involves the operation of a force (lb) over a specific distance (ft). The *amount of work* accomplished is measured in foot-pounds:

$$(ft)\ (lb) = ft\text{-}lb \qquad (2.6)$$

The *rate of doing work (power)* involves a time factor. Originally, the rate of doing work or power compared the power of a horse to that of a steam engine. The rate at which a horse could work was determined to be about 550 ft-lb/sec (or expressed as 33,000 ft-lb/min). This rate has become the definition of the standard unit called horsepower.

1. Horsepower

$$\textit{Horsepower, } hp = \frac{\textit{Power, } ft\text{-}lb/min}{33,000\ ft\text{-}lb/min/HP} \qquad (2.7)$$

When the major use of horsepower calculation is to determine the proper pumping station operation, the horsepower calculation can be modified as shown in the following section.

Area, Volume, and Density

To aid in performing these calculations, the following definitions are provided.

- *Area*—the area of an object, measured in square units
- *Base*—the term used to identify the bottom leg of a triangle, measured in linear units
- *Circumference*—the distance around an object, measured in linear units

When determined for other than circles, *circumference* may be called the *perimeter of the figure, object,* or *landscape.*

- *Cubic units*—measurements used to express volume, cubic feet, cubic meters, and so on
- *Depth*—the vertical distance from the bottom of the tank to the top

Depth is normally measured in terms of liquid depth and given in terms of SWD, measured in linear units.

- *Diameter*—the distance from one edge of a circle to the opposite edge passing through the center, measured in linear units
- *Height*—the vertical distance from one end of an object to the other, measured in linear units
- *Length*—the distance from one end of an object to the other, measured in linear units
- *Linear units*—measurements used to express distances: feet, inches, meters, yards, and so on
- *Pi, π*—a number in the calculations involving circles, spheres, or cones (π = 3.14)
- *Radius*—the distance from the center of a circle to the edge, measured in linear units
- *Sphere*—a container shaped like a ball
- *Square units*—measurements used to express area, square feet, square meters, acres, and so on
- *Volume*—the capacity of the unit, how much it will hold, measured in cubic units (cubic feet, cubic meters) or in liquid volume units (gallons, liters, million gallons)
- *Width*—the distance from one side of the tank to the other, measured in linear units

On occasion, it may be necessary to determine the distance around grounds or landscapes. To measure the distance around property, buildings, and basin-like structures, it is necessary to determine either perimeter or circumference. The *perimeter* is the distance around an object: a border or outer boundary. *Circumference* is the distance around a circle or circular object, such as a clarifier. Distance is linear measurement, which defines the distance (or length) along a line. Standard units of measurement like inches, feet, yards, and miles and metric units like centimeters, meters, and kilometers are used.

PERIMETER

The perimeter of a rectangle (a four-sided figure with four right angles) is obtained by adding the lengths of the four sides (see Figure 2.1).

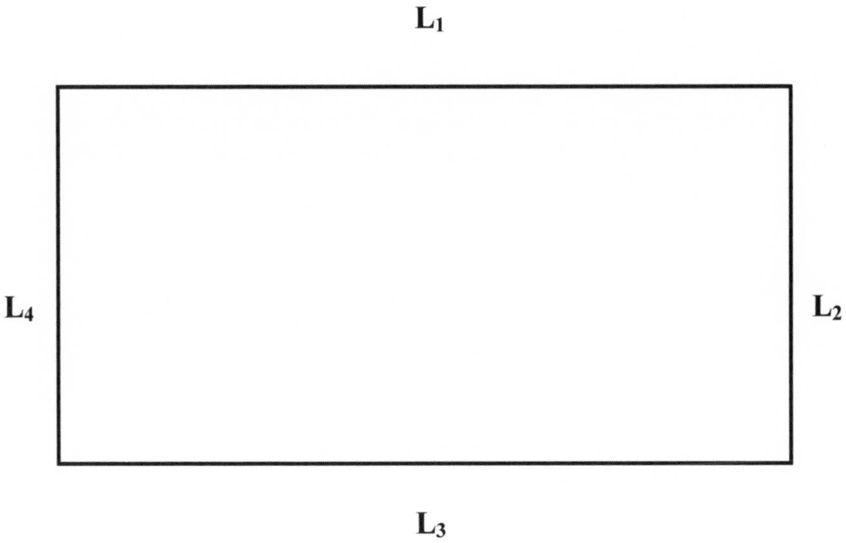

$$L_1$$

$$L_4 \qquad\qquad\qquad\qquad\qquad\qquad\qquad L_2$$

$$L_3$$

Figure 2.1.   Perimeter.

$$Perimeter = L_1 + L_2 + L_3 + L_4 \qquad\qquad (2.8)$$

Example 2.34

Problem:

Find the perimeter of the rectangle shown in Figure 2.2

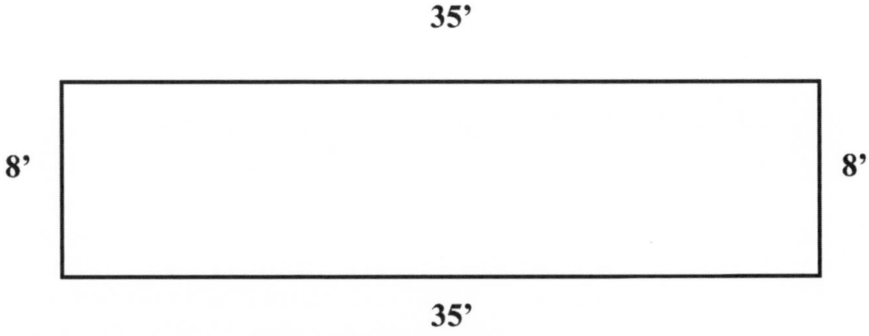

35'

8'   **8'**

35'

Figure 2.2.   Perimeter (for Example 2.34).

Solution:

$$P = 35' + 8' + 35' + 8'$$
$$P = 86'$$

CIRCUMFERENCE

The circumference is the length of the outer border of a circle. The circumference is found by multiplying pi (π) times the diameter (D) (diameter is a straight line passing through the center of a circle—the distance across the circle; see Figure 2.3).

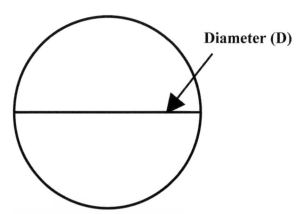

Figure 2.3.   Diameter of circle.

$$C = \pi D \qquad (2.9)$$

where

C = circumference
π = Greek letter pi = 3.1416
D = diameter

Use this calculation if, for example, the circumference of a circular tank must be determined.

Example 2.35

Problem:
    Find the circumference of a circle that has a diameter of 25 feet (π = 3.14)

Solution:

$$C = \pi \times 25'$$
$$C = 3.14 \times 25'$$
$$C = 78.5'$$

Example 2.36

Problem:
    A circular chemical holding tank has a diameter of 18 m. What is the circumference of this tank?

Solution:

$$C = \pi\ 18\ \text{m}$$
$$C = (3.14)\ (18\ \text{m})$$
$$C = 56.52\ \text{m}$$

AREA

For area measurements, three basic shapes are particularly important: circles, rectangles, and triangles. *Area* is the amount of surface an object contains or the amount of material it takes to cover the surface. The area on top of a chemical tank is called the *surface area*. The area of the end of a ventilation dust is called the *cross-sectional area* (the area at right angles to the length of ducting). Area is usually expressed in square units, such as square inches (in^2) or square feet (ft^2). Land may also be expressed in terms of square miles (sections) or acres (43,560 ft^2) or in the metric system as *hectares*.

    A *rectangle* is a two-dimensional box. The area of a rectangle is found by multiplying the length (L) times width (W); see Figure 2.4.

$$Area = L \times W \tag{2.10}$$

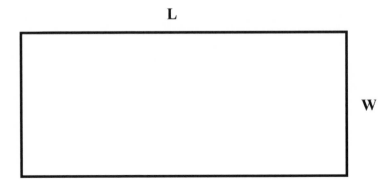

Figure 2.4.   Rectangle.

Example 2.37

Problem:

Find the area of the rectangle shown in Figure 2.5.

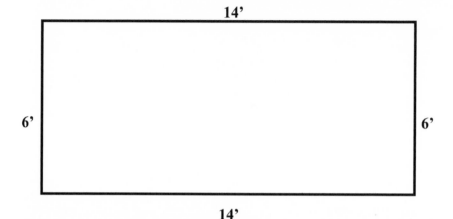

Figure 2.5.   Rectangle (for Example 2.37).

Solution:

$$Area = L \times W$$
$$= 14' \times 6'$$
$$= 84 \text{ ft}^2$$

AREA OF A CIRCLE

To find the area of a circle, we need to introduce one new term, the *radius,* which is represented by *r*. In Figure 2.6, we have a circle with a radius of 6 inches.

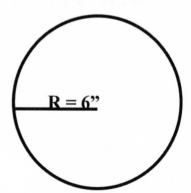

Figure 2.6.   Area of circle (for Example 2.38).

The radius is any straight line that radiates from the center of the circle to some point on the circumference. By definition, all radii (plural of radius) of the same circle are equal. The surface area of a circle is determined by multiplying π times the radius squared.

$$Area\ of\ Circle = \pi r^2 \tag{2.11}$$

where

A = area
π = pi (3.14)
r = radius of circle—radius is one-half the diameter

Example 2.38

Problem:
What is the area of the circle shown in Figure 2.6?

Solution:

$$\begin{aligned} Area\ of\ circle &= \pi r^2 \\ &= \pi 6^2 \\ &= 3.14 \times 36 \\ &= 113\ \text{ft}^2 \end{aligned}$$

AREA OF A CIRCULAR OR CYLINDRICAL TANK

If we were assigned to paint a water storage tank, we must know the surface area of the walls of the tank—we need to know how much paint is required. To determine the tank's surface area, we need to visualize the cylindrical walls as a rectangle wrapped around a circular base. The area of a rectangle is found by multiplying the length by the width; in this case, the width of the rectangle is the height of the wall, and the length of the rectangle is the distance around the circle, the circumference. Thus, the area of the side walls of the circular tank is found by multiplying the circumference of the base ($C = \pi \times D$) times the height of the wall (H):

$$\begin{aligned} A &= \pi \times D \times H \\ A &= \pi \times 20\ \text{ft} \times 25\ \text{ft} \\ A &= 3.14 \times 20\ \text{ft} \times 25\ \text{ft} \\ A &= 1570.8\ \text{ft}^2 \end{aligned} \tag{2.12}$$

To determine the amount of paint needed, remember to add the surface area of the top of the tank, which is 314 ft². Thus, the amount of paint needed must cover 1,570.8 ft² + 314 ft² = 1,884.8 or 1,885 ft². If the tank floor should be painted, add another 314 ft².

VOLUME

The amount of space occupied by or contained in an object, *volume* (see Figure 2.7), is expressed in cubic units, such as cubic inches (in³), cubic feet (ft³), acre feet (1 acre foot = 43,560 ft³), and so on. The volume of a rectangular object is obtained by multiplying the length times the width times the depth or height.

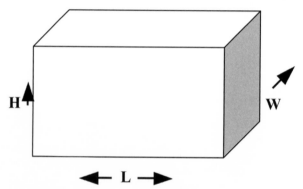

Figure 2.7.   Volume of rectangle.

$$V = L \times W \times H \tag{2.13}$$

where

L = length
W = width
D or H = depth or height

Example 2.39

Problem:
   A unit rectangular process basin has a length of 15', width of 7', and depth of 9'. What is the volume of the basin?

Solution:

$$\begin{aligned} V &= L \times W \times D \\ &= 15' \times 7' \times 9' \\ &= 945 \text{ ft}^3 \end{aligned}$$

Table 2.2.   Volume Formulas

Sphere volume = (π/6) (diameter)³
Cone volume = 1/3 (volume of a cylinder)
Rectangular tank volume = (area of rectangle) (D or H) = (LW) (D or H)
Cylinder volume = (area of cylinder) (D or H)
π² (D or H)
Triangle volume = (area of triangle) (D or H) = (bh/2) (D or H)

The practical volume formulas for determining representative surface areas of rectangles, triangles, circles, or a combination of these are given in Table 2.2.

$$Volume\ of\ cone = \frac{\pi}{12} \times Diameter \times Diameter \times Height \qquad (2.14)$$

$$\frac{\pi}{12} = \frac{3.14}{12} = 0.262$$

✓ **Key Point:** The diameter used in the formula is the diameter of the base of the cone.

Example 2.40

Problem:
The bottom section of a circular settling tank has the shape of a cone. How many cubic feet of water are contained in this section of the tank if the tank has a diameter of 120 ft and the cone portion of the unit has a depth of 6 ft?

Solution:

$$Volume = \frac{3.14}{6} \times Diameter \times Diameter \times Diameter \qquad (2.15)$$

$$\frac{\pi}{6} = \frac{3.14}{6} = 0.52$$

Example 2.41

Problem:
What is the volume of cubic feet of a gas storage container that is spherical and has a diameter of 60 ft?

Solution:

$$Volume, \text{ft}^3 = 0.524 \times 60 \text{ ft} \times 60 \text{ ft} \times 60 \text{ ft} = 113,184 \text{ ft}^3$$

VOLUME OF A CIRCULAR OR CYLINDRICAL TANK

A circular tank consists of a circular floor surface with a cylinder rising above it (see Figure 2.8). The volume of a circular tank is calculated by multiplying the surface area times the height of the tank walls.

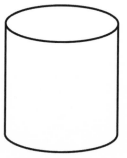

Figure 2.8.   Volume of circular tank (*Volume = Surface Area × Height*).

Example 2.42

Problem:
  If a tank is 20 feet in diameter and 25 feet deep, how many gallons of water will it hold?
  Hint: In this type of problem, calculate the surface area first, multiply by the height, and then convert to gallons.

Solution:

$$r = D \div 2 = 20 \text{ ft} \div 2 = 10 \text{ ft}$$
$$A = \pi \times r^2$$
$$A = \pi \times 10 \text{ ft} \times 10 \text{ ft}$$
$$A = 314 \text{ ft}^2$$
$$V = A \times H$$
$$V = 314 \text{ ft}^2 \times 25 \text{ ft}$$
$$V = 7,850 \text{ ft}^3 \times 7.5 \text{ gal/ft}^3 = 58,875 \text{ gal}$$

Force, Pressure, and Head

Before we study calculations involving force, pressure, and head, we must first define these terms.

- *Force*—the push exerted by water on any confined surface
  Force can be expressed in pounds, tons, grams, or kilograms.

- *Pressure*—the force per unit area
  The most common way of expressing pressure is in pounds per square inch (psi).

- *Head*—the vertical distance or height of water above a reference point
  Head is usually expressed in feet. In the case of water, head and pressure are related.

A cubical container measuring one foot on each side can hold one cubic foot of water. A basic fact of science states that one cubic foot of water weights 62.4 pounds and contains 7.48 gallons. The force acting on the bottom of the container would be 62.4 pounds per square foot. The area of the bottom in square inches is:

$$\frac{62.4 \text{ lb/ft}^2}{1 \text{ ft}^2} = \frac{62.4 \text{ lb/ft}^2}{144 \text{ in}^2/\text{ft}^2} = 0.433 \text{ lb/in}^2 \text{ (psi)}$$

If we use the bottom of the container as our reference point, the head would be 1 foot. From this we can see that 1 foot of head is equal to 0.433 psi—an important parameter to remember.

✓ **Important Point:** *Force* acts in a particular direction. Water in a tank exerts force down on the bottom and out on the sides. *Pressure,* however, acts in all directions. A marble at a water depth of 1 foot would have 0.433 psi of pressure acting inward on all sides.

Using the preceding information, we can develop Equations (2.16) and (2.17) for calculating pressure and head.

$$Pressure \text{ (psi)} = 0.433 \times Head \text{ (ft)} \qquad (2.16)$$

$$Head \text{ (ft)} = 2.31 \times Pressure \text{ (psi)} \qquad (2.17)$$

As mentioned, *head* is the vertical distance the water must be lifted from the supply tank or unit process to the discharge. The total head includes the vertical distance the liquid must be lifted (static head), the loss to friction (friction head), and the energy required to maintain the desired velocity (velocity head).

$$Total\ Head = Static\ Head + Friction\ Head + Velocity\ Head \qquad (2.18)$$

Static head is the actual vertical distance the liquid must be lifted.

$$Static\ Head = Discharge\ Elevation - Supply\ Elevation \qquad (2.19)$$

Example 2.43

Problem:

The supply tank is located at elevation 108 ft. The discharge point is at elevation 205 ft. What is the static head in feet?

Solution:

$$Static\ Head,\ \text{ft} = 205\ \text{ft} - 108\ \text{ft} = 97\ \text{ft}$$

Friction head is the equivalent distance of the energy that must be supplied to overcome friction. Engineering references include tables showing the equivalent vertical distance for various sizes and types of pipes, fittings, and valves. The total friction head is the sum of the equivalent vertical distances for each component.

$$Friction\ Head,\ \text{ft} = Energy\ Losses\ caused\ by\ Friction \qquad (2.20)$$

Velocity head is the equivalent distance of the energy consumed in achieving and maintaining the desired velocity in the system.

$$Velocity\ Head,\ \text{ft} = Energy\ Losses\ to\ Maintain\ Velocity \qquad (2.21)$$

The pressure exerted by water is directly proportional to its depth or head in the pipe, tank, or channel. If the pressure is known, the equivalent head can be calculated.

$$Head,\ \text{ft} = Pressure,\ \text{psi} \times 2.31\ \text{ft/psi} \qquad (2.22)$$

Example 2.44

Problem:

The pressure gauge on the discharge line from the influent pump reads 75.3 psi. What is the equivalent head in feet?

Solution:

$$Head, \text{ ft} = 75.3 \times 2.31 \text{ ft/psi} = 173.9 \text{ ft}$$

If the head is known, the equivalent pressure can be calculated by:

$$Pressure, \text{ psi} = \frac{Head, \text{ ft}}{2.31 \text{ ft/psi}} \qquad (2.23)$$

Example 2.45

Problem:

The tank is 15 feet deep. What is the pressure in psi at the bottom of the tank when it is filled with wastewater?

Solution:

$$Pressure, \text{ psi} = \frac{15 \text{ ft}}{2.31 \text{ ft/psi}}$$

$$= 6.49 \text{ psi}$$

Before we look at a few example problems dealing with force, pressure, and head, it is important to list the key points related to force, pressure, and head.

- By definition, water weighs 62.4 pounds per cubic foot.
- The surface of any one side of the cube contains 144 square inches (12 in × 12 in = 144 in²). Therefore, the cube contains 144 columns of water 1-foot tall and 1-inch square.
- The weight of each of these pieces can be determined by dividing the weight of the water in the cube by the number of square inches.

$$Weight = \frac{62.4 \text{ lbs}}{144 \text{ in}^2} = 0.433 \text{ lbs/in}^2 \text{ or } 0.433 \text{ psi}$$

- Because this is the weight of one column of water 1 foot tall, the true expression would be 0.433 pounds per square inch per foot of head, or 0.433 psi/ft.

✓ **Key Point:** 1 foot of head = 0.433 psi.

In addition to remembering the important parameter, 1 foot of head = 0.433 psi, it is important to understand the relationship between pressure and feet of head—in other words, how many feet of head 1 psi represents. This is determined by dividing 1 by 0.433.

$$Feet \text{ of head} = \frac{1 \text{ ft}}{0.433 \text{ psi}} = 2.31 \text{ ft/psi}$$

If a pressure gauge reads 12 psi, the height of the water necessary to represent this pressure is 12 psi × 2.31 ft/psi = 27.7 feet.

Example 2.46

Problem:
   Convert 40 psi to feet head.

Solution:

$$\frac{40 \text{ psi}}{1} \times \frac{ft}{0.433 \text{ psi}} = 92.4 \text{ feet}$$

Example 2.47

Problem:
   Convert 40 feet to psi.

Solution:

$$40 \frac{ft}{1} \times \frac{0.433 \text{ psi}}{1 \text{ ft}} = 17.32 \text{ psi}$$

As the previous examples demonstrate, when attempting to convert psi to feet, we divide by 0.433, and when attempting to convert feet to psi, we multiply by 0.433. This process can be most helpful in clearing up the confusion regarding whether to multiply or divide. There is another way, however—one that might be more beneficial and easier for many operators to use. Notice that the relationship between psi and feet is almost 2 to 1. It takes slightly more than 2 feet to make 1 psi. Therefore, when looking at a problem in which the data is in pressure and the result should be in feet, the answer will be at least twice as large as the starting number. For instance, if the pressure were 25 psi, we intuitively know that the head is over 50 feet. Therefore, we must divide by 0.433 to obtain the correct answer.

Example 2.48

Problem:
    Convert a pressure of 45 psi to feet of head.

Solution:

$$45 \, \frac{psi}{1} \times \frac{1 \text{ ft}}{0.433 \text{ psi}} = 104 \text{ ft}$$

Example 2.49

Problem:
    Convert 15 psi to feet.

Solution:

$$15 \, \frac{psi}{1} \times \frac{1 \text{ ft}}{0.433 \text{ psi}} = 34.6 \text{ ft}$$

Example 2.50

Problem:
    Between the top of a reservoir and the watering point, the elevation is 125 feet. What will the static pressure be at the watering point?

Solution:

$$125 \, \frac{psi}{1} \times \frac{1 \text{ ft}}{0.433 \text{ psi}} = 54.1 \text{ ft}$$

Example 2.51

Problem:

Find the pressure (psi) in a 12-foot-deep tank at a point 5 feet below the water surface.

Solution:

$$Pressure \text{ (psi)} = 0.433 \times 5 \text{ ft}$$
$$= 2.17 \text{ psi}$$

CHAPTER 3

Force and Motion

"I can calculate the movement of the stars, but not the madness of men."

—Sir Isaac Newton

Every one of the measurable quantities that we discuss in this text can be specified in terms of only three basic dimensions: mass, length, and time.

In geoscience we are interested in trying to understand the motion of objects. In starting our discussion of motion, we describe position in two and three dimensions and discuss the motion of objects in only one dimension. Later, after presenting and explaining foundational material, we discuss motion in regard to the real world; that is, we discuss the motion of objects in two dimensions (and three dimensions, for that matter).

Position

Position and time are two fundamental quantities that can be used to describe where an object is, where it is headed, and how long it will take to get there. The position of an object (its location in space; often indicated by the letter x; see Figure 3.1) along a straight line can be uniquely identified (we can measure it) by its distance from an origin (see Figure 3.1). Because the position shown in Figure 3.1 is specified by one coordinate, it is said to be a one-dimensional problem. Some examples of one-dimensional motion include a car moving on a straight road, dropping a pencil, throwing a ball straight up, and many others.

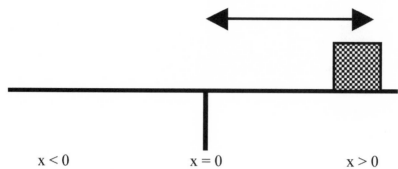

x < 0   x = 0   x > 0

Figure 3.1.   One-Dimensional Position.

Speed and Velocity

In routine conversation, the terms *speed* and *velocity* are often used interchange-ably—they are commonly thought to have the same meaning. In science, however, they are two distinct quantities. *Speed* is a scalar quantity that refers to how fast an object is moving (i.e., the rate of change of distance with time). Thus, if you travel 16 miles in 2 hours, the average speed is 8 miles per hour (mph). Even though most people are confused about the difference between speed and velocity (defined later in this chapter), most people do know the difference between two identical objects traveling at different speeds. Many know, for example, that a person moving faster (the one with the greater speed) will go farther than the one moving slower in the same amount of time. If we do not understand this concept, we certainly know that someone moving faster will get where he or she is going before someone who is moving slower. Intuitively, we all know that speed deals with both distance and time. We know that sooner means less time and that faster means greater distance (farther). Thus, it logically follows that doubling one's speed means doubling one's distance traveled in a given amount of time. Moreover, doubling one's speed would also mean having the time required to travel a given distance.

We can summarize our introductory presentation about speed by the fol-lowing: A fast-moving object has a high speed and covers a relatively large distance in a short amount of time. A slow-moving object has a low speed and covers a relatively small amount of distance in a short amount of time. An object with no movement at all has a zero speed.

✓ **Key Point:** Simply, how fast an object moves is its speed.

Speed is measured in meters per second (m/s), feet per second (ft/s), miles per hour, and knots—a *knot* is 1 nautical mph or 1.15 statute mph. The stan-dard formula for determining speed is:

$$Speed = Distance/Time \qquad (3.1)$$

or

$$s = d/t$$

Example 3.1

Problem:

A truck leaves Norfolk, Virginia, at 11:00 a.m. (EDT) and arrives in Washington, D.C., 200 miles away, at 2:00 p.m. (EDT) the same day. What is the truck's average speed?

Solution:

$$s = d/t$$
$$200 \text{ miles}/3 \text{ hours} = 67 \text{ mph}$$

Velocity is speed with direction. Velocity is a vector quantity that refers to the rate at which an object changes it position. To determine velocity, it is important to know if you are traveling 50 mph due east or 50 mph due south. These same two speeds (but with different velocities) will take you to two very different final locations. Problems in science generally involve velocities (represented by vectors that are simply measured quantities that have both a direction and magnitude [or size]) because the direction of motion is typically an important piece of information.

Again, velocity is a vector quantity and speed is a scalar quantity. Scalar quantities have magnitude, but no direction. Velocity is defined as a change in distance in a given direction divided by a change in time. The Greek letter delta (Δ) is used to represent the concept of change. Thus, the equation

$$\text{Velocity} = \Delta x/\Delta t \qquad (3.2)$$

is read as "velocity equals the change in x divided by the change in t."

To determine how far an object has traveled from an initial position after a set amount of time traveling at constant velocity, we need to derive equation (3.3) from equation (3.2).

$$x = x_o + vt \qquad (3.3)$$

where

x = distance traveled
x_o = initial position
t = a set amount of time
v = constant velocity

Example 3.2

Problem:

If you travel with an annual velocity of 60 mph, you drive for 12 hr on your second travel day, your starting point (x_o) was 400 miles beyond where you were the day before, then your total distance traveled at the end of the second day would be:

Solution:

$$x = (400 \text{ mi}) + (60 \text{ mph}) \times (12 \text{ h})$$
$$x = 1120 \text{ mi}$$

In typical geoscience problems involving the velocity of an object, there are sometimes several velocities involved, and the velocity that results from the sum of all the velocities described (resultant velocity) must be determined. The resultant velocity is the sum of all the velocity vectors. A *vector* is a quantity that has direction as well as magnitude (size). The resultant velocity is the sum of all the velocity vectors.

To gain understanding of vectors and vector math operations, imagine a speedboat crossing a river. If the speedboat were to point its bow straight toward the other side of the river, it would not reach the shore directly across from its starting point (see Figure 3.2). The river current influences the motion of the boat and carries it downstream. The speedboat may be moving with a velocity of 5 m/s directly across the river, yet the resultant velocity of the boat will be greater than 5 m/s and at an angle in the downstream direction. While the speedometer of the speedboat may read 5 m/s, its speed with respect to observers on the shore will be greater than 5 m/s. The resultant velocity of the motor boat can be determined by using vectors. The resultant velocity of the boat is the vector sum of the boat velocity and the river velocity. Because the speedboat heads across the river and

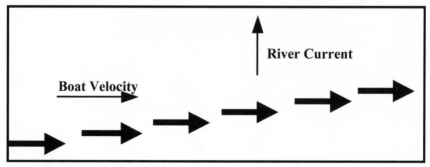

Figure 3.2. Motion of speed boat with current.

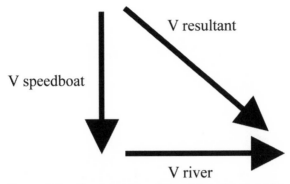

Figure 3.3. The Lengths of the sides of a right triangle are related according to the Pythagorean Theorem.

because the current is always directed straight downstream, the two vectors are at right angles to each other. The lengths of the sides of a right triangle are related by the Pythagorean theorem (see Figure 3.3), which states that in a right triangle the square of the length of the long side (hypotenuse) is equal to the sum of the squares of the other two sides. Generally, the Pythagorean theorem is written as $a^2 + b^2 = c^2$.

In our example of the speedboat crossing the river, the two velocity vectors form a right triangle, so that the resultant velocity can be computed with the formula:

$$v \text{ (resultant)} = (v_1^2 + v_2^2)^{1/2}$$

where v_1 is the velocity of the river, and v_2 is the velocity of the speedboat.

Example 3.3

Problem:

Suppose that the river in our previous example is moving with a velocity of 4 m/s north and the speedboat is moving with a velocity of 5 m/s east. What will be the resultant velocity of the speedboat (i.e., the velocity relative to observers on the shore)? The magnitude of the resultant can be found as follows:

Solution:

$$(5.0 \text{ m/s})^2 + (4.0 \text{ m/s})^2 = R^2$$
$$25 \text{ m}^2/s^2 + 16 \text{ m}^2/s^2 = R^2$$
$$41 \text{ m}^2/s^2 = R^2$$
$$\sqrt{(41 \text{ m}^2/s^2)} = R$$
$$6.4 \text{ m/s} = R$$

Acceleration

When an object's velocity increases, we say it *accelerates*. Acceleration shows the change in velocity in a unit time, or:

$$\text{Acceleration} = \Delta v / \Delta t$$

Because velocity is measured in meters per second (m/s) and time is measured in seconds (s), acceleration is measured in (m/s)/s, m/s^2, which can be both positive and negative and always varying; it is rarely constant.

Accelerations determine the final velocity of an object. To determine an object's final velocity (v), given that it started at some initial velocity (v_o) and experienced acceleration (a) over a period of time (t), use the equation derived from the definition of acceleration:

$$v = v_o + at$$

Example 3.4

Problem:

If a truck starts at rest (with an initial velocity $v_o = 0$) and it accelerates for 5 s at an acceleration rate of 12 m/s^2, then what will be the final velocity?

Solution:

$$v = v_o + at$$
$$v = 0 \text{ m/s} + (12 \text{ m/s}^2) \times (5 \text{ s})$$

or

$$v = 60 \text{ m/s}$$

It is important to note that a change in direction also constitutes acceleration. Remember, our definition of *velocity* is a speed in a given direction. Thus, a change in direction is a change in velocity, and any change in velocity is acceleration. Moreover, whenever we step on the brakes to slow our cars, we are experiencing another kind of acceleration called *deceleration*—a negative acceleration, or slowing down.

THE ACCELERATION OF GRAVITY

On Earth, a free-falling object has an acceleration of 9.8 m/s/s. The acceleration of gravity at the surface of Earth is often referred to as *1g*. Any object that is falling to the surface of Earth owing to the acceleration of gravity is in *free fall.* When the velocity and time for a free-falling object being dropped from a position of rest is tabulated, it displays the following pattern.

Time (s)	Velocity (m/s)
00	
1	9.8
2	19.6
3	29.4
4	39.2
5	49.0

The general acceleration equation is:

$$x = x_o + v_o t + \tfrac{1}{2} at^2$$

where

x = position of object
v_o = initial velocity (original speed)
t = time elapsed
a = constant acceleration

Example 3.5

Problem:
A water balloon is dropped from a five-story building. How long will it take to hit the street below? Ignoring air friction, the only acceleration involved is the acceleration of gravity (g), and the height of the building is 20 m.

Solution:
In the acceleration equation, we use x = 20 m (when the balloon hits the street), x_o = 0 m (we take the balloon's starting point at the top of the five-story building to be zero), v_o = O m/s (the balloon starts from rest), and g = 9.8 m/s² (only gravity is acting on the balloon). Inserting these values into the equation and solving for $t,$ we determine that:

$$x = x_o + v_o t + \tfrac{1}{2} at^2$$

or

$$20 \text{ m} = 0 \text{ m} + 0 \text{ m/s}(t) + 1/2 \ (9.8 \text{ m/s})t^2$$

or

$$40/9.8 \; s^2 = t^2$$
$$t = 2.0 \text{ s}$$

Force

We define *force* as a push or pull from the object's interaction with another object that can cause an object with mass to accelerate. Force has both magnitude (size) and direction, making it a vector quantity. When the interaction between two objects ceases, the objects no longer experience the force. All interactions (forces) between objects can be placed into two categories: contact forces (e.g., friction, normal, applied forces, etc.) and forces resulting from action-at-distance (e.g., magnetic, gravitational, or electrical force). Force is represented in units of newtons (abbreviated N). A *newton* is the force required to accelerate a 1-kg mass at a rate of 1 m/s^2.

Did You Know?

Truth is ever to be found in simplicity, and not in the multiplicity and confusion of things.

Newton's Laws of Motion

To this point in the text, we have presented foundational information important to grasping the concepts put forward by Sir Isaac Newton in his three laws of motion, which changed our understanding of the universe. Through careful observation and experimentation, Newton was able to describe the motion of objects by what are now called *Newton's laws of motion*. These laws include the *law of inertia* (Newton's first law of motion), the *law of constant acceleration* (Newton's second law of motion), and the *law of momentum* (Newton's third law of motion).

NEWTON'S FIRST LAW

According to Newton's first law,

> "An object at rest will remain at rest unless acted on by an unbalanced force. An object in motion continues in motion with the

same speed and in the same direction unless acted upon by an unbalanced force."

This law expresses what we mean when we say that an object has inertia. *Inertia* is that property of matter that causes matter to resist change in motion—it is the natural tendency for objects to keep on doing what they're doing.

✓ **Note**: Mass is a measure of how much inertia an object possesses.

NEWTON'S SECOND LAW

According to Newton's second law,

> "Acceleration is produced when a force acts on a mass. The greater the mass (of the object being accelerated) the greater the amount of force needed (to accelerate the object)."

We all know that heavier objects require more force to move the same distance as lighter objects. The second law does give us, however, an exact relationship between force, mass, and acceleration. It can be expressed as a mathematical equation:

$$F = MA$$

or

$$Force = Mass\ times\ Acceleration = newtons\ (\sim1/4\ pound)$$

Example 3.6

Problem:

A four-wheeled cart weights 1,200 kg and is at rest. A man tries to push the cart to a storeroom, and he makes the cart go 0.05 m/s/s. How much force is the man applying to the cart?

Solution:

$$F = MA$$
$$F = 1,200 \times 0.05$$
$$= 60\ newtons$$

NEWTON'S THIRD LAW

According to Newton's third law,

> "For every action there is an equal and opposite reaction."

Whenever an object pushes another object, it gets pushed back in the opposite direction equally hard.

CHAPTER 4

Work, Energy, and Momentum

Physics does not change the nature of the world it studies,
and no science of behavior can change the essential nature
of man, even though both sciences yield technologies with
a vast power to manipulate their subject matters.

—B. F. Skinner

Work

Work is defined as the product of the net force and the displacement through
which that force is exerted, or $W = F$ (Force) $\times d$ (displacement)—F and d are
vectors, but W is not. This corresponds to our everyday meaning of the word
in that when you lift a grocery bag 4 feet from the floor, you do twice as much
work as when your lift is 2 feet. Or suppose you lift 10 grocery bags to a height
of 3 feet. You do 30 times as much work as lifting one grocery bag that high.
Keep in mind that although work is a scalar quantity, both F and d are vector
quantities that must be in the same direction if they are to be multiplied to
obtain work.

When force is measured in pounds and distance in feet, the unit of work
is foot-pound (ft-lb). In SI units, recall that force is measured in newtons and
distance in meters. The unit of work is the newton-meter, called the *joule* (pronounced *"jewel"*).

To gain better understanding of this information, consider the force's action
on the box shown in Figure 4.1. The forces acting are represented by arrows, and
also show the acceleration of the box as an arrow.

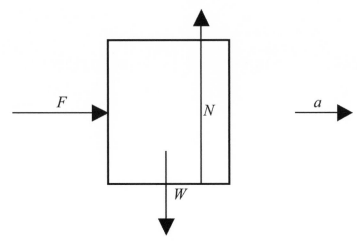

Figure 4.1.   Forces acting on a box.

As shown in Figure 4.1, different forces act on the box.

- W is the weight of the box.
- N is a force on the box arising from contact with the floor.
- F is the constant force applied to the box.
- a represents the acceleration of the box.

Example 4.1

Problem:

A force $F = 20\ N$ pushes a container across a frictionless floor for a distance $d = 10$ m. Determine how much work is done on the box.

Solution:

$$W = F \times d$$
$$= (20\ N) \times (10\ m) = 200\ \text{Joules (J)}$$

Energy

Energy (often defined as the ability to do work) is one of the most discussed topics today because of high prices for hydrocarbon products (gasoline and diesel fuel), electricity, and natural gas. These are all forms of energy that we are quite familiar with, but energy also comes in other forms—heat (thermal), light (radiant), mechanical, and nuclear energy. Energy is in everything. All things we do in life and death (biodegradation requires energy, too) are a result of energy. There are two types of energy—stored (potential) energy and working or moving (kinetic) energy.

POTENTIAL ENERGY

An object can have the ability to do work—to have energy—because of position. For example, a weight suspended high from a scaffold can be made to exert a force when it falls. Because gravity is the ultimate source of this energy, it is correctly called *gravitational potential energy* (GPE; *GPE = Weight × Height*), but we usually refer to this as *potential energy* (PE). Another type of PE is chemical PE—the energy stored in a battery or the gas in a vehicle's gas tank.

Consider Figure 4.2. When the suspended object is released, it will fall on top of the box and crush or squash it, exerting a force on the box over a distance. By multiplying the force exerted on the box by the distance the object falls, we could calculate the amount of work that is done.

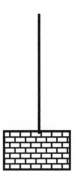

Figure 4.2. A box of bricks (gravitational potential energy, *GPE*) suspended above an empty cardboard box.

KINETIC ENERGY

Moving objects have energy (ability to do work). Kinetic energy (KE) of an object is related to its motion. Figure 4.3 shows the suspended box of bricks we used earlier to demonstrate PE, but now the box of bricks is free-falling—the PE is converted to KE because of movement. Specifically, KE of an object is defined as half its mass times its velocity squared, or:

$$KE = \tfrac{1}{2}\, mv^2$$

From this equation, it is apparent that the more massive an object, and the faster it is moving, the more KE it possesses. The units of KE are determined by taking the product of the units for mass (kg) and velocity squared (m²/s²)—the units of KE, like PE, are joules. KE can never be negative and only tells us about speed, not velocity.

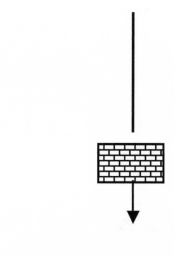

Figure 4.3.   Free-falling box of bricks (kinetic energy, KE) suspended above an empty cardboard box.

Momentum

An object in motion possesses *momentum*. Momentum, abbreviated with the letter **p**, is equal to the product of mass and velocity, $p = mv$. Thus, the SI unit of momentum is kg · m/s. Momentum has direction; therefore, it is a vector quantity. Newton's third law—for every action there is an equal and opposite reaction—is often referred to as the *law of conservation of momentum*.

The standard well-worn but appropriate example of someone jumping from a rowboat into a body of water to swim and cool off helps to explain momentum and the law of conservation. First we write total momentum as:

$$P_{total} = 0 = m_1 v_1 + m_2 v_2$$

or

$$m_1 v_1 = -m_2 v_2$$

and then designating the two objects as the boat, m_1, and the swimmer, m_2, we observe that each of these will move in opposite directions. If the mass of the boat (m_1) is 150 kg and the mass of the swimmer (m_2) is 50 kg, then the velocity of the boat (v_1) will be

$$v_1 = (-m_2/m_1) \, v_2$$

or

$$v_1 = -0.33 \, v_2$$

indicating that the boat will move away from the swimmer at one-third the speed of the swimmer swimming away from the boat. Keep in mind that if some force other than the swimmer jumping from the boat had caused the boat to move, the principle of conservation of momentum would not apply because the application of the principle of conservation of momentum is limited to systems isolated from other forces (basically, it must be a closed system).

CHAPTER 5

Circular Motion and Gravity

Kepler's First Law: All planets travel in elliptical orbits with the Sun at one focus.

Kepler's Second Law: Each planet travels in such a way that a line joining the planet and the Sun sweeps equal areas in equal times.

Kepler's Third Law: The square of the period is proportional to its distance from the Sun, squared. Stated differently, the farther a planet is from the Sun, the longer it takes to go around the Sun.

Circular motion, or more generally, *angular motion,* can include race cars whizzing around a track, rockets moving around planets, bees buzzing around a hive, children on a merry-go-round, a yo-yo moving up and down, planets orbiting the Sun, and the propeller of an airplane spinning. In the previous chapters we discussed concepts like *displacement, velocity,* and *acceleration* and how they relate to linear motion; now we describe the terms needed to describe and predict angular motion. Before discussing angular motions, qualitatively and quantitatively (and to aid in understanding the material presented later) a comparison of linear and angular motion terms is presented in Table 5.1.

Table 5.1. Terms Used for Linear and Angular Motion

Linear Motion	Units	Angular Motion	Units
Distance (x)	m	Angle (θ)	rad, °
Velocity (v)	m/s	Angular velocity (ω)	rad/s, or 1/s
Acceleration (α)	m/s^2	Angular acceleration (α)	rad/s^2 or 1/s^2
Force (F)	N	Torque (τ)	$N \cdot m$
$F = ma$	N	$N\tau = I\alpha$	$N \cdot m$
Mass (m)	kg	Moment of inertia (I)	$kg \cdot m^2$

Angular Motion

Earlier, when we discussed linear motion, we pointed out that linear distances are measured in meters and in other linear units of measure. These linear measurements were displacements from an arbitrary zero-point. Angular motion is measured differently. The different ways angular motion or angular distance is measured are shown in Figure 5.1 and explained later in this chapter.

Suppose the circular objects shown in Figure 5.1 are axle-mounted wheels wherein the top wheel is rotated through θ as shown. This rotation can be measured in three different ways. As you might expect, the two most common ways, though arbitrary, to measure the angle (θ) are with revolutions (one full rotation is a revolution) and degree units (one full rotation is defined as 360°). These are related by:

$$1 \text{ rev} = 360$$

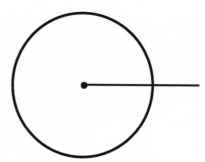

Figure 5.1.   Angular distance θ.

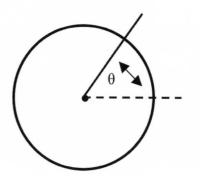

The third way to measure θ is not arbitrary. It is an angular measure scaled to the circle called a radian (rad). This is shown in Figure 5.2.

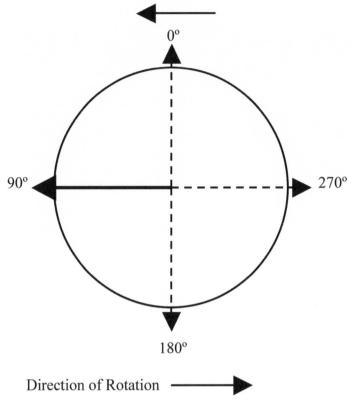

Figure 5.2. Degrees and radians in circular motion: 90° = π/2 radians; 360° = 2π radians.

Angular Velocity

Angular velocity is a measure of the rate of change in angular position. In an average day, we are exposed to many different examples of angular velocity. Consider, for example, the tachometer in your automobile. The automobile tachometer measures the number of revolutions per minute (rev/min) of the engine's crankshaft. When the auto is not moving (idling), the tachometer reads somewhere around 600–1000 rev/min. Actually, the car engine's crankshaft is idling but not stopped; it is revolving at what is known as *angular speed* (the number of revolutions per minute that the crankshaft is making). When you step on the gas pedal (accelerator) you feed more gas to the engine, causing the crankshaft to rotate at a higher angular velocity—the motion of the crankshaft is transferred to the wheels of the car manually or automatically by the transmission.

The average angular velocity of a rotating object is defined to be the angular distance divided by the time taken to turn through this angle, as defined in the following equation.

$$\text{Average angular velocity } (\omega\text{—overline}) = \theta/t$$

Where θ is the angle through which an object rotates in time, t. The units for angular velocity (ω—Greek letter "omega") are those of an angle divided by a time—units might be degrees per second, revolutions per minute, or radians per second.

Angular Acceleration

When a rotating object's angular velocity changes—it speeds up or slows down—it signifies the presence of angular acceleration. Because angular accelerations are changes in velocity, the units of angular acceleration are radians per second per second. Because radians are dimensionless units, angular acceleration is measured in units of $1/s^2$. Like angular velocity, ω, *angular acceleration, α,* (Greek "alpha") is a vector, meaning it has a magnitude and a direction. Angular acceleration is the rate of change of angular speed:

$$\alpha = \Delta\omega /\Delta t$$

Torque

Torque is a vector that measures the tendency of a force to rotate (or twist) an object about some axis. The SI unit for torque is newton meters (Nm). In U.S. customary units, it is measured in foot pounds (ft-lbs). The symbol for torque is τ, the Greek letter *tau.* Torque can be described simply as the application of a force at some distance from a pivot point (see Figure 5.3), or

$$\textit{Torque} = \textit{Force} \times \textit{Radius}$$

Or

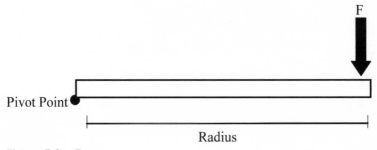

Figure 5.3. Torque.

Angular Momentum

The *angular momentum* of an object around a certain point is a vector quantity equal to the product of the distance from a point and its momentum measured with respect to the point. In equation form, we say that an object with a mass m moving in a circle of radius r at a velocity v has an angular momentum of

$$L = mvr$$

Gravity

Gravity refers to the force that is the cause of the attraction between objects. Newton, using foundational scientific explanations provided earlier by Kepler and Brahe, was the first to describe the force that makes the planets move in elliptical paths around the Sun. Newton claimed that the same physical laws that can explain how things move here on Earth can explain the motions of the planets. Newton asserted that all objects with mass exert an attractive force on all other objects with mass. Newton further posited that this force is directly proportional to the product of the masses of the two objects and inversely proportional to their distance squared.

Newton's law of gravity can be stated as follows:

$$F \sim \frac{m_1 \cdot m_2}{d^2}$$

The symbol $\sim$ means "is proportional to"; m_1 symbolizes the mass of one of the objects, and m_2 is the mass of the other; d is the distance between the centers of the objects.

In addition to the obvious attractive effects of gravity on the planets in the universe, upon earth and all objects with mass, in geoscience, gravity plays an important role in shaping Earth's structure; it is a main force of erosion. Whether it is a rock falling from a cliff, downhill creep, or massive landslide (mass wasting), gravity is the main force at work. Think of mass wasting as the ongoing process that works to reduce the highest levels on earth (mountains) to the lowest levels (ocean basins). We discuss gravity and mass wasting in greater detail later in the text.

CHAPTER 6

Thermal Properties and States of Matter

> This grand show is eternal. It is always sunrise somewhere; the dew is never all dried at once; a shower is forever falling; vapor is ever rising. Eternal sunrise, eternal sunset, eternal dawn and gloaming, on sea and continents and islands, each in its turn, as the round earth rolls.
>
> —John Muir

Thermal Properties

Thermal properties of chemicals and other substances are important in geoscience. Such knowledge is used in math calculations, in the study of the physical properties of materials, in hazardous materials spill mitigation, and in solving many other complex environmental problems. *Heat* is a form of energy—thermal energy that can be transferred between two bodies that are at different temperatures. Whenever work is performed, usually a substantial amount of heat is caused by friction. The conservation of energy law tells us the work done plus the heat energy produced must equal the original amount of energy available. That is:

$$\text{Total energy} = \text{work done} + \text{heat produced} \qquad (6.1)$$

A traditional unit for measuring heat energy is the *calorie* (cal), which is defined as the amount of heat necessary to raise one gram of pure liquid water by 1 degree Celsius at normal atmospheric pressure. In SI units

$$1 \text{ cal} = 4.186 \text{ J (Joule)}$$

The calorie we have defined should not be confused with the one used when discussing diets and nutrition. A kilocalorie is 1,000 calories as we have defined it—the amount of heat necessary to raise the temperature of one kilogram of water by 1° C.

In the British system of units, the unit of heat is the British thermal unit (BTU). One BTU is the amount of heat required to raise 1 pound of water 1 degree Fahrenheit at normal atmospheric pressure (1 atm).

Heat can be transferred in three ways: by conduction, convection, and radiation. When direct contact between two physical objects at different temperatures occurs, heat is transferred via **conduction** from the hotter object to the colder one. When a gas or liquid is placed between two solid objects, heat is transferred by **convection.** Heat is also transferred when no physical medium exists by **radiation** (for example, radiant energy from the sun).

Specific Heat

Earlier we pointed out that 1 kilocalorie of heat is necessary to raise the temperature of 1 kilogram of water 1 degree Celsius. Other substances require different amounts of heat to raise the temperature of 1 kilogram of the substance 1 degree. The *specific heat* of a substance is the amount of heat in kilocalories necessary to raise the temperature of 1 kilogram of the substance 1 degree Celsius.

The units of specific heat are Kcal/kg° C or, in SI units, J/kg° C. The specific heat of pure water, for example, is 1.000 kcal/kg° C, or 4186 J/kg° C.

The greater the specific heat of a material, the more heat is required. Also, the greater the mass of the material or the greater the temperature change desired, the more heat is required. If the specific heat of a substance is known, it is possible to calculate the amount of heat required to raise the temperature of that substance. In general, the amount of heat required to change the temperature of a substance is proportional to the mass of the substance and the change in temperature, according to the following relationship:

$$Q = mc\,\Delta T \tag{6.2}$$

where

Q = heat required
m = mass of the substance
c = specific heat of the substance
ΔT = change in temperature

The amount of heat necessary to change 1 kilogram of a solid into a liquid at the same temperature is called the *latent heat of fusion* of the substance. The temperature of the substance at which this change from solid to liquid takes place is known as the *melting point.* The amount of heat necessary to change 1 kilogram of a liquid into a gas is called the *latent heat of vaporization.* When this point is reached, the entire mass of substance is in the gas state. The temperature of the substance at which this change from liquid to gas occurs is known as the *boiling point.*

To compare the effect of different specific heats of different materials, one of the standard lab tests we have conducted in our environmental science and health undergraduate college courses is to heat equal amounts of different materials in boiling water. Then the students place those materials on squares of wax to see which heated material melts deeper into the wax. The students then measure the deepness of the melting; the deeper the melting, the more difficult the material is to cool, indicating which material has the higher specific heat.

States of Matter

The three common states (or phases) of matter (solid, liquid, gaseous) each have unique characteristics. In the *solid* state, the molecules or atoms are in a relatively fixed position. The molecules are vibrating rapidly, but about a fixed point. Because of this definite position of the molecules, a solid holds its shape. A solid occupies a definite amount of space and has a fixed shape.

When the temperature of a gas is lowered, the molecules of the gas slow down. If the gas is cooled sufficiently, the molecules slow down so much that they lose the energy needed to move rapidly throughout their container. The gas may turn into liquid. Common liquids are water, oil, and gasoline. A *liquid* is a material that occupies a definite amount of space, but that takes the shape of the container.

In some materials, the atoms or molecules have no special arrangement at all. Such materials are called *gases.* Oxygen, carbon dioxide, and nitrogen are common gases. A *gas* is a material that takes the exact volume and shape of its container.

Although the three states of matter discussed previously are familiar to most people, the change from one state to another is of primary interest to environmentalists. Changes in matter that include water vapor changing from the gaseous state to liquid precipitation, or a spilled liquid chemical changed to a semisolid substance by addition of chemicals, which aids in the cleanup effort, are two ways that changing from one state to another affects environmental concerns.

Did You Know?

During the phase change in water from solid to liquid (melting), 80 calories of heat are required for every gram. To boil into gas, liquid water requires 540 calories of that for every gram. It is interesting to note that during a phase change, the temperature will not change—no matter how hot the flame is under the container of water being heated.

The Gas Laws

The atmosphere is composed of a mixture of gases, the most abundant of which are nitrogen, oxygen, argon, carbon dioxide, and water vapor (gases and the atmosphere are addressed in greater detail later). The pressure of a gas is the force that the moving gas molecules exert on a unit area. A common unit of pressure is newton per square meter, N/m_2, called a pascal (Pa). An important relationship exists among the pressure, volume, and temperature of a gas. This relation is known as the *ideal gas law* and can be stated as:

$$PV = nRT$$

where P_1, V_1, and T_1 are pressure, volume, and absolute temperature at time 1, and P_2, V_2, and T_2 are pressure, volume, and absolute temperature at time 2. A gas is called *perfect* or *ideal*, when it obeys this law.

A temperature of 0° C (273 K) and a pressure of 1 atm have been chosen as *standard temperature and pressure (STP)*. At STP, the volume of 1 mole of ideal gas is 22.4 L.

Liquids and Solutions

The most common solutions are liquids. However, solutions, which are homogenous mixtures, can be solid, gaseous, or liquid. The substance in excess in a solution is called the *solvent*. The substance dissolved is the *solute*. Solutions in which water is the solvent are called *aqueous solutions*. A solution in which the solute is present in only a small amount is called a *dilute solution*. If the solute is present in large amounts, the solution is concentrated solution. When the maximum amount of solute possible is dissolved in the solvent, the solution is called a saturated solution.

The concentration (the amount of solute dissolved) is frequently expressed in terms of the molar concentration. The molar concentration, or *molarity*, is the number of moles of solute per liter of solution. Thus, a 1-molar solution, written *1.0 M,* has 1-gm formula weight of solute dissolved in 1 liter of solution. In general,

$$Molarity = \frac{moles\ of\ solute}{number\ of\ liters\ of\ solution} \qquad (6.4)$$

Note that the *number of liters of solution,* not the number of liters of solvent, is used.

Example: Exactly 40 gm of sodium chloride (NaCl), or table salt, were dissolved in water and the solution was made up to a volume 0.80 liter of solution. What was the molar concentration, M, of sodium chloride in the resulting solution?

Answer: First find the number of moles of salt.

$$Number\ of\ moles = \frac{40\ gm}{58.5\ gm/mole} = 0.68 \qquad (6.5)$$

$$Molarity = \frac{0.68\ mole}{0.80\ liter} = 0.85\ m$$

Part II

EARTH'S BREATH

CHAPTER 7

The Atmosphere

> This most excellent canopy, the air, look you, this brave
> o'erhanging firmament, this majestical roof fretted with
> golden fire. . . .
>
> —Shakespeare, *Hamlet*

> It suddenly struck me that that tiny pea, pretty and blue,
> was the earth. I put up my thumb and shut one eye, and
> my thumb blotted out the planet Earth. I didn't feel like a
> giant. I felt very, very small.
>
> —Neil Armstrong

Several theories of cosmogony attempt to explain the origin of the universe. Without speculating on the validity of any one theory, the following is simply the authors' view.

The time: 4,500 million years ago.

Before the universe there was time. Only time; otherwise, the vast void held only darkness everywhere.

Overwhelming darkness.

Not dim, not murky, not shadowy, not unlit. Simple nothingness—nothing but darkness, a shade of black so intense we cannot fathom or imagine it today. Light had no existence —this was the black of blindness, of burial in the bowels of the earth, the blackness of no other choice.

With time, eons of time, darkness came to a sudden, smashing, shattering, annihilating, scintillating, cataclysmic end, and there was light—light everywhere. This new force replaced darkness and lit up the expanse without end, creating a brightness fed by billions of glowing round masses so powerful as to renounce and overcome the darkness that had come before.

With the light was heat-energy that shone and warmed and transformed into mega-mega-mega trillions of super-excited ions, molecules, and atoms—heat of unimaginable proportions, forming gases, gases we don't even know how to describe, how to quantify, let alone how to name. But gases they were, and they were everywhere.

With light, energy, heat, and gases present, the stage was set for the greatest show of all time, anywhere—ever: the formation of the universe.

Over time—time in stretches we cannot imagine, so vast we cannot contemplate them meaningfully—the heat, the light, the energy, and the gases all came together and grew, like an expanding balloon, into one solid, glowing mass. But it continued to grow, with the pangs, sweating, and moans accompanying any birthing, until it had reached the point of no return: explosion level. And it did; it exploded with the biggest bang of all time.

The Big Bang sent masses of hot gases in all directions, to the farthest reaches of anything, everything, into the vast, wide, measureless void. Clinging together as they rocketed, soared, and swirled, forming galaxies that gradually settled into their arcs through the void, constantly propelled away from the force of their origin, these masses began their eternal evolution.

Two masses concern us: the Sun and Earth.

Forces well beyond the power of the Sun (beyond anything imaginable) stationed this massive gaseous orb approximately 93,000,000 miles from the dense molten core enveloped in cosmic gases and the dust of time that eventually became the insignificant mass we now call *Earth*.

Distant from the Sun, Earth's mass began to cool, slowly; the progress was slower than we can imagine, but cool it did. While the dust and gases cooled, Earth's inner core, mantle, and crust began to form—no more a quiet or calm evolution than the revolution that cast it into the void had been.

This transformation was downright violent. The cooling surface was only a facade for the internal machinations going on inside, out-gassing from huge, deep, destructive vents (we would call them *volcanoes* today) erupting continuously, never stopping, blasting away, delivering two main ingredients: magma and gas.

The magma worked to form the primitive features of Earth's early crust. The gases worked to form Earth's initial atmosphere, which is our point of interest. Without atmosphere, what is there?

About 4 billion years before the present, Earth's early atmosphere was chemically reducing, consisting primarily of methane, ammonia, water vapor, and hydrogen—for life as we know it today, an inhospitable brew.

Earth's initial atmosphere was not a calm, quiet, quiescent environment. To the contrary, it was an environment best characterized as dynamic and ever changing. Bombardment after bombardment by intense, bond-breaking, ultraviolet (UV) light, along with intense lightning and radiation from radionuclides, provided energy to bring about chemical reactions that resulted in the production of relatively complicated molecules, including amino acids and sugars, the building blocks of life.

About 3.5 billion years before the present, primitive life formed in two radically different theaters: on dry land and below the primordial seas near hydrothermal vents that spotted the wavering, water-covered floor. Initially, on Earth's unstable surface, these very primitive life forms derived their energy from fermentation of organic matter formed by chemical and photochemical processes, then gained the ability to produce organic matter (CH_2O) by photosynthesis. Thus, the stage was set for the massive biochemical transformation that resulted in the production of almost all the atmosphere's oxygen (O_2).

The oxygen initially produced was quite toxic to primitive life forms. However, much of this oxygen was converted to iron oxides by reaction with soluble iron. This process formed enormous deposits of iron oxides, the existence of which provides convincing evidence for the liberation of oxygen in the primitive atmosphere.

Eventually, enzyme systems developed that enabled organisms to mediate the reaction of waste-product oxygen with oxidizable organic matter in the sea. Later, the mode of waste gradient disposal was utilized by organisms to produce energy by respiration, which is now the mechanism by which nonphotosynthetic organisms obtain energy. In time, oxygen accumulated in the atmosphere. In addition to providing an abundant source of oxygen for respiration, the accumulated atmospheric oxygen formed an ozone (O_3) shield—the ozone shield absorbs bond-rupturing UV radiation.

With the ozone shield protecting tissue from destruction by high-energy UV radiation, the Earth, although still hostile to life forms we are familiar with, became a much more hospitable environment for life (self-replacing molecules), and life forms were enabled to move from the sea, where they flourished next to the hydrothermal gas vents, to the land. And from that point to the present, Earth's atmosphere became more life-form friendly.

Earth's Overhanging Roof

Shakespeare likened the sky to a majestic overhanging roof, constituting the transition between its surface and the vacuum of space; others have likened it to the skin of an apple. Both these descriptions of our atmosphere are fitting, as are descriptions of the Earth's envelope, veil, or gaseous shroud. The atmosphere is more like the apple skin, however. This thin skin, or layer, contains the life-sustaining 21% oxygen required by all humans and many other life forms; the carbon dioxide (0.03%) so essential for plant growth; the nitrogen (78%) needed for chemical conversion to plant nutrients; the trace gases such as methane, argon, helium, krypton, neon, xenon, ozone, and hydrogen; and varying amounts of water vapor and airborne particulate matter. Life on earth is supported by this atmosphere, solar energy, and other planets' magnetic fields.

Gravity holds about half the weight of a fairly uniform mixture of these gases in the lower 18,000 feet of the atmosphere; approximately 98% of the material in the atmosphere is below 100,000 feet.

Atmospheric pressure varies from 1,000 millibars (mb) at sea level to 10 mb 100,000 feet. From 100,000 to 200,000 feet, the pressure drops from 9.9 mb to 0.1 mb and so on.

The atmosphere is considered to have a thickness of 40–50 miles; however, here we are primarily concerned with the troposphere, the part of the earth's atmosphere that extends from the surface to a height of about 27,000 feet above the poles, about 36,000 feet in midlatitudes, and about 53,000 ft over the equator (Figure 7.1). Above the troposphere is the stratosphere, a region that increases in temperature with altitude (the warming is caused by absorption of the sun's radiation by ozone) until it reaches its upper limit of 260,000 feet.

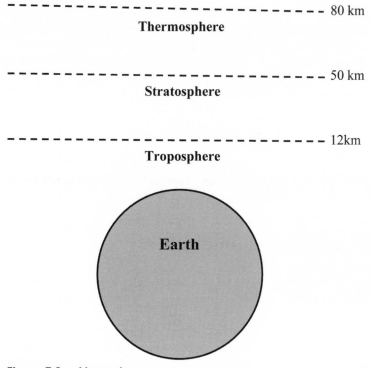

Figure 7.1. Atmosphere.

The Troposphere

Extending above earth approximately 27,000 feet, the troposphere is the focus of this text because people, plants, animals, and insects live here and depend on this thin layer of gases. Moreover, all of the Earth's weather takes place within the troposphere. The troposphere begins at ground level and extends 7.5 miles up into the sky where it meets with the second layer called the *stratosphere.*

Did You Know?

It was pointed out earlier that the gases that are so important to life on Earth are primarily contained in the troposphere. Also note that another important substance is contained in the troposphere: water vapor. Along with being the most remarkable of the trace gases contained in the troposphere, water vapor is also the most variable. Unlike the other trace gases in the atmosphere, water vapor alone exists in gas, solid, and liquid forms. It also functions to add and remove heat from the air when it changes form one form to another. Water vapor (in conjunction with airborne particles) is essential for the stability of earth's ecosystem. This water vapor–particle combination interacts with the global circulation of the atmosphere and produces the world's weather, including clouds and precipitation.

The Stratosphere

The stratosphere begins at the 7.2-mile point and reaches 21.1 miles into the sky. In the rarified air of the stratosphere, the significant gas is O_3 (life-protecting ozone, not to be confused with pollutant ozone), which is produced by the intense UV radiation from the Sun. In quantity, the total amount of ozone in the atmosphere is so small that if it were compressed to a liquid layer over the globe at sea level, it would have a thickness of less than 3/16 inch.

Did You Know?

The troposphere, stratosphere, mesosphere, and thermosphere act together as a giant safety blanket. They keep the temperature on the earth's surface from dipping to extreme icy cold that would freeze everything solid, or from soaring to blazing heat that would burn up all life.

Ozone contained in the stratosphere can also impact (add to) ozone in the troposphere. Normally, the troposphere contains about 20 parts per billion parts of ozone. On occasion, however, via the jet stream, this concentration can increase to 5–10 times higher than average.

"Revolutionary" Science

Today, almost every elementary school child can explain in basic terms the composition of air and water. Most young children understand that the air we breathe contains oxygen, nitrogen, and other gases. There was a time just a few hundred years ago, however, when the actual composition of air and water was nothing more than speculation—a mystery.

The French aristocrat Antoine Lavoisier (1743–1794) is universally regarded as the founder of modern chemistry. This lofty title was bestowed on Lavoisier for his great experiments and discoveries related to the major components that make up air (oxygen and nitrogen) and to a lesser degree for identifying the components of water (hydrogen and oxygen).

Most of Lavoisier's experiments and discoveries took place in the years just preceding the French Revolution. And even though he ranks with the great scientists of his time, Lavoisier was found guilty of trumped-up charges and guillotined during the French Revolution in 1794. Joseph Lagrange (1736–1813), the great French mathematician, said, "It required only a moment to sever his head, and probably one hundred years will not suffice to produce another like it."

Lagrange's eulogy concerning Lavoisier and his scientific accomplishments is quite fitting. You might ask why? What was so difficult about discovering the basic components of air, water, and the oxygen theory of combustion? Note that which seems so simple and elementary to us today was not so clear 200 years ago. Indeed, at that time Lavoisier's discoveries were quite difficult to arrive at. We must remember that in Lavoisier's time, so-called chemists had no clear idea of what a chemical element was, nor any understanding of the nature of gases.

Lavoisier's discoveries were built on the works of others who preceded him or who were working on similar experiments during his lifetime. Lavoisier's work also provided a foundation for scientific discoveries that followed. For example, Lavoisier experimented with the findings of the German chemist Georg Stahl (1660–1743) and disproved them. Stahl proposed a theory that a combustible material burned because it contained a substance called *phlogiston* (charcoal is a prime example). Stahl knew that metallurgists obtained some metals from their ores by heating them with charcoal, which seemed to support the phlogiston theory of combustion. However, in 1774, Lavoisier with the help of Joseph Priestley (1773–1804) proved that the phlogiston theory was wrong. Priestly heated a clax (in this particular case, the burned residue of oxide of mercury) in a closed apparatus and collected the gas liberated in the process. Priestley discovered that this gas supported combustion better than air.

Lavoisier repeated Priestley's experiments. He convinced himself of the presence in air of a gas that combined with substances when they burn and that it was the same gas given off when the oxide of mercury was heated. Thus, he proved that when a substance burned it combined with the oxygen in the air. He named this gas *oxygine,* or "acid former" (from the Greek), because he believed all acids contained oxygen.

In the meantime, Lavoisier had identified the other main component of air, nitrogen, which he named *azote,* from the Greek for "no life." He also demonstrated that when hydrogen, which chemists of the day called *inflammable air,* was burned with oxygen, water was formed.

Lavoisier restructured chemistry and gave it its modern form. His work provided a firm foundation for the atomic theory proposed by British chemist and physicist John Dalton, and his elements were later classified in the periodic table. Lavoisier's work set the stage for the discovery of the other gaseous constituents in air made later by other scientists.

✓ **Interesting Point:** Justus von Liebig, in *Letters on Chemistry*, No. 3, has this to say about Lavoisier:

> He discovered no new body, no new property, no natural phenomenon previously unknown; but all the facts established by him were the necessary consequences of the labors of those who preceded him. His merit, his immortal glory, consists in this—that he infused into the body of the science a new spirit; but all the members of that body were already in existence, and rightly joined together.

With nitrogen and oxygen already identified as the primary constituents in air, and later carbon dioxide, water vapor, helium, ozone, and particulate matter, it was some time before the other gaseous constituents were identified. Argon was discovered in 1894 by British chemists John Rayleigh and William Ramsay after all oxygen and nitrogen had been removed chemically from a sample of air. Ramsay, along with Englishman Morris Travers, discovered neon. They also discovered krypton and xenon in 1889.

Components of Atmospheric Air: Characteristics and Properties

It was pointed out that air is a combination of component parts: gases (see Table 7.1) and other matter (suspended minute liquid or particulate matter). In this section we discuss each of these components.

✔ **Note:** Much of the information pertaining to atmospheric gases that follows was adapted from the Compressed Gas Association's *Handbook of Compressed Gases* (1990) and *Environmental Science and Technology: Concepts and Applications* (2006).

ATMOSPHERIC NITROGEN

Nitrogen (N_2) makes up the major portion of the atmosphere (78.03% by volume, 75.5% by weight). It is a colorless, odorless, tasteless, nontoxic, and almost totally inert gas. Nitrogen is nonflammable, will not support combustion, and cannot support life. The obvious question becomes, if gaseous nitrogen does not support life, what is it doing in our atmosphere? What good is it? That is a logical question. In fact, nitrogen is indeed "good." Without nitrogen, we could not survive.

Nitrogen is part of earth's atmosphere primarily because, over time, it has simply accumulated in the atmosphere and remained in place and in balance. This nitrogen accumulation process has occurred because, chemically, nitrogen is not very reactive. When released by any process, it tends not to recombine with other elements and accumulates in the atmosphere. And this is a good thing, because we need nitrogen for life-sustaining processes other than breathing.

Let's take a look at a couple of reasons why gaseous nitrogen is so important to us. Although nitrogen in its gaseous form is of little use to us, after oxygen, carbon, and hydrogen, it is the most common element in living tissues. As a chief constituent of chlorophyll, amino acids, and nucleic acids—the "building blocks" of proteins, which are used as structural components in cells—nitrogen is essential to life. Nitrogen is dissolved in and is carried by the blood. Nitrogen does not appear to enter into any chemical combination as it is carried throughout the body. Each time we breathe, the same amount of nitrogen is exhaled as is inhaled. Animals cannot use nitrogen directly but only when it is obtained by eating plant or animal tissues; plants obtain the nitrogen they need when it is in the form of inorganic compounds, principally nitrate and ammonium. Gaseous nitrogen is converted to a form usable by plants (nitrate ions) chiefly through the process of nitrogen fixation via the nitrogen cycle, shown in simplified form in Figure 7.2.

Via the *nitrogen cycle,* aerial nitrogen is converted into nitrates mainly by microorganisms, bacteria, and blue-green algae. Lightning also converts some aerial nitrogen gas into forms that return to the earth as nitrate ions in rainfall and other types of precipitation. From Figure 7.2 it can be seen that ammonia plays a major role in the nitrogen cycle. Excretion by animals and anaerobic decomposition of dead organic matter by bacteria produce ammonia. Ammonia, in turn, is converted by *nitrite bacteria* (*Nitrosococcus* and *Nitrosomonas*), which are aerobic, into nitrites and then into nitrates. This process is known as *nitrification*. Although nitrite is toxic to many plants, it usually does not accumulate in the soil. Instead, other bacteria (such as *Nitrobacter*) oxidize the nitrite to form nitrate (NO_3^-), the most common biologically usable form of nitrogen.

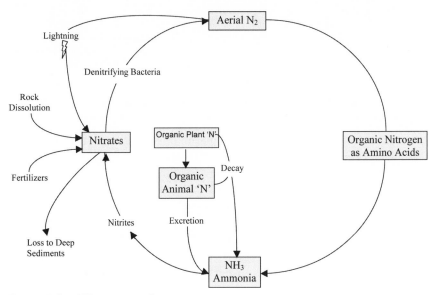

Figure 7.2. Nitrogen cycle.

Nitrogen reenters the atmosphere through the action of denitrifying bacteria, which are found in nutrient-rich habitats such as marshes and swamps. These bacteria break down nitrates into nitrogen gas and nitrous oxide (N_2O), which then reenter the atmosphere. Nitrogen also reenters the atmosphere from exposed nitrate deposits, emissions from electric power plants and automobiles, and from volcanoes.

Nitrogen: Physical Properties

The physical properties of nitrogen are noted in Table 7.1.

Nitrogen: Uses

In addition to being the preeminent (in regard to volume) component of Earth's atmosphere and providing an essential ingredient in sustaining life, nitrogen gas

Table 7.1.   Nitrogen: Physical Properties

Chemical Formula	N_2
Molecular weight	28.01
Density of gas at 70° F	0.072 lb/ft^3
Specific gravity of gas at 70° F & 1 atm (air = 1)	0.967
Specific volume of gas at 70° F & 1 atm	13.89 ft^3
Boiling point at 1 atm	–320.4° F
Melting point at 1 atm	–345.8° F
Critical temperature	–232.4° F
Critical pressure	493 psia
Critical density	19.60 lb/ft^3
Latent heat of vaporization at boiling point	85.6 BTU/lb
Latent heat of fusion at melting point	11.1 BTU/lb

atm, Atmosphere; *BTU*, British thermal unit; *psia*, pounds per square inch absolute.

has many commercial and technical applications. As a gas, it is used in heat-treating primary metals; the production of semiconductor electronic components, as a blanketing atmosphere; blanketing of oxygen-sensitive liquids and of volatile liquid chemicals; inhibition of aerobic bacteria growth; and the propulsion of liquids through canisters, cylinders, and pipelines.

Nitrogen Oxides

There are six oxides of nitrogen: nitrous oxide (N_2O), nitric oxide (NO), dinitrogen trioxide (N_2O_3), nitrogen dioxide (NO_2), dinitrogen tetroxide (N_2O_4), and dinitrogen pentoxide (N_2O_5). Nitric oxide, nitrogen dioxide, and nitrogen tetroxide are fire gases. One or more is generated when certain nitrogenous organic compounds (polyurethane) burn. Nitric oxide is the product of incomplete combustion, whereas a mixture of nitrogen dioxide and nitrogen tetroxide is the product of complete combustion.

The nitrogen oxides are usually collectively symbolized by the formula NO_x. The U.S. Environmental Protection Agency, under the Clean Air Act, regulates the amount of nitrogen oxides that commercial and industrial facilities may emit to the atmosphere. The primary and secondary standards are the same: The annual concentration of nitrogen dioxide may not exceed 100 µg/m^3 (0.05 parts per million [ppm]).

ATMOSPHERIC OXYGEN

Oxygen (O_2; Greek *oxys*—"acid" and *genes*—"forming") constitutes approximately one-fifth (21% by volume and 23.2% by weight) of the air in Earth's atmosphere. Gaseous oxygen is vital to life as we know it. On Earth, oxygen is the most abundant element. Most oxygen on Earth is not found in the free state, but in combination with other elements as chemical compounds. Water and carbon dioxide are common examples of compounds that contain oxygen, but there are countless others.

At ordinary temperatures, oxygen is a colorless, odorless, tasteless gas that not only supports life but also combustion. All the elements except the inert gases combine directly with oxygen to form oxides. However, oxidation of different elements occurs over a wide range of temperatures.

Oxygen is nonflammable but it readily supports combustion. All materials that are flammable in air burn more vigorously in oxygen. Some combustibles, such as oil and grease, burn with nearly explosive violence in oxygen if ignited.

Oxygen: Physical Properties

The physical properties of oxygen are noted in Table 7.2.

Table 7.2. Oxygen: Physical Properties

Chemical Formula	O_2
Molecular weight	31.9988
Freezing point	–361.12° F
Boiling point	–297.33° F
Heat of fusion	5.95 BTU/lb
Heat of vaporization	91.70 BTU/lb
Density of gas at boiling point	0.268 lb/ft³
Density of gas at room temperature	0.081 lb/ft³
Vapor density (air = 1)	1.105
Liquid-to-gas expansion ratio	875

BTU, British thermal unit.

Oxygen: Uses

The major uses of oxygen stem from its life-sustaining and combustion-supporting properties. It also has many industrial applications (when used with other fuel gases such as acetylene), including metal cutting, welding, hardening, and scarfing.

Ozone: Just Another Form of Oxygen

Ozone (O_3) is a highly reactive pale-blue gas with a penetrating odor. Ozone is an allotropic modification of oxygen. An allotrope is a variation of an element that possesses a set of physical and chemical properties significantly different from the "normal" form of the element. Only a few elements have allotropic forms; oxygen, phosphorous, and sulfur are some of them. Ozone is just another form of oxygen. It is formed when the molecule of the stable form of oxygen is split by UV radiation or electrical discharge; it has three atoms of oxygen per molecule instead of two. Thus, its chemical formula is represented by O_3.

Ozone forms a thin layer in the upper atmosphere, which protects life on Earth from UV rays, a cause of skin cancer. At lower atmospheric levels, it is an air pollutant and contributes to the greenhouse effect. At ground level, ozone, when inhaled, can cause asthma attacks, stunted growth in plants, and corrosion of certain materials. It is produced by the action of sunlight on air pollutants, including car exhaust fumes, and is a major air pollutant in hot summers. Ozone and the greenhouse effect is discussed later in the text.

ATMOSPHERIC CARBON DIOXIDE

Carbon dioxide (CO_2) is a colorless, odorless gas (although it is felt by some persons to have a slight pungent odor and biting taste); is slightly soluble in water and denser than air (one and a half times heavier than air); and is a slightly acid gas. Carbon dioxide gas is relatively nonreactive and nontoxic. It will not burn, and it will not support combustion or life.

Carbon dioxide is normally present in atmospheric air at about 0.035% by volume and cycles through the biosphere (carbon cycle) as shown in Figure 7.3. Carbon dioxide, along with water vapor, is primarily responsible for the absorption of infrared energy reemitted by the Earth and, in turn, some of this energy is reradiated back to the Earth's surface. It is also a normal end product of human and animal metabolism. The exhaled breath contains up to 5.6% carbon diox-

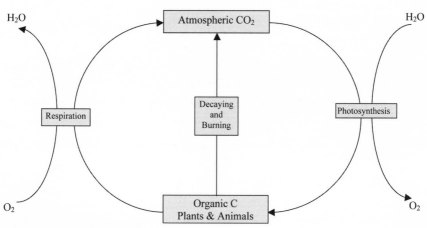

Figure 7.3. Carbon cycle.

ide. In addition, the burning of carbon-laden fossil fuels releases carbon dioxide into the atmosphere. Much of this carbon dioxide is absorbed by ocean water, some of it is taken up by vegetation through photosynthesis in the carbon cycle (see Figure 7.3), and some remains in the atmosphere. Today, it is estimated that the concentration of carbon dioxide in the atmosphere is approximately 350 ppm and is rising at a rate of approximately 20 ppm every decade. The increasing rate of combustion of coal and oil has been primarily responsible for this occurrence, which (as we will see later in this text) may eventually have an impact on global climate.

Carbon Dioxide: Physical Properties

The physical properties of carbon dioxide are noted in Table 7.3.

Table 7.3. Carbon Dioxide: Physical Properties

Chemical Formula	CO_2
Molecular weight	44.01
Vapor pressure at 70° F	838 psig
Density of the gas at 70° F and 1 atm	0.1144 lb/ft³
Specific gravity of the gas at 70° F and 1 atm (air = 1)	1.522
Specific volume of the gas at 70° F and 1 atm	8.741 ft³/lb
Critical temperature	−109.3° F
Critical pressure	1070.6 psia
Critical density	29.2 lb/ft³
Latent heat of vaporization at 32° F	100.8 BTU/lb
Latent heat of fusion at −69.9° F	85.6 BTU/lb

atm, Atmosphere; *BTU*, British thermal unit; *psia*, pounds per square inch absolute; *psig*, pounds per square inch gauge.

Carbon Dioxide: Uses

Solid carbon dioxide is used quite extensively to refrigerate perishable foods while in transit. It is also used as a cooling agent in many industrial processes, such as grinding, rubber work, cold-treating metals, vacuum cold traps, and so on.

Gaseous carbon dioxide is used to carbonate soft drinks, for pH control in water treatment, in chemical processing, as a food preservative, and in pneumatic devices.

ATMOSPHERIC ARGON

Discovered by Sir William Ramsey in 1894 Argon (*Ar*—Greek *argos,* "idle") is a colorless, odorless, tasteless, nontoxic, nonflammable gaseous element (noble gas). It constitutes almost 1% of the Earth's atmosphere and is plentiful compared with the other rare atmospheric gases. It is extremely inert and forms no known chemical compounds. It is slightly soluble in water.

Argon: Physical Properties

The physical properties of argon are noted in Table 7.4.

Table 7.4. Argon: Physical Properties

Chemical Formula	Ar
Molecular weight	39.95
Density of the gas at 70° F	0.103 lb/ft³
Specific gravity of the gas at 70° F	1.38
Specific volume of the gas at 70° F	9.71 ft³/lb
Boiling point at 1 atm	–302.6° F
Melting point at 1 atm	–308.6° F
Critical temperature	–188.1° F
Critical pressure	711.5 psia
Critical density	33.444 lb/ft³
Latent heat of vaporization at boiling point and 1 atm	69.8 BTU/lb
Latent heat of fusion	12.8 BTU/lb

atm, Atmosphere; *BTU,* British thermal unit; *psia,* pounds per square inch absolute.

Argon: Uses

Argon is used extensively in filling incandescent and fluorescent lamps and electronic tubes; to provide a protective shield for growing silicon and germanium crystals; and as a blanket in the production of titanium, zirconium, and other reactive metals.

ATMOSPHERIC NEON

Neon (*Ne*—Greek *neon,* "new") is a colorless, odorless, gaseous, nontoxic, chemically inert element. Air is about 2 parts per thousand neon by volume.

Neon: Physical Properties

The physical properties of neon are noted in Table 7.5.

Neon: Uses

Neon is used principally to fill lamp bulbs and tubes. The electronics industry uses neon singly or in mixtures with other gases in many types of gas-filled electron tubes.

Table 7.5.   Neon: Physical Properties

Chemical Formula	Ne
Molecular weight	20.183
Density of the gas at 70° F and 1 atm	0.05215 lb/ft^3
Specific gravity of the gas at 70° F and 1 atm	0.696
Specific volume of the gas at 70° F and 1 atm	19.18 ft^3/lb
Boiling point at 1 atm	–410.9° F
Melting point at 1 atm	–415.6° F
Critical temperature	–379.8° F
Critical pressure	384.9 psia
Critical density	30.15 lb/ft^3
Latent heat of vaporization at boiling point	37.08 BTU/lb
Latent heat of fusion	7.14 BTU/lb

atm, Atmosphere; *BTU,* British thermal unit; *psia,* pounds per square inch absolute.

ATMOSPHERIC HELIUM

Helium (*He*—Greek *helios,* "Sun") is inert (and as a result, does not appear to have any major effect on or role in the atmosphere), nontoxic, odorless, tasteless, nonreactive, and colorless; forms no compounds; and makes about 0.00005% (5 ppm) by volume of air in the Earth's atmosphere. Helium, as with neon, krypton, hydrogen, and xenon, is a noble gas. Helium is the second lightest element; only hydrogen is lighter. It is one-seventh as heavy as air. Helium is nonflammable and is only slightly soluble in water.

Helium: Physical Properties

The physical properties of helium are noted in Table 7.6.

Table 7.6.   Helium: Physical Properties

Chemical Formula	He
Molecular weight	4.00
Density of the gas at 70° F and 1 atm	0.0103 lb/ft^3
Specific gravity of the gas at 70° F and 1 atm	0.138
Specific volume of the gas at 70° F and 1 atm	97.09 ft^3/lb
Boiling point at 1 atm	–452.1° F
Critical temperature	–450.3° F
Critical pressure	33.0 psia
Critical density	4.347 lb/ft^3
Latent heat of vaporization at boiling point and 1 atm	8.72 BTU/lb

atm, Atmosphere; *BTU,* British thermal unit; *psia,* pounds per square inch absolute.

ATMOSPHERIC KRYPTON

Krypton (*Kr*—Greek *kryptos,* "hidden") is a colorless, odorless, inert gaseous component of Earth's atmosphere. It is present in very small quantities in the air (about 114 ppm).

Krypton: Physical Properties

The physical properties of krypton are noted in Table 7.7.

Table 7.7.   Krypton: Physical Properties

Chemical Formula	*Kr*
Molecular weight	83.80
Density of the gas at 70° F and 1 atm	0.2172 lb/ft^3
Specific gravity of the gas at 70° F and 1 atm	2.899
Specific volume of the gas at 70° F and 1 atm	4.604 ft^3/lb
Boiling point at 1 atm	–244.0° F
Melting point at 1 atm	–251° F
Critical temperature	–82.8° F
Critical pressure	798.0 psia
Critical density	56.7 lb/ft^3
Latent heat of vaporization at boiling point	46.2 BTU/lb
Latent heat of fusion	8.41 BTU/lb

atm, Atmosphere; *BTU*, British thermal unit; *psia*, pounds per square inch absolute.

Krypton: Uses

Krypton is used principally to fill lamp bulbs and tubes. The electronics industry uses it singly or in mixture in many types of gas-filled electron tubes.

ATMOSPHERIC XENON

Xenon (*Xe*—Greek *xenon,* "stranger") is a colorless, odorless, nontoxic, inert, heavy gas that is present in very small quantities in the air (about 1 part in 20 million).

Xenon: Physical Properties

The physical properties of xenon are noted in Table 7.8.

Xenon: Uses

Xenon is used principally to fill lamp bulbs and tubes. The electronics industry uses it singly or in mixtures in many types of gas-filled electron tubes.

Table 7.8.   Xenon: Physical Properties

Chemical Formula	Xe
Molecular weight	131.3
Density of the gas at 70° F and 1 atm	0.3416 lb/ft^3
Specific gravity of the gas at 70° F and 1 atm	4.560
Specific volume of the gas at 70° F and 1 atm	2.927 ft^3/lb
Boiling point at 1 atm	–162.6° F
Melting point at 1 atm	–168° F
Critical temperature	61.9° F
Critical pressure	847.0 psia
Critical density	68.67 lb/ft^3
Latent heat of vaporization at boiling point	41.4 BTU/lb
Latent heat of fusion	7.57 BTU/lb

atm, Atmosphere; BTU, British thermal unit; psia, pounds per square inch absolute.

ATMOSPHERIC HYDROGEN

Hydrogen (H_2—Greek *hydros* + *gen,* "water generator") is a colorless, odorless, tasteless, nontoxic, flammable gas. It is the lightest of all the elements and occurs on Earth chiefly in combination with oxygen as water. Hydrogen is the most abundant element in the universe, where it accounts for 93% of the total number of atoms and 76% of the total mass. It is the lightest gas known, with a density approximately 0.07 that of air. Hydrogen is present in the atmosphere, occurring in concentrations of only about 0.5 ppm by volume at lower altitudes.

Hydrogen: Physical Properties

The physical properties of hydrogen are noted in Table 7.9.

Table 7.9.   Hydrogen: Physical Properties

Chemical Formula	H_2
Molecular weight	2.016
Density of the gas at 70° F and 1 atm	0.00521 lb/ft^3
Specific gravity of the gas at 70° F and 1 atm	0.06960
Specific volume of the gas at 70° F and 1 atm	192.0 ft^3/lb
Boiling point at 1 atm	–423.0° F
Melting point at 1 atm	–434.55° F
Critical temperature	–399.93° F
Critical pressure	190.8 psia
Critical density	1.88 lb/ft^3
Latent heat of vaporization at boiling point	191.7 BTU/lb
Latent heat of fusion	24.97 BTU/lb

atm, Atmosphere; BTU, British thermal unit; psia, pounds per square inch.

Hydrogen: Uses

Hydrogen is used by refineries and petrochemical and bulk chemical facilities for hydro-treating, catalytic reforming, and hydro-cracking. Hydrogen is used in the production of a wide variety of chemicals. Metallurgical companies use hydrogen in the production of their products. Glass manufacturers use hydrogen as a protective atmosphere in a process whereby molten glass is floated on a surface of molten tin. Food companies hydrogenate fats, oils, and fatty acids to control various physical and chemical properties. Electronic manufacturers use hydrogen at several steps in the complex processes for manufacturing semiconductors.

ATMOSPHERIC WATER

Leonardo da Vinci understood the importance of water when he said, "Water is the driver of nature." Da Vinci was actually acknowledging what most scientists and many of the rest of us have come to realize: Water, propelled by the varying temperatures and pressures in Earth's atmosphere, allows life as we know it to exist on our planet (Graedel & Crutzen, 1995).

The water vapor content of the lower atmosphere (troposphere) is normally with a range of 1% to 3% by volume with a global average of about 1%. However, the percentage of water in the atmosphere can vary from as little as 0.1% or as much as 5% water, depending on altitude; water in the atmosphere decreases with increasing altitude. Water circulates in the atmosphere in the hydrologic cycle (discussed in detail later).

Water vapor contained in Earth's atmosphere plays several important roles: (1) it absorbs infrared radiation; (2) it acts as a blanket at night, retaining heat from the Earth's surface; and (3) it affects the formation of clouds in the atmosphere.

ATMOSPHERIC PARTICULATE MATTER

Significant numbers of particles (particulate matter) are suspended in the atmosphere, particularly the troposphere. These particles originate in nature from smokes, sea sprays, dusts, and the evaporation of organic materials from vegetation. There is also a wide variety of nature's living or semiliving particles—spores and pollen grains, mites and other tiny insects, spider webs, and diatoms. The atmosphere also contains a bewildering variety of anthropogenic (artificial) particles produced by automobiles, refineries, production mills, and many other human activities.

Atmospheric particulate matter varies greatly in size (colloidal-sized particles in the atmosphere are called *aerosols*—usually less than 0.1 µm in diameter). The smallest are gaseous clusters and ions and submicroscopic liquids and solids; somewhat larger ones produce the beautiful blue haze in distant vistas. Those two to three times larger are highly effective in scattering light. The largest consist of such things as rock fragments, salt crystals, and ashy residues from volcanoes, forest fires, or incinerators.

The numbers of particulates concentrated in the atmosphere vary greatly, ranging from more than 10,000,000/cc to less than 1/L (0.001/cc). Excluding the particles in gases as well as vegetative material, sizes range from 0.005 to 500 microns, a variation in diameter of 100,000 times.

The largest number of airborne particulates is always in the invisible range. These numbers vary from less than 1 L to more than a half million per cubic centimeter in heavily polluted air, and to at least 10 times more than that when a gas-to-particle reaction is occurring (Schaefer & Day, 1981).

Based on particulate level, there are two distinct regions in the atmosphere: very clean and dirty. In the clean parts there are so few particulates that they are almost invisible, making them hard to collect or measure. In the dirty parts of the atmosphere—the air of a large metropolitan area —the concentration of particles includes an incredible variety of particulates from a wide variety of sources.

Atmospheric particulate matter performs a number of functions and undergoes several processes, and is involved in many chemical reactions in the atmosphere. Probably the most important function of particulate matter in the atmosphere is their action as nuclei for the formation of water droplets and ice crystals. Much of the work of Vincent J. Schaefer (inventor of cloud seeding) involved using dry ice in early attempts, but later evolved around the addition of condensing particles to atmospheres supersaturated with water vapor and the use of silver iodide, which forms huge numbers of very small particles. Another important function of atmospheric particulate matter is that it helps determine the heat balance of the Earth's atmosphere by reflecting light. Particulate matter is also involved in many chemical reactions in the atmosphere such as neutralization, catalytic effects, and oxidation reactions. These chemical reactions are discussed in greater detail later.

References and Recommended Reading

Compressed Gas Association. 1990. *Handbook of compressed gases.* Chantilly, VA: Author.

Environmental Protection Agency. 2007. *Air pollution control orientation course: Air pollution.* Accessed 01/05/08 from www.epa.gov/air/oaqps/eog/course422/ap.1.html.

Graedel, T. E., & Crutzen, P. J. 1995. *Atmosphere, climate, and change.* New York: W. H. Freeman.

Schaefer, V. J., & Day, J. A. 1981. *Person field guides: Atmosphere.* New York: Houghton Mifflin.

Spellman, F. R., & Whiting, N. 2006. *Environmental science and technology: Concepts and applications,* 2nd edition. Rockville, Maryland: Government Institutes.

CHAPTER 8

Moisture in the Atmosphere

Hath the rain a father? or who hath begotten the drops of dew? Out of whose womb came the ice? and the hoary frost of heaven, who hath gendered it? . . . Can't thou lift up thy voice to the clouds, that abundance of water may cover thee?

—Job 38:28–29, 34

I wandered lonely as a cloud
That floats on high o'er vales and hills,
When all at once I saw a crowd,
A host, of golden daffodils;
Beside the lake, beneath the trees,
Fluttering and dancing in the breeze.

—William Wordsworth, 1804

Clouds.

On a hot day when clouds build up signifying that a storm is imminent, we do not always appreciate what is happening.

What is happening?

This cloud buildup actually signals one of the most vital processes in the atmosphere: the condensation of water as it is raised to higher levels and cooled within strong updrafts of air created either by convection currents, turbulence, or physical obstacles like mountains. The water originated from the surface— evaporated from the seas, from the soil, or transpired by vegetation. Once within the atmosphere, however, a variety of events combine to convert the water vapor (produced by evaporation) to water droplets. The air must rise and cool to its dew point, of course. At dew point, water condenses around minute airborne

121

particulate matter to make tiny cloud droplets, forming clouds from which precipitation occurs.

Whether created by the Sun heating up a hillside, by jet aircraft exhausts, or factory chimneys, there are actually only 10 major cloud types. The deliverers of countless millions of tons of moisture from the Earth's atmosphere, they form even from the driest desert air containing as little as 0.1% water vapor. They not only provide a visible sign of motion but also indicate change in the atmosphere, portending weather conditions that may be expected up to 48 hours ahead. In this chapter we take a brief look at the nature and consequences of these cloud-forming processes.

Cloud Formation

The atmosphere is a highly complex system, and the effects of changes in any single property tend to be transmitted to many other properties. The most profound effect on the atmosphere is the result of alternate heating and cooling of the air, which causes adjustments in relative humidity (RH) and buoyancy; this effect causes condensation, evaporation, and cloud formation.

The temperature structure of the atmosphere (along with other forces that propel the moist air upward) is the main force behind the form and size of clouds. Exactly how does temperature affect atmospheric conditions? For one thing, temperature (that is, heating and cooling of the surface atmosphere) causes vertical air movements. We next discuss what happens when air is heated.

Let's start with a simple parcel of air in contact with the ground. As the ground is heated, the air in contact with it warms also. This warm air increases in temperature and expands. Remember, gases expand on heating much more than liquids or solids, so this expansion is quite marked. In addition, as the air expands, its density falls (meaning that the same mass of air now occupies a larger volume). You've heard that warm air rises. Because of its lessened density, this parcel of air is now lighter than the surrounding air and tends to rise. Conversely, if the air cools, the opposite occurs—it contracts, its density increases, and it sinks. Alternate heating and cooling are intimately linked with the process of evaporation, condensation, and precipitation. But how does a cloud actually form? Let's look at another example.

On a sunny day, some patches of ground warm up more quickly than others because of differences in topography (soil, vegetation, etc.). As the surface temperature increases, heat passes to the overlying air. Later, by mid-morning, a bulbous mass of warm, moisture-laden air rises from the ground. This mass of air cools as it meets lower atmospheric pressure at higher altitudes. If cooled to its dew point temperature, condensation follows and a small cloud forms. This cloud breaks free from the heated patch of ground and drifts with the wind. If it passes over other rising air masses, it may grow in height. The cloud may encounter a mountain and be forced still higher into the air. Condensation continues as the cloud cools; if the droplets it holds become too heavy, they fall as rain.

Major Cloud Types

Earlier, it was mentioned that there are 10 major cloud types. These include:

STRATIFORM GENERA SPECIES

Cirrus
Cirrostratus
Cirrocumulus
Altostratus
Altocumulus
Stratus
Stratocumulus
Nimbostratus

CUMULIFORM GENERA SPECIES

Cumulus
Cumulonimbus

This list makes clear that the cloud groups are classified into a system that uses Latin words to describe the appearance of clouds as seen by an observer on the ground. Table 8.1 summarizes the four principal components of this classification system (Ahrens, 1994).

Table 8.1. Summary of Components of Cloud Classification System

Latin Root	Translation	Example
Cumulus	Heaped, puffy	Fair-weather cumulus
Stratus	Layered	Altostratus
Cirrus	Curl of hair, wispy	Cirrus
Nimbus	Ran	Cumulonimbus

Further classification identifies clouds by the height of the cloud base. For example, cloud names containing the prefix *cir-*, as in cirrus clouds, are located at high levels; cloud names with the prefix *alto-*, as in altostratus, are found at middle levels. This module introduces several cloud groups. The first three groups are identified based on their height above the ground. The fourth group consists of vertically developed clouds, while the final group consists of a collection of miscellaneous cloud types.

A *stratus* cloud is a featureless, gray, low-level cloud. Its base may obscure hilltops or occasionally extend right down to the ground, and, because of its low altitude, it appears to move very rapidly on breezy days. Stratus can produce drizzle or snow, particularly over hills, and may occur in huge sheets covering several thousand miles.

Cumulus clouds also seem to scurry across the sky, reflecting their low altitude. These small dense, white, fluffy, flat-based clouds are typically short lived, lasting no more than 10–15 minutes before dispersing. They are typically formed on sunny days, when localized convection currents are set up: These currents can form over factories or even brush fires, which may produce their own clouds.

Cumulus may expand into low-lying, horizontally layered, massive *stratocumulus*, or into extremely dense, vertically developed, giant *cumulonimbus* with a relatively hazy outline and a glaciated top up to 7 miles in diameter. These clouds typically form on summer afternoons; their high, flattened tops contain ice, which may fall to the ground in the form of heavy showers of rain or hail.

Rising to middle altitudes, the bluish-gray layered *altostratus* and rounded, fleecy, whitish-gray *altocumulus* appear to move slowly because of their greater distance from the observer.

Cirrus (meaning tuft of hair) clouds are made up of white, narrow bands of thin, fleecy parts and are relatively common over northern Europe, and generally ride the jet stream rapidly across the sky.

Cirrocumulus are high altitude clouds composed of a series of small, regularly arranged cloudlets in the form of ripples or grains; they are often present with cirrus clouds in small amounts. *Cirrostratus* are high-altitude, thin, hazy clouds, usually covering the sky and giving a halo effect surrounding the Sun or Moon.

Did You Know?

Contrails are clouds formed around the small particles (aerosols), which are in aircraft exhaust. When these persist after the passage of the plane they are indeed clouds, and are of great interest to researchers. Under the right conditions, clouds initiated by passing aircraft can spread with time to cover the whole sky.

Precipitation and Evaporation

The principal actions brought on by weather systems that affect land and sea and the humans, animals, and vegetation thereon are winds and precipitation. The latter comes in a variety of forms, as discussed later in this chapter. Most weather of consequence to people occurs in storms. These may be local in origin but more commonly are carried to locations in wide areas along pathways followed by active air masses consisting of *highs* and *lows*. The key ingredient in storms is water, either as a liquid or as a vapor. The vapor acts like a gas and thus contributes to the total pressure of the atmosphere, making up a small but vital fraction of the total (National Aeronautics and Space Administration [NASA], 2008).

Precipitation is found in a variety of forms. Which form actually reaches the ground depends on many factors, for example, atmospheric moisture content, surface temperature, intensity of updrafts, and method and rate of cooling.

Water vapor in the air varies in amount depending on sources, quantities, processes involved, and air temperature. Heat, mainly as solar irradiation but with some contributed by the earth and human activity and some from change of state processes, causes some water molecules either in water bodies (oceans, lakes, rivers) or in soils to be excited thermally and escape from their sources. This is called *evaporation;* the process in which water is released from trees and other vegetation is known as *evapotranspiration.* The evaporated water, or moisture, that enters the air is responsible for a state called *humidity.* Absolute humidity is the weight of water vapor contained in a given volume of air. The *mixing ratio* refers to the mass of the water vapor within a given mass of dry air. At any particular temperature, the maximum amount of water vapor that can be contained is limited to some amount; when that amount is reached, the air is said to be saturated for that temperature. If less than the maximum amount is present, then the property of air that indicates this is its RH, defined as the actual water vapor amount compared with the saturation amount at the given temperature; this is usually expressed as a percentage. RH also indicates how much moisture the air can hold above its stated level, which, after this point is attained, could lead to rain.

When a parcel of air attains or exceeds an RH of 100%, condensation occurs and water in some state begins to organize as some type of precipitation. One familiar form is dew, which occurs when the saturation temperature or some quantity of moisture reaches a temperature at the surface at which condensations sets in, leaving the moisture to coat the ground (especially obvious on lawns).

The term *dew point* has a more general use, being that temperature at which an air parcel must be cooled to become saturated. Dew frequently forms when the current air mass contains excessive moisture after a period of rain but the air is now clear; the dew precipitates out to coat the surface (noticeable on vegetation). Ground fog is a variant in which lowered temperatures bring on condensation within the near surface air as well as the ground.

The other types of precipitation are listed in Table 8.2 along with descriptive characteristics related to each type.

Evaporation and transpiration are complex processes that return moisture to the atmosphere. The rate of evapotranspiration depends largely on two factors: (1) how saturated (moist) the ground is and (2) the capacity of the atmosphere to absorb the moisture. In this chapter we discuss the factors responsible for both precipitation and evapotranspiration.

Precipitation

Earlier it was pointed out that if all the essentials are present, precipitation occurs when the dew point is reached. It was also pointed out that it is quite possible for an air mass or cloud containing water vapor to cool below the dew point without precipitation occurring. In this state the air mass is said to be *supercooled.*

How, then, are droplets of water formed? Water droplets form around microscopic foreign particles already present in the air. These particles on which the droplets form are called *hygroscopic nuclei.* They are present in the air primarily in the form of dust, salt from sea water evaporation, and from combustion residue. These foreign particles initiate the formation of droplets that eventually fall as precipitation. To have precipitation, larger droplets or drops must form. This may be brought about by two processes: (1) coalescence (collision) or (2) the Bergeron process.

COALESCENCE

Simply put, *coalescence* is the fusing together of smaller droplets into larger ones. The variation in the size of the droplets has a direct bearing on the efficiency of this process. Raindrops come in different sizes and can reach diameters up to 7 mm. Having larger droplets greatly enhances the coalescence process.

But what actually goes on inside a cloud to cause rain to fall? To answer this question, we must take a look inside a cloud to see exactly what processes occur to make rain—rain that actually falls as rain. Rainmaking is based on the essentials of the Bergeron process.

BERGERON PROCESS

Named after the Swedish meteorologist who suggested it, the *Bergeron process* is probably the more important process for the initiation of precipitation. To gain

Table 8.2.   Types of precipitation

Type	Approximate Size	State of Water	Description
Mist	0.005 to 0.05 mm	Liquid	Droplets large enough to be felt on face when air is moving 1 m/s. Associated with stratus clouds.
Drizzle	Less than 0.5 mm	Liquid	Small uniform drops that fall from stratus clouds, generally for several hours.
Rain	0.5 to 5 mm	Liquid	Generally produced by nimbostratus or cumulonimbus clouds. When heavy, size can be highly variable from one place to another.
Sleet	0.5 to 5 mm	Solid	Small, spherical to lumpy ice particles that form when raindrops freeze while falling through a layer of subfreezing air. Because the ice particles are small, any damage is generally minor. Sleet can make travel hazardous.
Glaze	Layers 1 mm to 2 cm thick	Solid	Produced when supercooled raindrops freeze on contact with solid objects. Glaze can form a thick covering of ice having sufficient weight to seriously damage trees and power lines.
Rime	Variable accumulation	Solid	Deposits usually consisting of ice feathers that point into the wind. These delicate, frostlike accumulations form as supercooled cloud or fog droplets that encounter objects and freeze on contact.
Snow	1 mm to 2 cm	Solid	The crystalline nature of snow allows it to assume many shapes, including six-sided crystals, plates, and needles. Produced in supercooled clouds where water vapor is deposited as crystals that remain frozen during their descent.
Hail	5 mm to 1 cm or larger	Solid	Precipitation in the form of hard, rounded pellets or irregular lumps of ice. Produced in large, convective cumulonimbus clouds, where frozen ice particles and supercooled water coexist.
Graupel	2 mm to 5 mm	Solid	Sometimes called "soft hail," graupel forms on rime and collects on snow crystals to produce irregular masses of "soft" ice. Because these particles are softer than hailstones, they normally flatten out on impact.

Source: NASA, 2008.

understanding on how the Bergeron process works, let's look at what actually goes on inside a cloud to cause rain.

Within a cloud made up entirely of water droplets there are a variety of droplet sizes. The air rises within the cloud anywhere from 10–20 cm per second (depending on the type of cloud). As the air rises, the drops become larger through collision and coalescence; many will reach drizzle size. Then the updraft intensifies up to 50 cm per second (and more), which reduces the downward movement of the drops, allowing them more time to become even larger. When the cloud becomes approximately 1 km deep, small raindrops of 700 μm diameter are formed.

The droplets, because of their small size, do not freeze immediately, even when the temperatures fall below 0° C. Instead, the droplets remain unfrozen in a supercooled state. However, when the temperature drops as low as –10° C, ice crystals might start to develop among the water droplets. This mixture of water and ice would not be particularly important but for a peculiar characteristic or property of water. Therefore, at –10° C, air saturated with respect to liquid water is supersaturated relative to ice by 10% and at –20° C by 21%. Thus, ice crystals in the cloud tend to grow and become heavier at the expense of the water droplets.

Eventually, the ice crystals sink to the lower levels of the cloud where temperatures are only just below freezing. When this occurs, they tend to combine (the supercooled droplets of water act as an adhesive) and form snowflakes. When the snowflakes melt, the resulting water drops may grow further by collision with cloud droplets before they reach the ground as rain. The actual rate at which water vapor is converted to raindrops depends on three main factors: (1) the rate of ice crystal growth, (2) supercooled vapor, and (3) the strength of the updrafts (mixing) in the cloud.

Types of Precipitation

We stated that for precipitation to occur, water vapor must condense, which occurs when water vapor ascends and cools. The three mechanisms by which air rises, allowing for precipitation to occur, are convectional, orographic, and frontal.

CONVECTIONAL PRECIPITATION

Convectional precipitation is the spontaneous rising of moist air caused by instability. This type of precipitation is usually associated with thunderstorms and occurs in the summer because localized heating is required to initiate the convection cycle. We have discussed that upward-growing clouds are associated with convection. Because the updrafts (commonly called a *thermal*) are usually strong, cooling of the air is rapid and lots of water can be condensed quickly; this is usually confined to a local area, and a sudden summer downpour might occur as a result.

Convectional thunderstorm clouds are also described as *supercells*. Convective thunderstorms are the most common type of atmospheric instability that produces lightning followed by thunder. Lightning is one of the most spectacular phenomena witnessed in storms.

Did You Know?

A lightning bolt can attain an electric potential up to 30 million volts and a current to as much as 10,000 amps. It can cause air temperatures to reach 10,000° C. But a bolt's duration is extremely short (fractions of a second). Although a bolt can kill people it hits, most can survive.

OROGRAPHIC PRECIPITATION

Orographic precipitation is a straightforward process, characteristic of mountainous regions; almost all mountain areas are wetter than the surrounding lowlands. This type of precipitation arises when air is forced to rise over a mountain or mountain range. The wind blowing along the surface of the Earth ascends along topographic variations. Where air meets this extensive barrier, it is forced to rise. This ascending wind usually gives rise to cooling and encourages condensation and thus orographic precipitation on the windward side of the mountain range.

FRONTAL PRECIPITATION

Frontal precipitation results when two different fronts (or the boundary between two air masses characterized by varying degrees of precipitation), at different temperatures, meet. Because the warm air mass is lighter, it moves up and over the colder air mass. The cooling is usually less rapid than in the vertical convection process because the warm air mass moves up at an angle, in more of a horizontal motion.

Evapotranspiration

Another important part or process of the hydrological cycle (though it is often neglected because it can rarely be seen) is *evapotranspiration*. More complex than precipitation, evaporation and transpiration is a land-atmosphere interface process whereby a major flow of moisture is transferred from ground level to the atmosphere. It returns moisture to the air, replenishing that lost by precipitation, and it also takes part in the global transfer of energy. The rate of evapotranspiration depends largely on two factors: (1) how moist the ground is and (2) the capacity of the atmosphere to absorb the moisture. Therefore, the greatest rates are over the tropical oceans, where moisture is always available and the long hours of sunshine and steady trade winds evaporate vast quantities of water.

Just how much moisture is returned to the atmosphere via transpiration? Table 8.3 makes clear that in the United States mainland alone, for example, about two-thirds of the average rainfall is returned via evaporation and transpiration.

Table 8.3. Water Balance in the United States (in bgd)

Precipitation	4,200
Evaporation and transpiration	3,000
Runoff	1,250
Withdrawal	310
Irrigation	142
Industry (utility cooling water)	142
Municipal	26
Consumed (irrigation loss)	90
Returned to streams	220

bgd, Billion gallons per day.

Source: National Academy of Sciences, 1962.

EVAPORATION

Evaporation is the process by which a liquid is converted into a gaseous state. Evaporation takes place (except when air reaches saturation at 100% humidity) almost on a continuous basis. It involves the movement of individual water molecules from the surface of Earth into the atmosphere, a process occurring whenever a vapor pressure gradient exists from the surface to the air (i.e., whenever the humidity of the atmosphere is less than that of the ground). Evaporation also requires energy (derived from the sun or from sensible heat from the atmosphere or ground): 2.48×10^6 Joules to evaporate each kilogram of water at 10° C.

TRANSPIRATION

A related process, *transpiration* is the loss of water from a plant by evaporation. Most water is lost from the leaves through pores known as *stomata,* whose primary function is to allow gas exchange between the plant's internal tissues and the atmosphere. Transpiration from the leaf surfaces causes a continuous upward flow of water from the roots via the xylem, which is known as the *transpiration stream*.

Transpiration occurs mainly by day, when the stomata open up under the influence of sunlight. Acting as evaporators, they expose the pure moisture (the plant's equivalent of perspiration) in the leaves to the atmosphere. If the vapor pressure of the air is less than that in the leaf cells, the water is transpired.

As you might guess, because of transpiration, far more water passes through a plant than is needed for growth. In fact, only about 1% or so is actually used in plant growth. Nevertheless, the excess movement of moisture through the plant is important to the plant because the water acts as a solvent, transporting vital nutrients from the soil into the roots and carrying them through cells of the plant. Obviously, without this vital process plants would die.

EVAPOTRANSPIRATION: THE PROCESS

Although evapotranspiration plays a vital role in cycling water over Earth's land masses, it is seldom appreciated. In the first place, distinguishing between evaporation and transpiration is often difficult. Both processes tend to be operating together, so the two are normally combined to give the composite term *evapotranspiration*.

Governed primarily by atmospheric conditions, energy is needed to power the process. Wind also plays an important role, which acts to mix the water molecules with the air and transport them away from the surface. The primary limiting factor in the process is lack of moisture at the surface (soil is dry). Evaporation can continue only so long as there is a vapor pressure gradient between the ground and the air.

References and Recommended Reading

Ahrens, D. 1994. *Meteorology today: An introduction to weather, climate and the environment,* 5th edition. St. Paul, Minnesota: West Publishing Company.

Environmental Protection Agency. 2007. Air pollution control—atmosphere. Accessed 12/28/07 from www.epa.gov/air/oaqpseog/course422/apl.html.

National Academy of Sciences. 1962. *Water balance in the U.S.* Washington, DC: National Research Council Publication 100-B.

National Aeronautics and Space Administration (NASA). 2008. *Observing cloud type.* Washington, DC: NASA.

National Oceanic and Atmospheric Administration (NOAA). 2007. Cloud types. Accessed 12/29/07 from www.gfdl.NOAA.gov/~01/weather/clouds.html.

Shipman, J. T., Adams, J. L., & Wilson, J. D. 1987. *An introduction to physical science.* Lexington, Massachusetts: D.C. Heath and Company.

Spellman, F. R. 2007. *The science of water,* 2nd edition. Boca Raton, Florida: CRC Press.

Spellman, F. R., & Whiting, N. 2006. *Environmental science and technology: Concepts and applications,* 2nd edition. Rockville, Maryland: Government Institutes.

US Geological Survey (USGS). 2008. *The water cycle: Evapotranspiration.* Accessed 1/07/08 from ga.water.usgs.gov/edu/watercycleevaportranspiration.html.

CHAPTER 9

Atmosphere in Motion

There are scientists and engineers out there in the real world who will tell us that perpetual motion in or for any machine is a pipedream—it's wishful thinking. It's impossible. Have you ever pondered the most dynamic perpetual motion machine of them all—Earth's atmosphere? Probably not—but that is exactly what Earth's atmosphere is. It must be in a state of perpetual motion because it constantly strives to eliminate the constant differences in temperature and pressure between different parts of the globe. How are these differences eliminated or compensated for? By its motion, winds and storms are produced. In this chapter, the horizontal movements that transfer air around the globe are considered.

Global Air Movement

Basically, winds are the movement of the Earth's atmosphere, which by its weight exerts a pressure on the Earth that we can measure using a barometer. Winds are often confused with air currents, but they are different. Wind is the horizontal movement of air or motion along the Earth's surface. Air currents, on the other hand, are vertical air motions collectively referred to as *updrafts* and *downdrafts*.

Throughout history, people have been both fascinated by and frustrated by winds. Humans have written about winds almost from the time of the first written word. For example, Herodotus (and later Homer and many others) wrote about winds in his *The Histories*. Wind has had such an effect on human existence that we have given winds names that describe a particular wind, specific to a particular geographical area. Table 9.1 lists some of these winds, their colorful names, and the region where they occur. Some of these names are more than just colorful—the winds are actually colored. For example, the *harmattan* blows across the Sahara filled with red dust; mariners called this red wind the "sea of darkness."

Table 9.1.   Assorted Winds of the World

Wind Name	Location
aajej	Morocco
alm	Yugoslavia
biz roz	Afghanistan
haboob	Sudan
imbat	North Africa
datoo	Gibraltar
nafhat	Arabia
besharbar	Caucasus
Samiel	Turkey
tsumuji	Japan
brickfielder	Australia
chinook	America
williwaw	Alaska

WIND INTENSITY

Winds are ranked by the intensity of the wind. In renewable energy practice whereby wind turbines are used to produce electricity, wind power varies with the intensity of the wind. This is usually measured by the Beaufort scale, which takes into account the effects of wind speed on the surrounding environment, trees, buildings, smoke, and so on. This scale ranges from 0 to 12, and the ranges are shown in Table 9.2.

Table 9.2.   Beaufort Wind Intensity Scale

Scale	Wind	Average Speed in Km/h	Description of Effects
0	Calm	1	Smoke rises vertically.
1	Breeze	3	Smoke shows wind.
2	Light breeze	9	Feel the wind in the face; the leaves move.
3	Gentle breeze	16	Leaves and twigs are in constant agitation.
4	Moderate breeze	24	Raises dust and loose papers.
5	Fresh breeze	34	Stirs shrubs with leaves.
6	Strong breeze	44	Large branches are in motion.
7	Moderate wind	56	Stirs whole trees.
8	Fresh wind	68	Breaks small tree branches.
9	Strong wind	81	Causes minor damage to buildings.
10	Temporal	95	Trees uprooted.
11	Storm	110	Widespread damage occurs.
12	Hurricane	Above 121	Devastation occurs.

EARTH'S ATMOSPHERE IN MOTION

To state that Earth's atmosphere is constantly in motion is to state the obvious. Anyone observing the constant weather changes is well aware of this phenomenon. Although the fact that the atmosphere is in motion is obvious, the *importance* of the dynamic state of our atmosphere is much less obvious.

As mentioned, the constant motion of Earth's atmosphere (air movement) consists of both horizontal (wind) and vertical (air currents) dimensions. The atmosphere's motion is the result of thermal energy produced from the heating of the Earth's surface and the air molecules above. Because of differential heating of the Earth's surface, energy flows from the equator poleward.

Wind and air currents are fundamental to how nature functions, but even though air movement plays the critical role in transporting the energy of the lower atmosphere, bringing the warming influences of spring and summer and the cold chill of winter, the effects of air movements on our environment are often overlooked. All life on Earth has evolved with mechanisms dependent on air movement: Pollen is carried by winds for plant reproduction; animals sniff the wind for essential information; wind power was the motive force that began the earliest stages of the industrial revolution. Now we see the effects of winds in other ways, too: Wind causes weathering (erosion) of the Earth's surface; wind influences ocean currents; air pollutants and contaminants such as radioactive particles transported by the wind affect our environment.

CAUSES OF AIR MOTION

In all dynamic situations, forces are necessary to produce motion and changes in motion—winds and air currents. The air (made up of various gases) of the atmosphere is subject to two primary forces: (1) gravity and (2) pressure differences from temperature variations.

Gravity (gravitational forces) holds the atmosphere close to the Earth's surface. Newton's law of universal gravitation states that each body in the universe attracts another body with a force equal to:

$$F = G \frac{m_1 m_2}{R_2} \qquad (9.1)$$

where:

F = Force
m_1 and m_2 = the masses of the two bodies
G = universal constant of 6.67×10^{-11} N $\times$ m^2/kg^2
R = distance between the two bodies

✓ **Important Point:** The force of gravity decreases as an inverse square of the distance between them.

Thermal conditions affect density, which in turn cause gravity to affect vertical air motion and planetary air circulation. This affects how air pollution is naturally removed from the atmosphere.

Although forces in other directions often overrule gravitational force, the ever-present force of gravity is vertically downward and acts on each gas molecule, accounting for the greater density of air near the Earth.

Atmospheric air is a mixture of gases, so the gas laws and other physical principles govern its behavior. The pressure of a gas is directly proportional to its temperature. Pressure is force per unit area ($P = F/A$), so a temperature variation in air generally gives rise to a difference in pressure of force. This difference in pressure resulting from temperature differences in the atmosphere creates air movement—on both large and local scales. This pressure difference corresponds to an unbalanced force, and when a pressure difference occurs, the air moves from a high- to a low-pressure region.

In other words, horizontal air movements (called *advective winds*) result from temperature gradients, which give rise to density gradients and, subsequently, pressure gradients. The force associated with these pressure variations (pressure gradient force) is directed at right angles (perpendicular) to lines of equal pressure and is directed from high to low pressure.

The pressures over a region are mapped by taking barometric readings at different locations. Lines drawn through the points (locations) of equal pressure are called *isobars*. All points on an isobar are of equal pressure, which means there is no air movement along the isobar. The wind direction is at right angles to the isobar in the direction of the lower pressure. Air moves down a pressure gradient toward a lower isobar like a ball rolls down a hill. If the isobars are close together, the pressure gradient force is large, and such areas are characterized by high wind speeds. If isobars are widely spaced, the winds are light because the pressure gradient is small.

Localized air circulation gives rise to thermal circulation (a result of the relationship based on a law of physics whereby the pressure and volume of a gas is directly related to its temperature). A change in temperature causes a change

Did You Know?

Air pressure at any location, whether it is on the Earth's surface or up in the atmosphere, depends on the weight of the air above. Imagine a column of air. At sea level, a column of air extending hundreds of kilometers above sea level exerts a pressure of 1,013 millibars (mb) (or 1.013 l). But if you travel up the column to an altitude of 4.4 km (18,000 feet), the air pressure would be roughly half, or approximately 506 mb.

in the pressure and volume of a gas. With a change in volume comes a change in density, since $P = m/V$, so regions of the atmosphere with different temperatures may have different air pressures and densities. As a result, localized heating sets up air motion and gives rise to thermal circulation.

Once the air has been set into motion, secondary forces (velocity-dependent forces) begin to act or react. These secondary forces are (1) Earth's rotation (Coriolis force) and (2) contact with the rotating Earth (friction). The *Coriolis force*, named after its discoverer, French mathematician Gaspard Coriolis (1772–1843), is the effect of rotation on the atmosphere and on all objects on the Earth's surface. In the northern hemisphere, it causes moving objects and currents to be deflected to the right; in the southern hemisphere, it causes deflection to the left because of the Earth's rotation. Air, in large-scale north or south movements, appears to be deflected from its expected path. That is, air moving poleward in the northern hemisphere appears to be deflected toward the east; air moving southward appears to be deflected toward the west.

Friction (drag) can also cause the deflection of air movements. This friction (resistance) is both internal and external. The friction of its molecules generates internal friction. Friction is also generated when air molecules run into each other. External friction is caused by contact with terrestrial surfaces. The magnitude of the frictional force along a surface is dependent on the air's magnitude and speed, and the opposing frictional force is in the opposite direction of the air motion.

Did You Know?

Friction, one of the major forces affecting the wind, comes into play near the Earth's surface and continues to be a factor up to altitudes of about 500 to 1,000 m. This section of the atmosphere is referred to as the *planetary* or *atmospheric boundary layer*. Above this layer, friction no longer influences the wind.

LOCAL AND WORLD AIR CIRCULATION

Air moves in all directions, and these movements are essential for those of us on Earth: Vertical air motion is essential in cloud formation and precipitation. Horizontal air movement near the Earth's surface produces winds.

Wind is an important factor in human comfort, especially affecting how cold we feel. A brisk wind at moderately low temperatures can quickly make us uncomfortably cold. Wind promotes the loss of body heat, which aggravates the chilling effect, expressed through wind chill factors in the winter (Table 9.3) and the heat index in the summer (Table 9.4). These two scales describe the cooling effects of wind on exposed flesh at various temperatures.

Table 9.3. Wind Chill Chart

Wind MPH	Temperature (Degrees Fahrenheit)											
	30	25	20	15	10	5	0	-5	-10	-15	-20	-25
5	25	19	13	7	1	-5	-11	-16	-22	-28	-34	-40
10	21	15	9	3	-4	-10	-16	-22	-28	-35	-41	-47
15	19	13	3	0	-7	-13	-19	-26	-32	-39	-45	-51
20	17	11	4	-2	-9	-15	-22	-29	-35	-42	-48	-55
25	16	9	3	-4	-11	-17	-24	-31	-37	-44	-51	-58
30	15	8	1	-5	-12	-19	-26	-33	-39	-46	-53	-60
35	14	7	0	-7	-14	-21	-27	-34	-41	-48	-55	-62
40	13	6	-1	-8	-15	-22	-29	-36	-43	-50	-57	-64
45	12	5	-2	-9	-16	-23	-30	-37	-44	-51	-58	-65
50	12	4	-3	-19	-17	-24	-31	-38	-45	-52	-60	-67
55	11	4	-3	-11	-18	-25	-32	-39	-46	-54	-61	-68
60	10	3	-4	-11	-19	-26	-33	-40	-48	-55	-62	-69

Note: Grey cells indicate wind chill temperatures at which frostbite occurs in 15 minutes or less.

Source: USA Today: http://www.usatoday.com/weather/resources/basics/windchill/wind-chill-chart.htm

Table 9.4. Heat Index Chart (Temperature and Relative Humidity)

RH (%)	Temperature (Degrees Fahrenheit)															
	90	91	92	93	94	95	96	97	98	99	100	101	102	103	104	105
90	119	123	128	132	137	141	146	152	157	163	168	174	180	186	193	199
85	115	119	123	127	132	136	141	145	150	155	161	166	172	178	184	190
80	112	115	119	123	127	131	135	140	144	149	154	159	164	169	175	180
75	109	112	115	119	122	126	130	134	138	143	147	152	156	161	166	171
70	106	109	112	115	118	122	125	129	133	137	141	145	149	154	158	163
65	103	106	108	111	114	117	121	124	127	131	135	139	143	147	151	155
60	100	103	105	108	111	114	116	120	123	126	129	133	136	140	144	148
55	98	100	103	105	107	110	113	115	118	121	124	127	131	134	137	141
50	96	98	100	102	104	107	109	112	114	117	119	122	125	128	131	135
45	94	96	98	100	102	104	106	108	110	113	115	118	120	123	126	129
40	92	94	96	97	99	101	103	105	107	109	111	113	116	118	121	123
35	91	92	94	95	97	98	100	102	104	106	107	109	112	114	116	118
30	89	90	92	93	95	96	98	99	101	102	104	106	108	110	112	114

RH, Relative humidity.

Note: Exposure to full sunshine can increase heat index values by up to 15° F.

Source: Weather Images. *Heat Index Charts.* Accessed from http://www.weatherimages.org/data/heatindex.html.

Local winds are the result of atmospheric pressure differences involved with thermal circulations because of geographic features. Land areas heat up more quickly than do water areas, giving rise to a convection cycle. As a result, during the day, when land is warmer than the water, we experience a lake or sea breeze.

At night, the cycle reverses. Land loses its heat more quickly than water, so the air over the water is warmer. The convection cycle sets to work in the opposite direction and a land breeze blows.

In the upper troposphere (above 11 to 14 km, west to east flows) are very narrow, fast-moving bands of air called *jet streams*. Jet streams have significant effects on surface airflows. When jet streams accelerate, divergence of air occurs at that altitude. This promotes convergence near the surface and the formation of cyclonic motion. Deceleration causes convergency aloft and subsidence near the surface, causing an intensification of high-pressure systems. Jet streams are thought to result from the general circulation structure in the regions where great high- and low-pressure areas meet.

Wind Farms

Though wind energy production (via wind turbines) is generally considered unusually environmentally clean, serious environmental issues do exist. For species protection, wind farm placement should be carefully studied. Wind farms put stresses on already fragmented and reduced wildlife habitats. Another serious factor is the avian mortality rate. Just as high-rise buildings, power lines, towers, antennas, and other artificial structures are passive killers of many birds, badly positioned wind farms put a heavy toll on bird populations, especially on migratory birds. The Altamont Pass wind farms near San Francisco are badly placed, and since their construction in the 1980s have killed many golden eagles and other species as well. Golden eagles lock on to a prey animal and dive for it, totally blocking out the threat of the wind turbine. They can see the propellers under normal circumstances, but their instinctive prey focus is so strong that when they stoop over a kill, they see only their prey.

Six to ten different companies, including U.S. Wind Power, Kenetech Wind power, and Green Mountain Energy, own the turbines at the Altamont Pass wind farms—over 7,000 of them. Another wind facility in Tehachapi Pass near Los Angeles poses little threat to bird populations.

In an interview with a reporter from the *San Francisco Chronicle,* conservationist Stan Moore states,

> It is estimated that 40 to 60 golden eagles are killed annually, plus 200 red-tailed hawks and smaller numbers of American kestrels, crows, burrowing owls and other birds. Those numbers are conservative. . . .

> I'm in favor of renewable energy when it is sited appropriately, but Altamont Pass is one of the worst places to put a wind farm on planet Earth, because it is adjacent to one of the densest breeding populations of golden eagles in the world. It's a unique place for raptors because of the abundant food source in ground squirrels. . . .
> Altamont Pass is not an appropriate place for wind turbines. What we have there is world-class golden-eagle habitat.

The California Energy Commission financed a 5-year study conducted by Dr. Grainger Hunt, a world authority on birds of prey who works with the Santa Cruz Predatory Bird Research Group. The study detected no population-level effects for golden eagles in Altamont Pass; however, the local eagles could provide source population for all of California if the wind farms deaths were halted. Instead, the local eagles are an at-risk population: If other pressures disturbed the Altamont Pass golden eagle population—an outbreak of West Nile virus, for example—catastrophic population losses would occur, because the wind turbines have removed much of the buffer population.

Because control guidelines are voluntary, not mandatory, the energy industry essentially polices itself on this issue. When the U.S. Fish and Wildlife Service (practicing what Service officials themselves call "discretionary" enforcement of Service laws) chooses not to enforce the Migratory Bird Treaty Act and the Bald and Golden Eagle Protection Act, and when California officials fail to enforce their own decrees (a state designation of the golden eagle as a "fully protected species" and a "species of special concern"), the protections supposedly provided by federal and state laws become a farce (Pellissier, 2003).

References and Recommended Reading

Anthes, R. A. *Meteorology,* 7th ed. Upper Saddle River, New Jersey: Prentice Hall, 1996.

Anthes, R. A., Cahir, J. J., Fraizer, A. B., & Panofsky, H. A. *The atmosphere,* 3rd ed. Columbus, Ohio: Charles E. Merrill, 1984.

Ingersoll, A. P. The Atmosphere. *Scientific American* 249(33):162–174, 1983.

Lutgens, F. K., & Tarbuck, E. J. *The atmosphere: An introduction to Meteorology.* Englewood Cliffs, New Jersey: Prentice-Hall, 1982.

Miller, G. R., Jr. *Environmental science,* 10th ed. Australia: Thompson-Brooks/Cole, 2004.

Moron, J. M., Morgan, M. D., & Wiersma, J. H. *Introduction to environmental science,* 2nd ed. New York: W. H. Freeman, 1986.

Pellissier, H. Golden Eagle Eco-Atrocity at Altamont Pass. Special to *SF Gate: San Francisco Chronicle*; Xcel Energy Ponnequin wind farm in northeastern Colorado. Accessed 1/07/08 from http://telosnet.com/wind/.

Shipman, J. T., Adams, J. L., & Wilson, J. D. *An Introduction to physical science,* 5th ed. Lexington, Massachusetts: D.C. Heath & Company, 1987.

Spellman, F. R., & Whiting, N. E. *Environmental science and technology: Concepts and applications,* 2nd ed. Rockville, Maryland: CRC Press.

CHAPTER 10

Meteorology: The Science of Weather

The Pharisees also with the Sadducees came and tempting desired him that he would show them a sign from heaven. He answered and said unto them, When it is evening ye say, It will be fair weather today for the sky is red and lowering. Oh ye hypocrites, ye can discern the face of the sky, but can ye not discern the signs of the times?

—Matthew 16:1–4

Mean Weather
Intermittent rain, I've learned,
Which forecasts tell about,
Is rain that stops when I go in
And starts when I come out.

—Elizabeth Dolan
The Breeze 2(8):6, 1945.

An eminent meteorologist once said, "A butterfly flapping its wings in Brazil can cause a tornado in Texas." What the meteorologist was implying is true to a point (and in line with what some critics might say): Because of tiny nuances in Earth's weather patterns, making accurate, long-range weather predictions is extremely difficult.

What is the difference between weather and climate? Some people get these two confused, believing they mean the same thing, but they do not. In this chapter you will gain a clear understanding of the meaning of and difference between the two and also gain basic understanding of the role weather plays in air pollution.

Meteorology is the science concerned with the atmosphere and its phenomena. The atmosphere is the media into which all air pollution is emitted. The meteorologist observes atmospheric processes such as temperature, density, winds (air), clouds, precipitation, and other characteristics, and endeavors to account for its observed structure and evaluation (weather, in part) in terms of external influence and the basic laws of physics. *Air pollution meteorology* is the study of how these atmospheric processes affect the fate of air pollutants.

Because the atmosphere serves as the medium into which air pollutants are released, the transport and dispersion of these releases are influenced significantly by meteorological parameters. Understanding air pollution meteorology and its influence in pollutant dispersion is essential in air-quality planning activities. Planners use this knowledge to help locate air pollution monitoring stations and to develop implementation plans to bring ambient air quality into compliance with standards. Meteorology is used in predicting the ambient effect of a new source of air pollution and to determine the effect on air quality from modifications to existing sources (Environmental Protection Agency [EPA], 2005).

Weather is the state of the atmosphere, mainly with respect to its effect on life and human activities; as distinguished from *climate* (the long-term manifestations of weather), weather consists of the short-term (minutes or months) variations of the atmosphere. Weather is defined primarily in terms of heat, pressure, wind, and moisture.

At high levels above the Earth, where the atmosphere thins to near vacuum, there is no weather; instead, weather is a near-surface phenomenon. This is evidenced clearly on a day-by-day basis where you see the ever-changing, sometimes dramatic, and often violent weather display.

In the study of air science and in particular of air quality, several determining factors are directly related to the dynamics of the atmosphere, resulting in local weather. These factors include strength of winds, the direction they are blowing, temperature, available sunlight (needed to trigger photochemical reactions, which produce smog), and the length of time since the last weather event (strong winds and heavy precipitation) cleared the air.

Weather events (such as strong winds and heavy precipitation) that work to clean the air we breathe are beneficial. However, few people would categorize the weather events such as tornadoes, hurricanes, and typhoons as beneficial.

The Sun: The Weather Generator

The Sun is the driving force behind weather. Without the distribution and re-radiation to space of solar energy, we would experience no weather (as we know it) on Earth. The Sun is the source of most of the Earth's heat. Of the gigantic amount of solar energy generated by the Sun, only a small portion bombards Earth. Most of the Sun's solar energy is lost in space. A little over 40% of the Sun's radiation reaching Earth hits the surface and is changed to heat. The rest stays in the atmosphere or is reflected back into space.

Like a greenhouse, the Earth's atmosphere admits most of the solar radiation. When solar radiation is absorbed by the Earth's surface, it is reradiated as heat waves, most of which is trapped by carbon dioxide and water vapor in the atmosphere, which work to keep the Earth warm in the same way a greenhouse traps heat.

By now you are aware of the many functions performed by the Earth's atmosphere. You should also know that the atmosphere plays an important role in regulating the Earth's heating supply. The atmosphere protects the Earth from too much solar radiation during the day and prevents most of the heat from escaping at night. Without the filtering and insulating properties of the atmosphere, the Earth would experience severe temperatures similar to other planets.

On bright, clear nights the Earth cools more rapidly than on cloudy nights because cloud cover reflects a large amount of heat back to Earth, where it is reabsorbed. The Earth's air is heated primarily by contact with the warm Earth. When air is warmed, it expands and becomes lighter. Air warmed by contact with Earth rises and is replaced by cold air, which flows in and under it. When this cold air is warmed, it too rises and is replaced by cold air. This cycle continues and generates a circulation of warm and cold air, which is called *convection*.

At the Earth's equator, the air receives much more heat than the air at the poles. This warm air at the equator is replaced by colder air flowing in from north and south. The warm, light air rises and moves poleward high above the Earth. As it cools, it sinks, replacing the cool surface air that has moved toward the equator.

The circulating movement of warm and cold air (convection) and the differences in heating cause local winds and breezes. Different amounts of heat are absorbed by different land and water surfaces. Soil that is dark and freshly plowed absorbs much more than grassy fields. Land warms faster than does water during the day and cools faster at night. Consequently, the air above such surfaces is warmed and cooled, resulting in production of local winds.

Winds should not be confused with air currents. Wind is primarily oriented toward horizontal flow. Air currents, on the other hand, are created by air moving upward and downward. Wind and air currents have a direct effect on air pollution. Air pollutants are carried and dispersed by wind. An important factor in determining the areas most affected by an air pollution source is wind direction. Because air pollution is a global problem, wind direction on a global scale is important.

Along with wind, another constituent associated with Earth's atmosphere is water. Water is always present in the air. It evaporates from the Earth, two-thirds of which is covered by water. In the air, water exists in three states: solid, liquid, and invisible vapor.

The amount of water in the air is called *humidity*. The *relative humidity* is the ratio of the actual amount of moisture in the air to the amount needed for saturation at the same temperature. Warm air can hold more water than cold. When air with a given amount of water vapor cools, its relative humidity increases; when the air is warmed, its relative humidity decreases.

Air Masses

An air mass is a vast body of air (a macroscale phenomena that can have global implications) in which the condition of temperature and moisture are much the same at all points in a horizontal direction. An air mass takes on the temperature and moisture characteristics of the surface over which it forms and travels, though its original characteristics tend to persist. The processes of radiation, convection, condensation, and evaporation condition the air in an air mass as it travels. Also, pollutants released into an air mass travel and disperse within the air mass. Air masses develop more commonly in some regions than in others. Table 10.1 summarizes air masses and their properties.

When two different air masses collide, a *front* is formed. A front is not a sharp wall but a zone of transition that is often several miles wide. Four frontal patterns—warm, cold, occluded, and stationary—can be formed by air of different temperatures. A *cold front* marks the line of advance of a cold air mass from below, as it displaces a warm air mass. A *warm front* marks the advance of a warm air mass as it rises up over a cold one.

When cold and warm fronts merge (the cold front overtaking the warm front), *occluded fronts* form. Occluded fr   can be called *cold front* or *warm front occlusions*. But in either case, a cold   mass takes over an air mass that is not as cold.

The last type of front is the *stationary front*. As the name implies, the air masses around this front are not in motion. A stationary front can cause bad weather conditions that persist for several days.

Thermal Inversions and Air Pollution

Earlier, it was pointed out that during the day the Sun warms the air near the Earth's surface. Normally, this heated air expands and rises during the day, diluting low-lying pollutants and carrying them higher into the atmosphere. Air

Table 10.1. Classification of Air Masses

Name	Origin	Properties	Symbol
Aortic	Polar regions	Low temperatures, low specific but high summer relative humidity, the coldest of the winter air masses	A
Polar continental*	Subpolar continental areas	Low temperatures (increasing with southward movement), low humidity, remaining constant	cP
Polar maritime	Subpolar area and arctic region	Low temperatures increasing with movement, higher humidity	mP
Tropical continental	Subtropical high-pressure land areas	High temperatures, low moisture content	cT
Tropical maritime	Southern borders of oceanic subtropical, high-pressure areas	Moderate high temperatures, high relative and specific humidity	mT

*The name of an air mass, such as *polar continental*, can be reversed to *continental polar*, but the symbol, *cP*, is the same for either name.

Source: EPA 2005.

from surrounding high-pressure areas then moves down into the low-pressure area created when the hot air rises. This continual mixing of the air helps keep pollutants from reaching dangerous levels in the air near the ground.

Sometimes, however, a layer of dense, cool air is trapped beneath a layer of less dense, warm air in a valley or urban basin. This is called a *thermal inversion*. In effect, a warm-air lid covers the region and prevents pollutants from escaping in upward-flowing air currents. Usually these inversions trap air pollutants (i.e., plume dispersion is inhibited) at ground level for a short period. However, sometimes they last for several days when a high-pressure air mass stalls over an area, trapping air pollutants at ground level where they accumulate to dangerous levels.

The best known location in the United States where thermal inversions occur almost on a daily basis is in the Los Angeles Basin. The Los Angeles Basin is a valley with a warm climate and light winds, surrounded by mountains located near the Pacific Coast. Los Angeles is a large city with a large population of people and automobiles and possesses the ideal conditions for smog, which is worsened by frequent thermal inversions.

References and Recommended Reading

Environmental Protection Agency. 2005. *Basic air pollution meteorology.* APTI Course SI: 409. Accessed 1/08/08 from www.epa.gov/apt.

Spellman, F. R., & Whiting, N. 2006. *Environmental science and technology: Concepts and applications,* 2nd edition. Rockville, Maryland: Government Institutes.

CHAPTER 11

Climate Change

> Humanity is conducting an unintended, uncontrolled, globally pervasive experiment whose ultimate consequences could be second only to nuclear war. The Earth's atmosphere is being changed at an unprecedented rate by pollutants resulting from human activities, inefficient and wasteful fossil fuel use and the effects of rapid population growth in many regions. These changes are already having harmful consequences over many parts of the globe.
>
> —Toronto Conference statement, June 1988

Setting the Stage

Time: 10,312 BCE

He sat on the ground leaning against a deadfall, his leather-wrapped legs pulled tight under him, and watched the swamp. He felt disoriented, detached from the world around him. Even the air around him felt strange. It was different; this place was unusually warm. A possum waddled from a copse of vine-maple below him, and Yurk watched the possum move off to the left—the possum, in a hurry, constantly jerking his head to the right, over his shoulder. The possum darted toward the marshy bank, and stopped to sniff the ground. Some noise or an odor carried on the wind seemed suddenly to startle the possum into attention, and he looked back toward Yurk, then moved off into the tall marsh grass, where he disappeared from view.

The rain was coming down in a fine drizzle. The wind sighed through the fir boughs and the afternoon was redolent with the smell of tree-perfumed air. Even with the light rain and wind, though, Yurk was warm—warmer than he ever remembered being before. He had never been so warm, his whole body at the same time. By a fire, only what faces the fire is warm.

Yurk's weathered face wore a mesmerized look as he chewed on a piece of bark, resting against the decaying trunk of the fallen tree, almost as if he was unaware of his surroundings. His eyes glazed over, as if he was there in shell form only—an empty one at that. Maybe his blank state and hypnotized appearance was the result of the view in front of those blank eyes: Great truncated tree trunks blackened by fire stood above the surface of the swamp water, stark remnants of a very ancient past. A misty pall hung over the swamp as the blackness lowered over the forest and the swamp took on an eerie, forbidding, spectral quality with the coming of night.

The cry of an owl drifted through the dark forest as Yurk stood (an effort that required much exertion from his tired, ancient body). Carefully, he stretched and yawned—careful not because of his frailty or from a sense of impending danger, but because of instinct—not fear exactly, just instinct. A lifetime, generations of lifetimes of vigilance for survival (both conscious and unconscious) had taught Yurk to be vigilant at all times. He was leg-weary and footsore, but that really didn't concern him. He knew his ending time was near—that was why he had traveled better than 200 miles to this place. This place he had come to had been familiar to him years before, but in a very different form. He wanted to see the wonderment of the swampy terrain that lay before him now.

Yurk was viewing something he had heard about from other clan members but something he had never witnessed before: a swamp.

Yes, a swamp with blackened, truncated tree remnants. In all his years (Yurk was unusually old for his time and circumstances, well more than 60), Yurk had never seen such a sight. Before—up until now—the landscape he had been familiar with had been covered in snow and ice. He had visited this place many times in the past—what seemed a bare plain of ice and snow. He had not been on this journey in many years, but the last time he had come, he had simply trudged through the open area (the swamp) over a bridge of thick ice and snow. He (and no one else) had any idea that the swamp lay below the thick layers of ice and snow. In his absence from this place, he had heard the tales from the younger clan hunters, and had decided to take his last journey to see such a place, such a site before he died.

It was so warm.

As he stood, wiping his wet brow and looking out upon the swamp, 30 feet to Yurk's left, working toward the top of the steep, craggy ledge on the sheer cliff edge climbed the cat.

Like Yurk, the cat had come to this place many times in the past; although she could not cognitively determine the exact difference between the past and the present, she, too, knew this place had changed.

It was so warm.

In the past 15 or so years, the cat, along with her running mates (these cats almost always ventured into the wilderness accompanied—to hunt and to kill required help, sometimes lots of help) had, like Yurk and his clan members, crossed the swamp using the ice-bridge. But now things were different; the cat knew this. She also knew that something else was different: It was so warm.

The cat (known today as *Smilodon* or saber-toothed tiger) continued slowly, inexorably up the steep slope of the stony ridge. In the past, she had allowed for the slipperiness of the ice sheet that covered the ledge, but at this time of year she should have had very little difficulty climbing the high terminal edge overlooking the swamp. But now things were different—much different. She was on her last legs, in all ways. But her difficulty was even more than that; even though the going was easier now without the ice and snow, she still struggled her way up to the terminal point. It was so hot. She labored even to breathe.

Yurk and the cat were aware of each other. Each knew the other was there—have no doubt about that. Yurk was probably more fearful of the cat than she was of him. But how could anyone tell? They had been bitter enemies throughout their lives. The cat preferred feasting on mammoths and mastodons (Yurk liked that kind of meat himself), but when confronted with her "only" threat, her only true enemy, the cat knew she was wise to be ready. Life of any sort was difficult enough; not being alert and wary at all times was certainly an invitation to disaster—for both of them.

But now things were different. It was so warm. Neither the cat nor Yurk was attentive to each other; they were not as alert, as wary of each other as they had been in the past. Each knew, in their own way, that the days of hunting and protecting themselves were behind them—food certainly wasn't a consideration with either. No, food was not a problem; they were not hungry. Afraid? No, not really.

The cat continued her climb and finally reached the summit. She stood looking out on the swamp, with one eye semifocused on Yurk. Yurk stood below, looking out on the swamp, aware of her presence as well.

They both knew in their own way that things were different. Hell, they could feel the difference; it was so warm.

Warm. Yes, it was warm. For their entire lives, they had never known such warmth, had never seen the snow and ice melt, had never witnessed the swampy landscape now before them. Their world was different, fearfully and wonderfully changed.

The warming trend had actually begun about 2 or 3 years earlier, though Yurk and the cat had barely been aware of it, because the increase in temperatures had been subtle, just about a half a degree Fahrenheit every 3 months or so. But now, now the difference was obvious. The temperature was a least 10 degrees warmer than they had ever experienced; thus the melt, the freshly uncovered swamp, the rock-strewn ledge—and the warmth, of course.

The cat and Yurk stood for a time, gazing out at the swamp. What this change would mean to their clan and mates—those to follow—they were not capable of determining. What this change would bring to their world, they were not capable of speculating. So they stood, until Yurk sat back down on the ground, his back against the deadfall, and the cat lay down on the heated rocks of the ledge; they were both exhausted, tired, worn out—old, so old. And warm—too warm.

About an hour later, as darkness fell total upon the blackened, spectral landscape before them, they both went to sleep, the sleep of the dead, and their own warmth turned cold.

The ambient temperature continued to rise, even now that it was dark, night. A night that when ended would bring the dawn of a new day and the dawn of a new era.

It was so warm and getting warmer.

The question is, Can't we do it with all the wastes we produce?

We think that goal is possible, especially if the scientific, political, social, and monetary commitment that needs to be made is made—and it will be.

Why?

What other choice do we have? (Spellman, 1999)

Are we headed for warmer times or colder times? Is global climate change actually happening, and if so, do we really need to worry about it? Are the tides rising? Does the ozone hole portend disaster right around the corner?

These days, many people are beginning to ask a variety of questions related to climate change. Such questions seem reasonable when you consider the constant barrage of newspaper headlines, magazine articles, and television news reports we have been exposed to in recent years. Recently, for example, El Niño (and its devastation of the west coast of the United States, Peru, and Ecuador) has received a lot of attention in the media as well as scientific attention. On the other side of the coin, it has helped reduce the usual number, magnitude, and devastation of hurricanes that annually blast the east coast of the United States—though the people affected by the ice storms in upstate New York and in Canada and the tornado victims in Florida in the winter of 1998 probably didn't consider themselves very lucky.

What's going on? What does all this mean? We have plenty of theories and doomsayers out there, but are they correct? Does anyone really know the answers? Is there anything we can do?

Not really.

Should we be concerned?

Yes.

Should we panic?

No.

Should we take decisive action? Should we practice feel-good science instead of good science?

Not exactly.

Is there anything we can do?

Yes. We can study the facts, the issues, the possible consequences—we can do all this, but we need to let scientific fact, common sense, and cool-headedness prevail. Shooting from the hip is not called for here, makes little sense, and could have Titanic consequences for us all.

One thing is certain: We cannot abandon ship. The SS Earth is the only ship around and whether we like it or appreciate it doesn't really matter; we are all passengers headed in the same direction. We hope our journey will continue in such a manner that we and SS Earth can continue to make headway with fair winds and following seas.

Will we be able to accomplish this?

Maybe a better question is whether there is anything we can do about it.

The answer, of course, is yes. We can do something—and we will because what other choice do we have?

The only question that really has any merit here is whether we will take the correct action before it is too late. The key words are "correct action."

In this chapter, global climate change related to our atmosphere and its problems, actual and potential, are discussed. Consider this: Any damage we do to our atmosphere affects the other three mediums—water, soil, and biota. Thus, the endangered atmosphere (if it is endangered) is a major concern to all of us.

The Past

Before we begin our discussion of the past, we need to define the era we refer to when we say "the past." Table 11.1 gives the entire expanse of time from Earth's beginning to present. Table 11.2 provides the sequence of geological epochs over the past 65 million years, as dated by modern methods. The Paleocene through Pliocene together make up the Tertiary Period; the Pleistocene and the Holocene compose the Quaternary Period.

When most people think about climatic conditions in the ancient past, they generally think of two eras: the Ice Age and the period of the dinosaurs. Of course, those two ages take up only a tiny fraction of the time Earth has been spinning around the Sun. Let's look at what we know about the past. In the first place, geological history has shown periods when the normal climate of the Earth was so warm that subtropical weather reached to 60° north and south latitude, and there was a total absence of polar ice.

Table 11.1. Geologic Eras and Periods

Era	Period	Millions of Years before Present
Cenozoic	Quaternary	2.5–present
	Tertiary	65–2.5
Mesozoic	Cretaceous	135–65
	Jurassic	190–135
	Triassic	225–190
Paleozoic	Permian	280–225
	Pennsylvanian	320–280
	Mississippian	345–320
	Devonian	400–345
	Silurian	440–400
	Ordovician	500–440
	Cambrian	570–500
	Precambrian	4600–570

Most people are unaware that glaciers have advanced and reached as far south as what is now the temperate zone of the northern hemisphere during less than about 1% of Earth's history. The latest such advance (which started about 1,000,000 years ago) was marked by geological upheaval and, arguably, the beginning of man. During this time, vast ice sheets advanced and retreated over the continents. The following takes a closer look at these Ice Ages.

A TIME OF ICE

The oldest known glacial epoch occurred nearly 2 billion years ago. In southern Canada, extending east to west about 1,000 miles is a series of deposits of glacial origin. Within the last billion years or so, the earth has experienced at least six major phases of massive, significant climatic cooling and subsequent glaciation, which apparently occurred at intervals of about 150 million years. Each may have lasted as long as 50 million years.

Table 11.2. Geological Epochs

Epochs	Million Years Ago
Holocene	0.01–0
Pleistocene	1.6–0.01
Pliocene	5–1.6
Miocene	24–5
Oligocene	35–24
Eocene	58–35
Paleocene	65–58

In more recent times (Pleistocene epoch to present), examination of land and oceanic sediment core samples clearly indicate that numerous alterations between warmer and colder conditions have occurred over the last 2 million years (during the middle and early Pleistocene epoch). At least eight such cycles have occurred in the last million years, with the warm part of the cycle lasting only a relatively short time.

During the Pleistocene Epoch (what we commonly call the Great Ice Age), a series of ice advances began, at times covering over one-fourth of the earth's land surface with great sheets of ice thousands of feet thick. Glaciers moved across North America many times, reaching as far south as the Great Lakes, and an ice sheet thousands of feet thick spread over northern Europe, sculpting the land and leaving behind its giant footprints in the form of numerous lakes and swamps and its toe prints in the form of terminal moraines as far south as Switzerland. Evidence appears to indicate that each succeeding glacial advance was more severe than the previous one. The most severe began about 50,000 years ago and ended about 10,000 years ago. Several glacial advances were separated by interglacial stages, during which the ice melted and temperatures were on average higher than today.

"Temperatures were higher than today?" Yes. Keep this important point in mind as we proceed.

Although scientists consider the Earth still to be in a glacial stage (because one-tenth of the globe's surface is still covered by glacial ice), ever since the climax of the last glacial advance, the ice sheet has been in a stage of retreating and world climates, although fluctuating, are slowing warming.

How do we know the ice sheet is in a retreating stage? We know from our observations and well-kept records that clearly show that the last 100 years have seen marked worldwide retreat of ice. Swiss resorts built during the early 1900s to offer scenic views of glaciers now have no ice in sight. Glacier National Park in Montana, world famous for its 50 glaciers and 200 lakes, is not quite the same place it was a hundred years ago. In 1937, a 10-foot pole was driven into the ground at the terminal edge of one of the main glaciers. Today this pole is still in place, but the glacier has retreated several hundred feet back up the slope of the mountain. If this glacial retreat continues and all the ice melts, sea levels will rise more than 200 feet, flooding many of the world's major cities. New York and Boston would then become aquariums.

What causes an Ice Age? The cause of these periodic ice ages is a deep enigma of Earth history. Scientists have advanced many theories ranging from changing ocean currents to sunspot cycles. One fact that is absolutely certain, however, is that an Ice Age event occurs because of a change in Earth's climate. But what brings about such a drastic change? To answer this question we need to take a closer look at some factors related to climate and climate change.

Climate results from the uneven distribution of heating over the surface of the Earth caused by the Earth's tilt. This tilt is the angle between the earth's rotational axis and its orbital plane around the sun. Currently this angle is 23.5 degrees, but it has not always been at that angle. The angle has changed; we will discuss this in greater detail shortly.

Long-term climate is also affected by the heat balance of the earth, which is driven mostly by the concentration of carbon dioxide (CO_2) in the atmosphere. Climate change can result if the pattern of solar radiation is changed or if the amount of carbon dioxide changes. Abundant evidence that the earth does undergo climatic change exists. Climatic change can be a limiting factor for the evolution of many species.

As stated previously, there is abundant evidence (primarily from soil core samples and topographical formations) that the earth undergoes climatic change. Climate change includes events such as periodic Ice Ages, characterized by glacial and interglacial periods. Major glacial periods lasted up to 100,000 years, with a temperature decrease of about 9° F, and most of the planet was covered with ice. Minor periods lasted up to 12,000 years with about a 5° F decrease in temperature, and ice covered 40° latitude and above. Smaller periods such as the "Little Ice Age" occurred from about 1000–1850 BCE, when there was about a 3.8° F drop in temperature. Despite its name, "Little Ice Age" was not a true glacial period, but rather a time of severe winters and violent storms.

We are presently in an interglacial stage that may be reaching its apogee. The Earth has gone through a series of glacial periods, and from the best information available to us, these periods are cyclical. What does that mean in the long run? No one knows for sure, but let's look at the effects of Ice Ages.

Ice Ages bring about changes in sea levels. A full-blown Ice Age can change sea level by about 100 meters, which would expose the continental shelves. The exposed continental shelves' composition would be changed because of increased deposition during melt. The hydrological cycle would change because less evaporation would occur. Significant changes in landscape would occur, such as the creation of huge formations on the scale of the Great Lakes. Drainage patterns throughout most of the world would change, with possible massive flooding episodes. Topsoil characteristics would change—glaciers deposit rock and grind away soil.

Are these changes significant? Many areas would be devastatingly affected by the changes that an Ice Age could bring. In particular, northern Europe, Canada, Seattle, Washington, around the Great Lakes, and coastal regions would be affected.

Let's go back to the main question: What causes Ice Ages? The answer is that we are not sure, but there are some theories. Scientists point out, for example, in order to generate a full-blown Ice Age (a massive ice sheet covering most of the globe), certain periodic or cyclic events or happenings would have to take place. The periodic fluctuations referred to would have to affect the solar cycle, for instance. However, we have no definitive evidence that this has ever occurred. Another theory speculates that periods of widespread volcanic activity generating masses of volcanic dust block or filter heat from the sun, thus cooling down the Earth.

Some speculate that the carbon dioxide cycle would have to be periodic or cyclic to bring about periods of climate change. References have been made to a so-called factor 2 reduction, causing a 7° F temperature drop worldwide.

Others speculate that another global Ice Age could be brought about by increased precipitation at the poles, caused by changing orientation of continental land masses. And still others theorize that a global Ice Age would result if changes in mean temperatures of ocean currents occurred. But the question is how? By what mechanism?

So what are the most probable causes of Ice Ages on Earth? According to the *Milankovitch hypothesis,* the occurrence of an Ice Age is governed by a combination of factors: (1) the Earth's change of altitude in relation to the sun: the way it tilts in a 41,000-year cycle and at the same time wobbles on its axis in a 22,000-year cycle, making the time of its closest approach to the sun come at different seasons; and (2) the 92,000-year cycle of eccentricity in its orbit round the sun, changing it from an elliptical to a near-circular orbit, with the severest period of an Ice Age coinciding with the approach to circularity.

What this means is that we have a lot of speculation about Ice Ages and their causes and their effects. We do not have any speculation about the fact that they actually occurred and that they caused things to occur (formation of the Great Lakes, etc.), but there is a lot we do not know—the old "we don't know what we don't know" paradox.

Many possibilities exist. At this point, no single theory is sound, and, doubtless, many factors are involved. But we should keep in mind that we are possibly still in the Pleistocene Ice Age. It may reach another maximum in another 60,000 years or so. This issue will be revisited again in the next section.

WARM WINTER

Maybe you've seen the headlines: "1997 Was the Warmest Year on Record," "Scientists Discover Ozone Hole Is Larger Than Ever," "Record Quantities of Carbon Dioxide Detected in Atmosphere," or "January 1998 Was the Third Warmest January on Record." Have you seen other reports about research that indicates that we are undergoing a cooling trend?

In the previous section, we discussed several possible causes of glaciation and subsequent climatic cooling. In most scenarios discussed, we were left with the old paradox: We do not know what we don't know. Now it's time to discuss how we know what we think we know about climatic change.

WHAT WE THINK WE KNOW ABOUT GLOBAL CLIMATE CHANGE

Two large-scale environmentally significant events took place in 1997: the return of El Niño and the Kyoto Conference: Summit on Global Warming and Climate Change. As to El Niño, 1997 and 1998 news reports have blamed this phenomenon for just about everything and anything that has to do with weather conditions throughout the world. Some of these occurrences are indeed El Niño–related or El Niño–generated: the out-of-control fires, droughts, floods; the stretches of dead coral; the lack of fish in the water; and the reduction in the number of birds around certain Pacific atolls are examples. Few would argue that the devastating storms that struck the west coasts of South America, Mexico, and California were not related to El Niño. Additionally, few argue against El Niño's affect on the 1997 hurricane season, one of the mildest on record. However, other anomalies or occurrences blamed on El Niño, such as lower or higher plant growth in certain regions of the globe and other absurdities like the appearance of a double rainbow in certain areas, certainly are suspect if not totally ridiculous.

On December 7, 1997, the Associated Press reported that, while delegates at the global climate conference in Kyoto haggled over greenhouse gases and emission limits, a compelling—and so far unanswered—question has emerged: "Is global warming fueling El Niño?"

Nobody is really sure because we need more data than we have today. One thing seems certain (based on our paltry amount of recorded data): El Niño is getting stronger and more frequent.

Some scientists fear that the increasing frequency and intensity of El Niño (based on records showing that two of this century's three worst El Niños have come recently, in 1982 and 1997) may be linked to global warming. Experts at the Kyoto Conference say the hotter atmosphere is heating up the world's oceans, which could set the stage for more frequent and extreme El Niños.

We have little doubt that weather-related phenomena seems to be intensifying throughout the globe. Can we be sure that this is related to global warming yet? No. The jury is still out. We need more data, more science, more time.

Is there cause for concern? Yes, there is. According to the Associated Press coverage of the Kyoto Conference, scientist Richard Fairbanks reported that he found some startling evidence of our need for concern. During two months of scientific experiments on Christmas Island (the world's largest atoll in the Pacific Ocean) conducted in autumn 1997, he discovered a frightening scene. The water surrounding the atoll was 7° F higher than average for that time of year, throwing the environmental system out of balance. According to Fairbanks, 40% of the coral was dead, the warmer water had killed off or driven away fish, and the atoll's normally plentiful bird population was almost completely gone.

Few would argue with the affect that El Niño is having on the globe. However, we are not certain that it is caused or intensified because of global warming. The natural question now shifts to what we know about global warming and climate change.

USA Today (December 1997) reported on the results of a report issued by the Intergovernmental Panel on Climate Change and an interview with Jerry Mahlman of the National Oceanic and Atmospheric Administration and Princeton University in which the following information about what most scientists agree on was obtained (p. A-2):

- There is a natural "greenhouse effect" (first discovered by Joseph Fourier in 1824). Scientists know how it works, and without it, Earth would freeze.
- The Earth undergoes normal cycles or warming and cooling on grand scales. Ice ages occur every 20,000 to 100,000 years.
- Globally, average temperatures have risen 1 degree in the past 100 years, within the range that might occur normally.
- The level of artificial carbon dioxide in the atmosphere has risen 30% since the beginning of the Industrial Revolution in the 19th century, and is still rising.
- Levels of artificial carbon dioxide will double in the atmosphere over the next 100 years and generate a rise in global average temperatures of about 3.5° F (larger than the natural swings in temperature that have occurred over the past 10,000 years).
- By 2050, temperatures will rise much higher in northern latitudes than the increase in global average temperatures. Substantial amounts of northern sea ice will melt, and snow and rain in the northern hemisphere will increase.
- As the climate warms, the rate of evaporation will rise, further increasing warming. Water vapor also reflects heat back to Earth.

WHAT WE THINK WE KNOW ABOUT GLOBAL WARMING

What is global warming? To answer this question we need to discuss the greenhouse effect. We know that water vapor, carbon dioxide, and other atmospheric gases (greenhouse gases; Table 11.3) help to warm the Earth. Without this

Table 11.3.   Greenhouse Gases

Greenhouse Gas	Percent of Total Greenhouse Gases	Sources and Percent of Total Greenhouse Gases
Carbon dioxide	50	Energy from fossil fuels (35) Deforestation (10) Agriculture (3) Industry (2)
Methane	16	Energy from fossil fuels (4) Deforestation (4) Agriculture (8)
Nitrous oxide	6	Energy from fossil fuels (4) Agriculture (2)
Chlorofluorocarbons (CFCs)	20	Industry (20)
Ozone	8	Energy from fossil fuels (6) Industry (2)

EPA (2005).

greenhouse effect, the Earth's average temperature would be closer to zero than its actual 60°. As gases are added to the atmosphere, the average temperature could increase, changing orbital climate.

Greenhouse Effect

Here is an explanation of Earth's greenhouse effect most people, especially gardeners, are familiar with. In a garden greenhouse, the glass walls and ceilings are largely transparent to shortwave radiation from the sun, which is absorbed by the surfaces and objects inside the greenhouse. Once absorbed, the radiation is transformed into longwave (infrared) radiation (heat), which is radiated back from the interior of the greenhouse. But the glass does not allow the longwave radiation to escape, instead absorbing the warm rays. With the heat trapped inside, the interior of the greenhouse becomes much warmer than the air outside.

The earth's atmosphere allows much the same greenhouse effect to take place. The shortwave and visible radiation that reaches Earth is absorbed by the surface as heat. The long heat waves are then radiated back out toward space, but the atmosphere instead absorbs many of them. This is a natural and balanced process and, indeed, is essential to life as we know it on Earth. The problem comes when changes in the atmosphere radically change the amount of absorption, and, therefore, the amount of heat retained. Scientists speculate that this may have been happening in recent decades as various air pollutants have caused the atmosphere to absorb more heat. This phenomenon takes place at the local level with air pollution, causing heat islands in and around urban centers.

As pointed out earlier, the main contributors to this effect are the greenhouse gases: water vapor, carbon dioxide, carbon monoxide, methane, volatile organic compounds, nitrogen oxides, chlorofluorocarbons, and surface ozone. These gases delay the escape of infrared radiation from the Earth into space, causing a general climatic warming. Note that scientists stress that this is a natural process. Indeed, the Earth would be 33° C cooler than it is presently if the "normal" greenhouse effect did not exist (Hansen et al., 1986).

The problem with Earth's greenhouse effect is that human activities are now rapidly intensifying this natural phenomenon, which may lead to global warming. Debate, confusion, and speculation about this potential consequence is rampant. Scientists are not entirely sure whether the recently perceived worldwide warming trend is because of greenhouse gases or some other cause or whether it is simply a wider variation in the normal heating and cooling trends they have been studying. If it continues unchecked, however, the process may lead to significant global warming, with profound effects. Human influence on the greenhouse effect is real; it has been measured and detected. The rate at which the greenhouse effect is intensifying is now more than five times what it was during the last century (Hansen & Lebedeff, 1989).

Greenhouse Effect and Global Warming

Those who support the theory of global warming base their assumptions on human altering of the Earth's normal greenhouse effect, which provides necessary warmth for life. They blame human activities (burning of fossil fuels, deforestation, and use of certain aerosols and refrigerants) for the increased amounts of greenhouses gases. These gases have increased the amounts of heat trapped in the Earth's atmosphere, gradually increasing the temperature of the whole globe.

Many scientists note that recent, short-term observation indicates that the last decade has been the warmest since temperature recordings began in the late 19th century, and that the more general rise in temperature in the last century has coincided with the Industrial Revolution, with its accompanying increase in the use of fossil fuels. Other evidence supports the global warming theory. For example, in the Arctic and Antarctica, places that are synonymous with ice and snow, we see evidence of receding ice and snow cover.

Taking a long-term view, scientists look at temperature variations over thousands or even millions of years. Having done this, they cannot definitively show that global warming is anything more than a short-term variation in Earth's climate. They base this assumption on historical records that show that the Earth's temperature does vary widely, growing colder with ice ages and then warming again. On another side of the argument, some people point out that the 1980s saw 9 of the 12 warmest temperatures ever recorded, and the Earth's average surface temperature has risen approximately 0.6° C (1° F) in the last century (Environmental Protection Agency [EPA], 1995). At the same time, still others offer as evidence that the same decade also saw three of the coldest years: 1984, 1985, and 1986.

So what is really going on? We are not certain, but let's assume that we are indeed seeing long-term global warming. If this is the case, we must determine what is causing it. But here we face a problem. Scientists cannot be sure of the greenhouse effect's causes. Global warming may simply be part of a much longer trend of warming since the last Ice Age. Though much has been learned in the past two centuries of science, little is actually known about the causes of the worldwide global cooling and warming that have sent the Earth through a succession of major and lesser ice ages. We simply don't have the long-term data to support our theories.

Factors Involved with Global Warming and Cooling

Right now, scientists are able to point to six factors that could be involved in long-term global warming and cooling:

1. Long-term global warming and cooling could result if changes in the Earth's position relative to the Sun occur (i.e., the Earth's orbit around the Sun changes), with higher temperatures when the two are closer together and lower when farther apart.
2. Long-term global warming and cooling could result if major catastrophes (meteor impacts or massive volcanic eruptions) occur that throw solar-radiation blocking pollutants into the atmosphere.
3. Long-term global warming and cooling could result if changes in albedo (reflectivity of the Earth's surface) occur. If the Earth's surface were more reflective, for example, the amount of solar radiation radiated back toward space instead of being absorbed would increase, lowering temperatures on Earth.
4. Long-term global warming and cooling could result if the amount of radiation emitted by the sun changes.
5. Long-term global warming and cooling could result if the shape and relationship of the land and oceans change.
6. Long-term global warming and cooling could result if the composition of the atmosphere changes.

This last possibility, the potential change in the composition of the atmosphere, of course, relates directly to our present concern: Have human activities had a cumulative effect large enough to affect the total temperature and climate of Earth? We are not certain right now, but we are somewhat concerned and alert to the problem.

So what does all this mean? If global warming is occurring, we can expect winters to be shorter, summers to be longer and warmer, and the sea level to rise on the order of a foot or so in the next 100 years and to continue to do so for many hundreds of years.

We have routine global temperature measurements for only about 100 years, and these are not too reliable because of changes in instrumentation and methods of observation. The only conclusion to be drawn about our climate is that we do not know whether it is changing drastically. The key word is *drastically*. Geologically, we may be at the end of an Ice Age. Evidence indicates that, during interglacial cycles, there is a period when temperatures increase before they plunge. Are we ascending the peak temperature range? How about human effects on climate? Have they become so marked that we cannot be sure that the natural cycle of Ice Ages (which has lasted for the last 5 million years or so) will continue? Or are we just having a breathing spell of a few centuries before the next glacial advance? No one knows for sure. What do you think?

✓ **Interesting Point:** Views on the effects of global climate change and subsequently on global warming are many and varied. It is interesting to note that in my undergraduate and graduate-level air pollution courses at Old Dominion University, we spend a lot of time discussing and researching global climate change. I have found (surprisingly, to me, at least) that many students do not believe that global warming is a problem. Instead, they argue that global warming is better than global cooling—no argument there. Enduring warm winters is better than enduring cold winters—in my old age, this seems sensible. Because energy supplies are limited, it is better that the globe is warming than cooling simply because it will be difficult to keep us all warm with a limited, dwindling amount of available energy. Some might say that these are simple statements that have some semblance to that rare substance that is not so common: common sense.

References and Recommended Reading

Associated Press. 1997. Does warming feed El Niño? *Virginian-Pilot* (Norfolk, Virginia), December 7, p. A-15.

Environmental Protection Agency (EPA). 2005. *Basic air pollution meteorology.* Accessed 1/11/08 from www.epa.gov/apti.

Global warming: Policies and economics further complicate the issue. *USA Today,* pp. A-1, 2, December 1, 1997.

Hansen, J. E., et al. 1986. Climate Sensitivity to Increasing Greenhouse Gases. *Greenhouse Effect and Sea Level Rise: A Challenge for this Generation.* M. C. Barth & J. G. Titus, editors, New York: Van Nostrand Reinhold.

Hansen, J. E., & Lebedeff, 1989. Greenhouse effect of chlorofluorocarbons and other trace gases. *Journal of Geophysical Research* 94 (November), pp. 16, 417–416, 421.

Spellman, F. R. 1999. *The science of environmental pollution.* Boca Raton, Florida: CRC Press.

Spellman, F. R., & Whiting, N. 2006. *Environmental science and technology: Concepts and applications,* 2nd ed. Rockville, Maryland: Government Institutes.

Toronto Conference. 1988.

Part III

EARTH'S GEOSPHERE (LITHOSPHERE)

CHAPTER 12

Earth's Structure

Nobody hurries geology.

—Mark Twain

Earth, Planet Earth, the *World,* and *Terra,* the third planet from the sun, the place we live—what could be more important to us? Yet, how little most of us really know about the composition and history of the planet that is our home.

—Jennifer Blue

These statements set the stage for what is to follow in this and subsequent chapters in this part of the text. As Mark Twain points out, Earth's structure, or geology, is all about time—lots of time, eons of time; time that is endless (we hope). The fact is that because of the passage of time, we do not know what we do not know about our own planet, Earth. How could we? None of us has lived the 4.54 billion years from the time the scientific evidence suggests that Earth was formed to the present. Well, we might think, to find out about Earth—its beginning and its subsequent history—couldn't we just refer back to the written record? That depends on the *written record* being referred to. In regard to the human written record, keep in mind that prior to 5,000–5,500 years ago there is no written record (that we are aware of) that has preserved all the knowledge of facts or events that occurred in the past. There simply is not a continuous narrative and systematic analysis of past events of importance to the human race because on the scale of longevity, the time humankind has spent on Earth is measured as a small drop in an enormous ocean; likewise, on a comparative scale, the period during which humankind has been able to write is so small it cannot be perceptively measured. Events that occurred before the advent of written communication are dubbed *prehistory*. In writing—the written historical record of Earth—there is little that is known or written about prehistorical times.

The existing *written historical record* is defined as a written document preserving knowledge of facts and events. As discussed previously, the written historical record as recorded by humankind is extremely limited. This is not to say, however, that there is no record of the history of the Earth available for our perusal. No one has "written" more about the history of past events of importance to Earth and the human race than Mother Nature. All we need do is to look at Earth (rocks) and its formations to see its past. It is true, however, that all Mother Nature has written in Earth's past is not that clear to us; again, the heavy hand of time has molded, shaped, reshaped, retooled, and reshaped again and again Earth's physical features in so many different ways and in so many directions that it is literally impossible for anyone to decipher exactly what has occurred to any given region on Earth at any given time in the ancient past. Compounding the problem brought about by the effects of time is that not many of us are able or qualified to decipher what it is we are looking at in regard to Earth and its history. To do this to any extent, we must know geology.

Some geology facts are common knowledge; we know them despite not having any detailed geological knowledge. We know about the products of soils that have been formed by weathered rock, oil formed from the remains of prehistoric plants and animals, and the beauty and value of precious stones. These are the basics; they constitute only a small fraction of the useful materials with which the Earth provides us.

Think about how impossible it would be for modern industry to have developed as we know it today without other Earth products. Mineral resources such as coal, iron, lead, and petroleum derived from the Earth have been made readily available through the application of basic geology, geological engineering, global positioning systems, and generic mapping tools.

Earth also provides us with areas of exceptional beauty. One need not be a geologist to marvel at the breathtaking vastness of the Grand Canyon; the enriching warmth of the sandstone formations and the very color of the Earth at Zion, Bryce Canyon, Arches National Parks, and Monument Valley Navajo Tribal Park all in Utah; the mystery of Luray Caverns; or the natural wonders of Yosemite or Yellowstone. All of these and many more are the results of geologic processes that are dynamic and are still at work today. They are the same geologic processes that began to shape the Earth more than 4.54 billion years ago!

What Is Geology?

It is important to point out that geology is not only the study of the structure of Earth as we see it today, but the history of the Earth as it has evolved to its present condition. In attempting to pin down a definition of *geology* (to explain what it is and what it is about), consider the definition provided by Press and Siever (2001):

> Earth is a unique place, home to more than a million life forms, including ourselves. No other planet yet discovered has the same delicate balance necessary to sustain life. Geology is the science that studies the Earth—how it was born, how it evolved, how it works, and how we can help it.

Did You Know?

- The Earth has evolved (changed) throughout its history, and will continue to evolve.
- The Earth is about 4.6 billion years old; human beings have been around for only the past 2 million years.

The term *geology* is derived from the Greek *geo,* "Earth," plus *logos,* "discourse or study of." Paraphrasing Press and Siever's (2001) definition of *geology* in simplistic terms, we can define it as the science that deals with the origin, structure, and history of the Earth and its inhabitants as recorded in the rocks.

Geology consists of the sciences of mathematics, physics (geophysics and seismology), and chemistry (petrology and geochemistry); these are the sciences that set the general principles for all the other sciences. Additionally, geology is composed of or interrelated to the sciences that describe the great systems that make up the universe: astronomy (planetary geology) and biology (paleontology). Some geologists are specialists who deal with very narrow parts of geology including mineralogy, meteorology, botany, zoology, and others. Geological science also includes exposure to the principles of sociology and psychology. In a nutshell and based on personal experience, it can be said with some certainty that the practicing geologist, whether a specialist or not, should and must be a generalist. That is, to be effective in practice, the geologist must have a well-rounded exposure to and some knowledge of just about everything related to the other scientific disciplines.

Professional geologists work to locate geologic resources. They also perform geological and mining engineering, site studies, and land-use planning. In regard to environmental protection, they conduct environmental impact studies and perform groundwater and waste management functions. Geologists also conduct basic research to maintain the science on the cutting edge of fundamental knowledge for current and future applications.

Scientific Principles and Geology

As with the other sciences, the science of geology is based on scientific principles. In the following, the geological principles of parsimony, superposition, and uniformitarianism are defined.

- Parsimony is a principle that states that, in arriving at a hypothesis or course of action, less is better (i.e., the least complex explanation for an observation is better). The meaning of this approach is best summed up by the acronym KISS ("Keep it simple, stupid"). The KISS principle states that simplicity is the key to understanding complex issues.
- Superposition, the basic principle to understand geologic history, states that in undisturbed sedimentary layers, rocks are deposited in a time sequence, with the oldest on the bottom and youngest on top. Note that in areas where the rocks have been greatly disturbed, it is necessary to determine the tops and bottoms of beds before the normal sequence can be established.
- Uniformitarianism is one of the most basic principles of modern geology, the observation that fundamentally the same geological processes that operate today also operated in the distant past. Or, more simply stated, *the present is the key to the past.*

Geology: Major Divisions

Because the scope of geology is so broad, it has been divided into two major divisions: *physical geology* and *historical geology.* Each of these divisions has been subdivided into a number of more specialized branches.

PHYSICAL GEOLOGY

Physical geology deals with the study of Earth materials as well Earth's structure, composition, the movements within and on the Earth's surface, and the geologic processes by which the Earth's surface is (or has been) changed.

To perform detailed studies and gain knowledge in all phases of geology (Earth science), the broad division of physical geology includes such basic sub-branches as geochemistry, the use of chemical principles to study and understand the composition of rocks and environmental quality. The study of the physical properties to infer internal structures is called geophysics. The sub-branches of mineralogy, the study and identification of minerals, and petrology, the study of rocks, provide much-needed information about the composition of the Earth. In addition, sedimentology studies sedimentary deposits. Another important branch of physical geology is structural geology, the study of the deformation of rocks. Economic geology, the exploration and exploitation of geological re-sources, is also an important branch.

HISTORICAL GEOLOGY

Historical geology is the study of the origin and evolution of Earth, its environ-ments, and its life (inhabitants) through time. It examines geologic history as a pattern and as process. Like physical geology, historical geology covers such a variety of fields that it has been subdivided into several branches. For example, the geologist utilizes stratigraphy, which is concerned with the origin, composi-tion, proper sequence, and correlation of the rock strata. Paleontology, the study of ancient organisms as revealed by their fossils, provides a background of the development of life on Earth, and paleogeography affords a means of studying geographic conditions of past times. Each of these branches, and others, is a sci-ence within itself that makes it possible to reconstruct the relations of ancient lands and seas and the organisms that inhabited them.

Did You Know?

Because the physical and historical geologist studies the same rocks, the unification of these two important divisions leads ultimately to a better understanding of the composition and history of Earth.

Earth's Geological Processes

Earlier it was pointed out that geology is about patterns and processes. We can also say that geology is about the materials that make up the Earth. The materials that make up the Earth, of course, are mainly rocks (including dust, silt, sand, and soil). Rocks in turn are composed of minerals and minerals are composed of atoms.

Earth materials and patterns are important and are discussed in detail later in the book, but for now it is Earth processes that are of interest to us. Earth processes are constantly acting upon and within the Earth to change it. Examples of these on-going processes include formation of rocks; chemical cementation of sand grains together to form rock; construction of mountain ranges; and erosion of mountain ranges. These are internal processes that get their energy from the interior of the Earth—most from radioactive decay (nuclear energy). Other examples of ongoing Earth processes include those that are more apparent to us (external processes) because they occur relatively quickly and are visible. These include volcanic eruptions, dust storms, mudflows, and beach erosion. The energy source for these processes is solar and gravitational.

It is important to point out that many of these processes are cyclical in nature. The two most important cyclical processes are the hydrological (water) cycle and the rock cycle.

HYDROLOGICAL CYCLE

Simply, the water cycle describes how water moves through the environment and identifies the links between groundwater, surface water, and the atmosphere (Figure 12.1). As illustrated, water is taken from the Earth's surface to the atmosphere by evaporation from the surface of lakes, rivers, streams, and oceans. This evaporation process occurs when the sun heats water. The sun's heat energizes surface molecules, allowing them to break free of the attractive force binding them together, and then evaporate and rise as invisible vapor in the atmosphere.

Water vapor is also emitted from plant leaves by a process called *transpiration.* Every day, a growing plant transpires 5 to 10 times as much water as it can hold at once. As water vapor rises, it cools and eventually condenses, usually on tiny particles of dust in the air. When it condenses, it becomes a liquid again or turns directly into a solid (ice, hail, or snow). These water particles then collect and form clouds. The atmospheric water formed in clouds eventually falls to Earth as precipitation. The precipitation can contain contaminants from air pollution. The precipitation may fall directly onto surface waters, be intercepted by plants or structures, or fall onto the ground. Most precipitation falls in coastal areas or in high elevations. Some of the water that falls in high elevations becomes runoff water, the water that runs over the ground picking up the sand, silt, and clay from the soil, and carries particles to lower elevations to form streams, lakes, and alluvial fertile valleys.

The water we see is known as *surface water.* Surface water can be broken down into five categories: oceans, lakes, rivers and streams, estuaries, and wetlands.

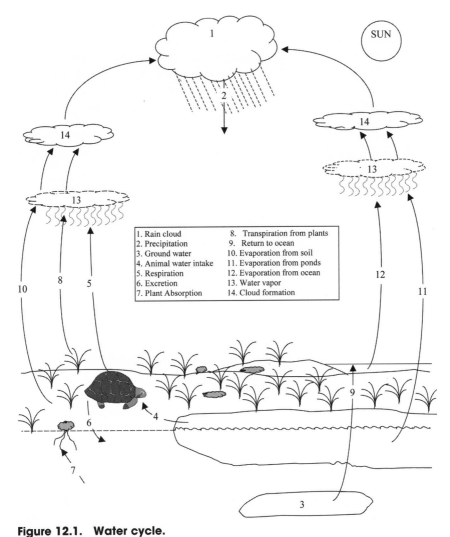

1. Rain cloud
2. Precipitation
3. Ground water
4. Animal water intake
5. Respiration
6. Excretion
7. Plant Absorption
8. Transpiration from plants
9. Return to ocean
10. Evaporation from soil
11. Evaporation from ponds
12. Evaporation from ocean
13. Water vapor
14. Cloud formation

Figure 12.1. Water cycle.

Modified from Carolina Biological Supply Co., 1966.

The health of rivers and streams is directly linked to the integrity of habitat (and geology) along the river corridor and in adjacent wetlands. Stream quality will deteriorate if activities damage vegetation along riverbanks and in nearby wetlands. Trees, shrubs, and grasses filter pollutants from runoff and reduce soil erosion. Removal of vegetation also eliminates shade that moderates stream temperature. Stream temperature, in turn, affects the availability of dissolved oxygen in the water column for fish and other aquatic organisms.

ROCK CYCLE

With time and changing conditions, the igneous, sedimentary, and metamorphic rocks of Earth are subject to alteration by the processes of weathering (erosion), volcanism, and tectonism. Known as the rock cycle (Figure 12.2), this series of

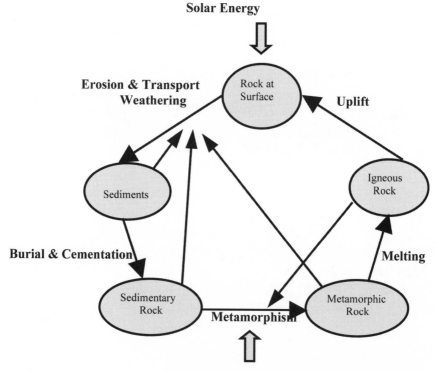

Figure 12.2. Rock cycle—if uninterrupted, will continue completely around the outer margin of the diagram. However, as shown by the arrows, the cycle may be interrupted or "short-circuited" at various points in its course.

events (or group of changes) represents a response of Earth materials to various forms of energy. As shown in Figure 12.2, most surface rocks started out as igneous rocks (rocks produced by crystallization from a liquid). When igneous rocks are exposed at the surface they are subject to weathering. Erosion moves particles into rivers and oceans where they are deposited to become sedimentary rocks.

Sedimentary rocks can be buried or pushed to deeper levels in the Earth, when changes in pressure and temperature cause them to become metamorphic rocks. At high temperatures, metamorphic rocks might melt to become magmas. Magmas rise to the surface, crystallize to become igneous rocks, and the processes starts over. Keep in mind that the cycles are not always completed, for there can be many short circuits along the way, as indicated in Figure 12.2.

Did You Know?

All rock on Earth (except for meteorites) today is made of the same stuff as the rocks that dinosaurs and other ancient life forms walked, crawled, or swam over. While the stuff that rocks are made from stays the same, the rocks do not. Over millions of years, rocks are recycled into other rocks. Moving tectonic plates help to destroy and form many types of rocks.

Planet Earth

The Earth has a radius of about 6,370 km, although it is about 22 km larger at the equator than at the poles. The circumference of the Earth is about 24,874 miles, and the surface area comprises roughly 197 million square miles (about 29%). The remaining 71% of the Earth's surface is covered by water.

The Earth, like the other planets in our solar system, revolves around the Sun within its own orbit and period of revolution. The Earth also rotates on its own axis. The Earth rotates from west to east and makes one complete rotation each day. It is this rotating motion that gives us the alternating periods of daylight and darkness that we know as *day* and *night*.

The Earth also precesses (or wobbles) as it rotates on its axis, much as a top wobbles as it spins. This single wobble has to do with the fact that the Earth's axis is tilted at an angle of 23½ degrees. The tilting of the axis is also responsible for the seasons. Over the years, various hypotheses have been put forward in attempts to identify the force, or excitation mechanism, that propels the wobble, such as atmospheric phenomena, continental water storage (changes in snow cover, river runoff, lake levels, or reservoir capacity), interaction at the boundary of Earth's core and its surrounding mantle, and earthquakes. In an explanation by the National Aeronautics and Space Administration (2000), the principal cause of the wobble is fluctuating pressure on the bottom of the ocean caused by temperature and salinity changes and wind-driven changes in the circulation of the oceans.

The Earth revolves around the sun in a slightly elliptical orbit approximately once every 365¼ days. During this solar year, the Earth travels at a speed of more than 60,000 miles per hour, and on the average it remains about 93 million miles from the sun.

Principal Divisions of the Earth

The Earth consists of four interconnected *geospheres:* the atmosphere, a gaseous envelope surrounding the Earth; the hydrosphere, the waters filling the depressions and covering almost 71% of the Earth; and the lithosphere, the solid part of the Earth that underlies the atmosphere and hydrosphere; and the biosphere, composed of all living organisms.

THE ATMOSPHERE

As mentioned, the atmosphere is the body of air that surrounds our planet. But what is air? Air is a mixture of gases that constitutes the Earth's atmosphere. The atmosphere is a thin shell or veil, an envelope of gases that surrounds Earth like the skin of an apple—it is thin but vital. The approximate composition of dry air is, by volume at sea level, nitrogen 78%, oxygen 21% (necessary for life as we know it), argon 0.93%, and carbon dioxide 0.03%, together with very small amounts of numerous other constituents (Table 12.1). The water vapor content

Table 12.1. Composition of Earth's Atmosphere (Air)

Gas	Chemical Symbol	Volume(%)
Nitrogen	N_2	78.08
Oxygen	O_2	20.94
Carbon dioxide	CO_2	0.03
Argon	Ar	0.093
Neon	Ne	0.0018
Helium	He	0.0005
Krypton	Kr	trace
Xenon	Xe	trace
Ozone	O_3	0.00006
Hydrogen	H_2	0.00005

is highly variable and depends on atmospheric conditions. Air is said to be pure when none of the minor constituents is present in sufficient concentration to be injurious to the health of human beings or animals, to damage vegetation, or to cause loss of amenity (e.g., through the presence of dirt, dust, or odors, or by diminution of sunshine).

Where does air come from? Genesis 1:2 states that God separated the water environment into the atmosphere and surface waters on the second day of creation. Many scientists state that 4.6 billion years ago a cloud of dust and gases forged the Earth and also created a dense molten core enveloped in cosmic gases. This was the proto-atmosphere or proto-air, composed mainly of carbon dioxide, hydrogen, ammonia, and carbon monoxide, but it did not last long before it was stripped away by a tremendous outburst of charged particles from the Sun. As the outer crust of Earth began to solidify, a new atmosphere began to form from the gases outpouring from gigantic hot springs and volcanoes. This created an atmosphere of air composed of carbon dioxide, nitrogen oxides, hydrogen, sulfur dioxide, and water vapor. As the Earth cooled, water vapor condensed into highly acidic rainfall, which collected to form oceans and lakes.

For much of Earth's early existence (the first half), only trace amounts of free oxygen were present. But then green plants evolved in the oceans and they began to add oxygen to the atmosphere as a waste gas and later oxygen increased to about 1% of the atmosphere and with time to its present 21%.

How do we know for sure about the evolution of air on Earth? Are we guessing or using "voodoo" science? There is no guessing or voodoo involved with the historical geological record. Consider, for example, geological formations that are dated to 2 billion years ago. In these early sediments there is a clear and extensive band of red sediment ("red bed" sediments)—sands colored with oxidized (ferric) iron. Earlier ferrous formations show no oxidation. But there is more evidence. We can look at the time frame of 4.5 billion years ago, when carbon dioxide in the atmosphere was beginning to be lost in sediments. The vast amount of carbon deposited in limestone, oil, and coal indicate that carbon dioxide concentrations must once have been many times greater than today, which currently stands at only 0.03%. The first carbonated deposits appeared about 1.7 billion years ago, the first sulfate deposits about 1 billion years ago. The decreasing carbon dioxide was balanced by an increase in the nitrogen content of the air. The forms of respiration practiced advanced from fermentation 4 billion years ago to anaerobic photosynthesis 3 billion years ago to aerobic photosynthesis 1.5 billion years ago. The aerobic respiration that is so familiar today only began to appear about 500 million years ago. The atmosphere itself continues to evolve, but human activities—with their highly polluting effects—have now overtaken nature in determining the changes. In fact, human beings and their effect on planet Earth is one of the overriding themes of this text.

The atmosphere is an important geological agent and is responsible for the processes of weathering that are continually at work on the Earth's surface.

THE HYDROSPHERE

The hydrosphere includes all the waters of the oceans, lakes, and rivers, as well as groundwater that exists within the lithosphere. Approximately 40 million cubic miles of water covers or resides within the Earth. The oceans contain about 97% of all water on Earth. The other 3% is freshwater: (1) snow and ice on the surface of Earth contains about 2.25% of the water; (2) usable groundwater is approximately 0.3%; and (3) surface freshwater is less than 0.5%.

In the United States, for example, average rainfall is approximately 2.6 feet (a volume of 5,900 cubic kilometers). Of this amount, approximately 71% evaporates (about 4,200 cubic cm), and 29% goes to stream flow (about 1,700 cubic km).

Beneficial freshwater uses include manufacturing, food production, domestic and public needs, recreation, hydroelectric power production, and flood control. Stream flow withdrawn annually is about 7.5% (440 cubic km). Irrigation and industry use almost half of this amount (3.4% or 200 cubic km per year). Municipalities use only about 0.6% (35 cubic km per year) of this amount.

Historically, in the United States, water usage is increasing (as might be expected). For example, in 1990, 40 billion gallons of freshwater were used. In 1975, the total increased to 455 billion gallons. The use was about 780 billion gallons in 2008.

The primary sources of freshwater include the following:

1. Captured and stored rainfall in cisterns and water jars
2. Groundwater from springs, artesian wells, and drilled or dug wells
3. Surface water from lakes, rivers, and streams
4. Desalinized seawater or brackish groundwater
5. Reclaimed wastewater

In the United States, current federal drinking water regulations actually define three distinct and separate sources of freshwater. They are surface water, groundwater, and groundwater under the direct influence of surface water (GUDISW). This last classification is the result of the surface water treatment rule. The definition of the conditions that constitute GUDISW, while specific, is not obvious. This classification is discussed in detail later.

THE LITHOSPHERE

The lithosphere is of prime importance to the geologist. This, the solid, inorganic, rocky crust portion of the Earth, is composed of rocks and minerals that, in turn, comprise the continental masses and ocean basins. The rocks of the lithosphere are of three basic types: igneous, sedimentary, and metamorphic.

THE BIOSPHERE

Geologist Eduard Suess (1875) coined the term biosphere and defined it as: "The place on earth's surface where life dwells." In the broadest sense; biosphere is the sum total of all ecosystems; that is, it is the global ecological system integrating all living beings and their relations, including their interaction with the elements of the lithosphere, hydrosphere, and atmosphere. Postulated to have evolved, the biosphere evolved via biogenesis (i.e., the production of new living organisms) or biopoesis (i.e., formation of life from self-replicating molecules), at least some 3.5 billion years ago (Campbell et al., 2006). Others have defined biosphere as that zone of life on Earth; a closed and self-regulating system.

Key Terms

After reviewing the characteristics of life listed previously, it should be obvious that we need to accede to Voltaire's request: "If you wish to converse with me, please define your terms." The following list defines many of the terms used to describe the characteristics of life or that are related to the characteristics:

- **Aerobic**—occurring or living only in the presence of oxygen
- **Anabolism**—the utilization of energy and materials to build and maintain complex structures from simple components
- **Anaerobic**—active, living, occurring, or existing in the absence of free oxygen
- **Amino acids**—building blocks of proteins
- **Asexual reproduction**—requires one parent cell
- **Autotrophic**—using light energy (photosynthesis) or chemical energy (chemosynthesis)
- **Catabolism**—the breaking down of complex materials into simpler ones using enzymes and releasing energy
- **Carbohydrates**—main source of energy for living things such as sugar and starch
- **Cellular basis of life**—two types of cells: prokaryotes and eukaryotes

- **Cold-blooded animals**—body temperatures change with the environment
- **Compound**—two or more elements chemically combined
- **Digestion**—process by which food is broken down into simpler substances
- **DNA**—the double helix of DNA is the unifying chemical of life; its linear sequence defines the diversity of living things
- **Elements**—pure substance that cannot be broken down into simpler substances
- **Enzymes**—special types of proteins that regulate chemical activities
- **Excretion**—process of getting rid of waste materials
- **Eukaryotes**—cells with a nucleus; they are found in human and other multi-cellular organisms (plants and animals) also algae, protozoa
- **Evolution**—the modification of species; the core theme of biology
- **Food**—needed by living things to grow, develop, and repair body parts
- **Heterotrophic**—obtaining materials and energy by the breaking down of other biological material using digestive enzymes and then assimilating the usable byproducts
- **Ingestion**—taking in food or producing food
- **Inorganic compounds**—may or may not contain carbon
- **Life span**—maximum length of time an organism can be expected to live
- **Lipids**—energy-rich compounds made of C, O, and H
- **Metabolism**—chemical reactions that occur in living things
- **Movement**—nonliving material moves only as a result of external forces, whereas living material moves as a result of internal processes at cellular level or at organism level (locomotion in animals and growth in plants)
- **Nucleic acids**—store information that helps the body make proteins it needs
- **Organic compounds**—found in living things and contain carbon
- **Organism**—any living thing
- **Prokaryotes**—cells without a nucleus, including bacteria and cyanophytes (blue-green algae); genetic material is a single circular DNA and is contained in the cytoplasm, since there is no nucleus
- **Proteins**—used to build and repair cells; made of amino acids
- **Respiration**—taking in oxygen and using it to produce energy
- **Response**—action, movement, or change in behavior caused by a stimulus
- **Sexual reproduction**—requires two parent cells
- **Stimulus**—signal that causes an organism to react
- **Structure and function**—at all levels of organization, biological structures are shaped by natural selection to maximize their ability to perform their functions
- **Unity in diversity**—explained by evolution: all organisms linked to their ancestors through evolution; scientists classify life on Earth into groups related by ancestry; related organisms descended from common ancestor and have certain similar characteristics
- **Warm-blooded animals**—maintain a constant body temperature

Did You Know?

All four divisions of Earth can be and often are present in a single location. For example, a piece of soil of course has mineral material from the lithosphere. Additionally, elements of the hydrosphere are present as moisture within the soil, the biosphere is present as insects and plants, and even the atmosphere is present as pockets of air between soil pieces.

Earth's Skeleton

Earth is made up of three main compositional layers: crust, mantle, and core (Figure 12.3). The crust has variable thickness and composition: Continental crust is 10–50 km thick whereas the oceanic crust is 8–10 km thick. The elements silicon, oxygen, aluminum, and ion make up the Earth's crust. Like the

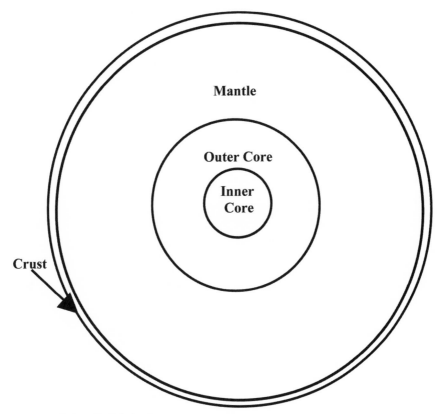

Figure 12.3. Earth's layers.

shell of an egg, the Earth's crust is brittle and can break. Earth's continental crust is 35 kilometers thick. The oceanic crust is 7 kilometers thick.

Based on seismic (earthquake) waves that pass through the Earth, we know that below the crust is the mantle, a dense, hot layer of semisolid (plastic-like liquid) rock approximately 2,900 km thick. The mantle, which contains silicon, oxygen, aluminum, and more iron, magnesium, and calcium than the crust, is hotter and denser because temperature and pressure inside the Earth increase with depth. According to the U.S. Geological Survey (1999), as a comparison, the mantle might be thought of as the white of a boiled egg. The 30-km thick transitional layer between the mantle and crust is called the *Moho layer*. The temperature at the top of the mantle is 870° C. The temperature at the bottom of the mantle is 2,200° C.

At the center of the Earth lies the core, which is nearly twice as dense as the mantle because its composition is metallic (iron–nickel alloy) rather than stony. Unlike the yolk of an egg, however, the Earth's core is actually made up of two distinct parts: a 2,200-km thick liquid outer core and a 1,250-km thick solid inner core. As the Earth rotates, the liquid outer core spins, creating the Earth's magnetic field.

References and Recommended Reading

Air Quality Criteria. 1968. Staff report, Subcommittee on Air and Water Pollution, Committee on Public Works, U.S. Senate, 94-411.

American heritage dictionary of the English language, 4th ed. 2000. New York: Houghton Mifflin Company.

Bergman, J. 2005. *Rocks and the rock cycle.* Accessed 05/23/08 from www.windows .ucar.edu/ tour/ling=/earth/geology/rocks_intro.html.

Blue, J. 2007. Descriptor Terms. *Gazetteer of planetary nomenclature.* U.S. Geological Service. Accessed 05/20/2008 from http://planetarynames.wr.usgs.gov/jsp/apped5.jsp.

Campbell, N. A. 2004. *Biology: Concepts & connections,* 4th ed. on CD-ROM. Harlow, Essex, England: Benjamin-Cummings.

Campbell, N. A., Williamson, B., and Heyden, R. J. (2006). *Biology: Exploring life.* Boston: Person Prentice Hall.

Goshorn, D. 2006. Proceedings—DELMARVA Coastal Bays Conference III: Tri-State approaches to preserving aquatic resources. Washington, D.C.: U.S. Environmental Protection Agency.

Huxley, T. H. 1876. *Science & education, collected essays,* vol 3. Boston: D. Appleton & Company.

Jones, A. M. 1997. *Environmental biology.* New York: Routledge.

Keeton, W. T. 1996. *Biological science.* W. W. Norton.

King, R. M. 2003. *Biology made simple.* New York: Broadway Books.

Koch, R. 1882. *Uber die Atiologie der Tuberkulose.* In *Verhandlungen des Knogresses fur Innere Medizin.* Erster Kongress, Wiesbaden.

Koch, R. 1884. *Mitt Kaiser Gesundh* 2, 1–88.

Koch R. 1893. *J Hyg Inf* 14, 319–333.

Larsson, K. A. 1993. Prediction of the pollen season with a cumulated activity method., *Grana*, 32, 111–114.

Med Net. 2006. *Definition of Koch's postulates.* www.medicinenet.com.

National Aeronautics and Space Administration (NASA). 2000. *A mystery of Earth's wobble solved: It's the ocean.* Accessed 05/24/08 from www.jpl.nasa.gov/releases/2000/chandlerwobble.html.

Press, R., and Siever, F. 2001. *Earth,* 3rd ed. New York: W. H Freeman & Co.

Seuss, E. (1875). *Die Entstechung Der Alpen* [The origin of the Alps]. Vienna: W. Braunmuller.

Society of General Microbiology. 2006. *The scientific method, fish health and pfiesteria.* College Park: University of Maryland, National Oceanic and Atmospheric Administration.

Spellman, F. R. 1998. *Environmental science and technology: Concepts and applications.* Lancaster, PA: Technomic Publishing Company.

Spellman, F. R. 2008. *The science of air: Concepts and applications.* Boca Raton, FL: CRC Press.

Spellman, F. R., & Whiting, N. E. 2006. *Environmental science and technology,* 2nd ed. Rockville, MD: Government Institutes.

Spieksma, F. T. 1991. Aerobiology in the Nineties: Aerobiology and pollinosis. *International Aerobiology Newsletter,* 34, 1–5.

U.S. Environmental Protection Agency. 2006. *What is the scientific method?* Accessed from http://www.epa.gov/maia/html/scientific.html.

USGS. 1999. *Inside the earth.* Accessed 5/10/11 at http://pubs.usgs.gov/g:p/dynamic/inside.html.

CHAPTER 13

Minerals

The Earth is just too small and fragile a basket for the human race to keep all its eggs in.

—Robert Heinlein

The geologist is primarily interested in the Earth's rocky crust. Because rocks are aggregates of minerals, in order to study and understand the rocks that make up Earth's crust, it is necessary to know something about minerals. Minerals are solid chemical elements or compounds that occur naturally within the crust of the Earth. They are inorganic, have a definite chemical composition or range of composition, a specified orderly internal arrangement of atoms (crystalline structure), and certain other distinct physical properties.

Again, the Earth is composed of rocks. Rocks are aggregates of minerals. Minerals are composed of atoms. To understand rocks, we must first have an understanding of minerals. To understand minerals, we must have some basic understanding of atoms—what they are and how they interact with one another to form minerals.

Matter

Although a detailed discussion of chemistry is not within the scope of this book,* an introduction to chemical terminology and atoms is necessary if we are to understand the chemical composition of minerals.

---

* For an easy-to-understand introduction to general chemistry, read Spellman, F. R. 2006. *Chemistry for Non-Chemists,* Lanham, MD: Government Institutes Press.

Just as we use several adjectives to describe an object (color of object, how tall or short, how bulky or thin, how light or heavy, etc.), several **properties,** or characteristics, must be used in combination to adequately describe a kind of matter. We also distinguish one form of matter from another by its properties. However, simply saying that something is a colorless solid isn't enough to identify it as diamond. A lot of solids are colorless as, for example, some quartz and feldspar, as well as many rock aggregates. More details are needed before we can zero in on the identity of a substance. Chemists, therefore, determine several properties, both chemical and physical, to characterize a particular sample of matter—to distinguish one form of matter from another. The following discussion describes the differences between the two kinds of properties, chemical and physical.

DEFINING MATTER

A thorough understanding of matter—how it consists of elements that are built from atoms—is critical for grasping chemistry. *Matter* is anything that occupies space and has weight (mass). Matter (or mass–energy) is neither created nor destroyed during chemical change.

As mentioned, along with the properties of having weight and taking space, matter has two other distinct properties, **chemical** and **physical**. These properties are actually used to describe **substances**, which are definite varieties of matter. Copper, gold, salt, sugar, and rust are all examples of substances. All of these substances are uniform in their makeup. However, if we pick up a common rock from our garden, we cannot call the rock a substance because it is a mixture of several different substances.

✓ **Key Point:** Matter is measured by making use of its two properties. Anything that has the properties of having weight and taking up space *must* be matter. A *substance* is a definite variety of matter, all specimens of which have the same properties.

Physical Properties

Substances have two kinds of physical properties: intensive and extensive. **Intensive physical properties** include those features that definitely distinguish one substance from another. Intensive physical properties do not depend on the amount of the substance (matter present). Some of the important intensive physical properties are color, taste, melting point, boiling point, density, luster, and hardness. It is important to note that it takes a combination of several intensive properties to identify a given substance. For example, a certain substance may have a particular color that is common to it but not necessarily unique to it. A white diamond is, as its name implies, white. However, is another gemstone that is white and faceted to look like a diamond really a diamond? Remember that a diamond is one of the hardest known substances. To determine if the white faceted gemstone is really a diamond we would also have to test its hardness and not rely on its appearance alone.

Some of the important intensive physical properties are defined as follows:

DENSITY

Density is the mass per unit volume of a substance (mass of a substance divided by its volume). Suppose we had a cube of lard and a large box of crackers, each having a mass of 400 gm. The density of the crackers would be much less than the density of the lard because the crackers occupy a much larger volume than the lard occupies. The density of an object or substance can be calculated by using the formula:

$$Density = \frac{mass}{volume} \qquad (13.1)$$

Starting with water as one example, consider that perhaps the most common measures of density are pounds per cubic foot (lb/ft^3) and pounds per gallon (lb/gal). However, the density of a dry substance, such as sand, lime, or soda ash, is usually expressed in pounds per cubic foot. The density of a gas, such as methane or carbon dioxide, is usually expressed in pounds per cubic foot. As shown in Table 13.1, the density of a substance like water changes slightly as the temperature of the substance changes. This happens because substances usually increase in volume (size), as they become warmer. Because of this expansion with warming, the same weight is spread over a larger volume, so the density is lower when a substance is warm than when it is cold.

SPECIFIC GRAVITY

Specific gravity (SG) is a unitless measure of the weight of a substance compared with the weight of an equal volume of water—a substance having an SG of 2.5 weighs two and a half times more than water. This relationship is easily seen when a cubic foot of water, which weighs 62.4 pounds, is compared with

Table 13.1.   Temperature and Density Relationship

Temperature (° F)	Specific Weight (lb/ft3)	Density (slugs/ft3)
32	62.4	1.94
40	62.4	1.94
50	62.4	1.94
60	62.4	1.94
70	62.3	1.94
80	62.2	1.93
90	62.1	1.93
100	62.0	1.93
110	61.9	1.92
120	61.7	1.92
130	61.5	1.91
140	61.4	1.91
150	61.2	1.90
160	61.0	1.90
170	60.8	1.89
180	60.6	1.88
190	60.4	1.88
200	60.1	1.87
210	59.8	1.86

a cubic foot of aluminum, which weighs 178 pounds. Aluminum is 2.7 times as heavy as water.

Gold can easily be distinguished from "fool's gold" by SG alone because it is not that difficult to find the SG of a piece of metal. All we need do is to weigh the metal in air, then weigh it under water. Its loss of weight is the weight of an equal volume of water. To find the SG, divide the weight of the metal by its loss of weight in water.

$$SG = \frac{Weight\ of\ a\ substance}{Weight\ of\ equal\ volume\ of\ water} \tag{13.2}$$

Example 13.1

Problem:

Suppose a piece of tin weighs 110 pounds in air and 74 pounds under water. What is the SG?

Solution:

Step 1: 110 lb subtract 74 lb = 36 lb loss of weight in water

Step 2:

$$SG = \frac{110 \text{ lb}}{36 \text{ lb}} = 3.1$$

✓ **Key Point:** In a calculation of SG, it is *essential* that the densities be expressed in the same units.

The SG of water is 1, which is the standard, the reference for which all other substances (i.e., liquids and solids) are compared. Any object that has an SG greater than 1 will sink in water. Considering the total weight and volume of a ship, its SG is less than 1; therefore, it can float.

In geology, SG has application. For example, it has application in regard to the Earth's crust, which is composed mostly of the minerals calcite, feldspar, and quartz. These minerals have SGs around 2.75 and that is close to the average SG of the rocks on the outer surface of the Earth's crust. Thus, the SG of most rocks that are initially picked up and later sorted out is familiar to the geologist; they have an SG of approximately 2.75.

HARDNESS

Hardness is commonly defined as a substance's relative ability to resist scratching or indentation. Hardness is one of the easiest ways to distinguish one mineral from another. Actual hardness testing involves measuring how far an "indenter" can be pressed into a given material under a known force. A substance will scratch or indent any other substance that is softer. Table 13.2 is used for comparing the hardness of mineral substances.

Table 13.2. Moh's Hardness Scale

Hardness (H)	Mineral	
1	Talc	
2	Gypsum	(fingernail: H = 2.5)
3	Calcite	(penny: H = 3)
4	Fluorite	
5	Apatite	
6	Feldspar	(glass plate: H = 5.5)
7	Quartz	
8	Topaz	
9	Corundum	
10	Diamond	(hardest)

Do not confuse mineral hardness for water hardness. In water treatment, *hardness* is a characteristic of water, caused primarily by calcium and magnesium ions. Water hardness can cause many maintenance problems, especially with piping and process components where scale buildup can occur.

STREAK

Streak is the color of a crushed mineral's powder that is left when the mineral is rubbed across a piece of unglazed tile, leaving a line similar to a pencil or crayon mark. This line is composed of the powdered minerals. The color of this powdered material is known as the *streak of the mineral,* and the unglazed tile used in such a test called a *streak plate.* Most light-colored, nonmetallic minerals have a white or colorless streak, as do most silicates, carbonates, and most transparent minerals. The streak test is most useful for identifying dark-colored minerals, especially metals.

COLOR

Unless we are color-blind, probably one of the first things we notice about a mineral is its color. Most of us are familiar with the colors of various substances. Pure water is usually described as colorless. Water takes on color when foreign substances such as organic matter from soils, vegetation, minerals, and aquatic organism are present. Like water, the same mineral may vary greatly in color from one specimen to another, and with certain exceptions, color is of limited use in mineral identification. When using color in mineral identification, we must consider whether the specimen is being examined in natural or artificial light, whether the surface being examined is weathered or fresh, and whether the mineral is wet or dry.

LUSTER

The appearance of the surface of a mineral as seen in reflected light is called *luster.* Terms such as *shining* (bright), *glistening* (sparkling brightness), *splendent* (glossy brilliance), and *dull* (lacking luster) are commonly used to indicate the degree of luster present. Some minerals shine like the metals gold and silver. These are said to have metallic luster. A mineral does not have to be a metal to have luster. Other lusters are called *nonmetallic.* Some common examples are shown in the following list:

- Adamantine—brilliant glossy luster (diamonds)
- Vitreous—glassy, looks like glass (topaz or quartz)
- Resinous—the luster of resin (sphalerite)
- Greasy—like an oily surface (nepheline)
- Pearly—like mother-of-pearl (talc)
- Silky—luster of silk or rayon (asbestos)
- Dull—as the name implies (chalk or clay)

MALLEABILITY

When a mineral, such as gold or copper, can be hammered into thin sheets it is said to be *malleable*.

DUCTILITY

When a substance (e.g., copper, gold, or silver) can be drawn into thin wires, it is said to be *ductile*.

CONDUCTIVITY

The ability of a substance (copper, aluminum, gold, silver, etc.) to conduct electricity is called *conductivity*.

BOILING POINT

The temperature at which the vapor pressure of a liquid is equal to the pressure on the liquid (e.g., atmospheric pressure) is known as the *boiling point*.

MELTING/FREEZING POINT

The temperature at which the solid and liquid phases of a substance are in equilibrium at atmospheric pressure is know as the *melting/freezing point*.

Extensive physical properties are such features as *mass, weight, volume, length,* and *shape*. Extensive physical properties are dependent upon the amount of the substance (matter present).

Chemical Properties

The nonchemist often has difficulty in distinguishing the physical versus the chemical properties of a substance. One test that can help is to ask the question, "Are the properties of a substance determined without changing the identity of the substance?" If we answer *yes,* then the substance is distinguished by its physical properties. If we answer *no,* then we can assume the substance is defined by its chemical properties. Simply, the **chemical properties** of a substance describe its ability to form new substances under given conditions.

To determine the chemical properties of certain substances we can observe how the substance reacts in the presence of air, acid, water, a base, and other chemicals. We can also observe what happens when the substance is heated. If we observe a change from one substance to another, we know that a chemical change, or a chemical reaction, has taken place.

✓ **Key Point:** The chemical properties of a substance may be considered to be a listing of all the chemical reactions of a substance and the conditions under which the reactions occur.

Another example can be used to demonstrate the difference between physical and chemical change. When a carpenter cuts pieces of wood from a larger piece of wood to build a wooden cabinet, the wood takes on a new appearance. The value of the crafted wood is increased as a result of its new look. This kind of change, in which the substance remains the same, but only the appearance is different, is called a **physical change**. When this same wood is consumed in a fire, however, ashes result. This change of wood into ashes is called a **chemical change**. *In a chemical change a new substance is produced.* Wood has the property of being able to burn. Ashes cannot burn.

KINETIC THEORY OF MATTER

There are three states of matter—solids, liquids, and gases. All matter is made of molecules. Matter is held together by attractive forces that prevent substances from coming apart. The molecules of a solid are packed more closely together and have little freedom of motion. In liquids, molecules move with more freedom and are able to flow. The molecules of gases have the greatest degree of freedom and their attractive forces are unable to hold them together.

The **Kinetic Theory of Matter** (German: kinetic = *motion*; theory = *idea*) is a statement of how we believe atoms and molecules behave and how that behavior relates to the ways we have to look at the things around us. Essentially, the theory states that all molecules are always moving. More specifically, the theory says:

- All matter is made of atoms, the smallest bit of each element. A particle of a gas could be an atom or a group of atoms.
- Atoms have energy of motion that we feel as temperature. At higher temperatures, the molecules move faster.

Matter changes its state from one form to another. Examples of how matter changes its state include the following:

- Melting is the change of a solid into its liquid state.
- Freezing is the change of a liquid into its solid state.
- Condensation is the change of a gas into its liquid form.
- Evaporation is the change of a liquid into its gaseous state.
- Sublimation is the change of a gas into its solid state and vice versa (without becoming liquid).

Crystal Structure

Smooth, angular shapes known as *crystals* normally form when crystalline minerals solidify and grow without interference. Packing of atoms in a crystal structure requires an orderly and repeated atomic arrangement. The planes that form the outside of the crystals are known as *faces*. The shape of crystals and the angles between related sets of crystal faces are important in mineral identification.

Each mineral has been assigned to one of six crystal systems. The classes have been established on the basis of the number, position, and relative length of the crystal axes—imaginary lines extending through the center of the crystal.

Geologists recognize the following crystal systems:

Isometric: Crystals belonging to the isometric system have three axes of equal length at right angles to one another. Examples of minerals that crystallize in the isometric system are halite, magnetite, and garnet.

Tetragonal: Tetragonal crystals are characterized by having all three axes at right angles. The horizontal axes are of equal length but are longer or shorter than the vertical axis. Mineral species that crystallize in the tetragonal crystal system are zircon and cassiterite.

Hexagonal: The hexagonal system has crystals marked by three horizontal axes of equal length that interest at angles of 120 degrees, and a vertical axis at right angles to these. Example species that crystallize in the rhombohedral division are calcite, dolomite, low quartz, and tourmaline.

Orthorhombic: Crystals assigned to the orthorhombic system have all three axes all at right angles and each of different length. Species that belong to this system are olivine and barite.

Monoclinic: Crystals from the monoclinic system have three unequal axes, two of which intersect at right angles. Mineral species that adhere to the monoclinic crystal system include pyroxene, amphibole, orthoclase, azurite, and malachite, among others.

Triclinic: Crystals of the triclinic system are characterized by three axes of unequal length that are all oblique to one another. Mineral species of this system include plagioclase and axinite.

Rock-Forming Minerals

Rocks consist of aggregates of minerals. Minerals are made up of one or a number of chemical elements with definite chemical compositions. Of the 3,000 or so minerals that are known to be present in the Earth's crust, only about 20 are very common and relatively few (only 9) are major constituents of the more common rocks. These 9 minerals are all silicates. Those minerals that do make up a large part of the more common types of rocks are called the *rock-forming minerals*. Some of the more important rock-forming minerals are briefly discussed in the following list:

Quartz: Quartz is the most common mineral on the face of the Earth and is one of the most widely distributed. It is a glassy-looking, transparent or translucent mineral that forms an important part of many igneous rocks and is common in many sedimentary and metamorphic rocks. Found in nearly every geological environment, it is at least a component of almost every rock type.

Feldspars: Minerals belonging to the feldspar group constitute the most important group of rock-forming minerals. Feldspar, like quartz, is a very common, light-colored rock-forming mineral. Unlike quartz, feldspar is not glassy, but instead is generally dull to opaque with a porcelain-like appearance. Feldspars are found in almost all igneous rocks as well as in many sedimentary and metamorphic rocks. Feldspar is hard but can be scratched by quartz. Feldspars are split into two principal groups: orthoclase (potash feldspar) and plagioclase (soda-lime feldspar).

Mica: Minerals of the mica group are easily distinguished by characteristic peeling into many flat, paper-thin, smooth sheets or flakes. These sheets are chemically inert, dielectric, elastic, flexible, hydrophilic, insulating, lightweight, platy, reflective, refractive, resilient, and transparent to opaque

(US Geological Survey, 2008). Mica may be white and pearly (muscovite) or dark and shiny (biotite).

Pyroxenes: The pyroxene group is composed of complex silicates, and is among the most common of all rock-forming minerals. The most common pyroxene mineral is augite, which is generally dark green to black in color. Augite is a common constituent of many of the dark-colored igneous rocks. Augite forms short, stubby crystals that have square or rectangular cross-sections. Pyroxenes are also commonly found in certain metamorphic rocks.

Amphiboles: The amphiboles are closely related to the pyroxenes. The most common amphibole is hornblende, a common constituent of igneous and metamorphic rocks. *Hornblende* is not a mineral and the name is used as a general field term. Hornblende is quite similar to augite in that both are dark minerals; however, hornblende crystals are generally longer, thinner, and shinier than augite and the mineral cross-sections are diamond shaped.

Calcite: Calcite (calcium carbonate) is a very common mineral in sedimentary rocks. It is commonly white to grey in color. Individual crystals are generally clear and transparent. It occurs in many sedimentary and metamorphic rocks, and is the primary constituent of most limestone.

Dolomite: Dolomite is a common sedimentary rock-forming mineral that can be found in massive beds several hundred feet thick. It may occur in association with many ore deposits and in cavities in some igneous rocks and is also found in metamorphic marbles, hydrothermal veins, and replacement deposits.

Aragonite: Like calcite, aragonite is composed of calcium carbonate, but it differs from calcite it that it is a polymorph of calcite, which means that that it has the same chemistry of calcite but it has a different structure. Aragonite is also less stable than calcite and crystallizes in the orthorhombic system.

Gypsum: Gypsum, calcium sulfate, is one of the more common minerals in sedimentary environments. It is a major rock forming mineral that produces massive beds, usually from precipitation out of highly saline waters. Gypsum is a mineral of great economic importance; it is commonly used in the manufacture of sheet rock and plaster of paris.

Anhydrite: Although chemically similar to gypsum, anhydrite does not contain water and is harder and heavier than gypsum. Anhydrite is a relatively common sedimentary mineral that forms massive rock layers.

Halite: Composed of sodium chloride, halite is commonly called *rock salt.* Large occurrences of halite have been formed as a result of the evaporation of prehistoric seas, and in some areas (e.g., Michigan and New York) the salt occurs in thick beds.

Kaolinite: Kaolinite occurs in soft, compact, earthy masses with a dull luster and greasy feel, and is typical of the other minerals commonly found in clay.

Kaolinite is important to the production of ceramics and porcelain. The greatest demand for Kaolinite is in the paper industry to produce a glossy paper such as is used by most magazines.

Serpentine: Serpentine, a complex group of hydrous magnesium silicates, is a major rock-forming mineral and is found as a constituent in many metamorphic and weathered igneous rocks. It often colors many of these rocks to a green color and most rocks that have a green color probably have serpentine in some amount.

Chlorite: Chlorite, a complex silicate of aluminum, magnesium, an iron in combination with water, is the general name for several minerals that are difficult to distinguish by ordinary methods.

Metallic Minerals

The metallic minerals have always been of interest to geologists because of their intrinsic value. These minerals are found in ore deposits—rock masses from which metals may be obtained commercially. Using U.S. Geological Survey (2008) descriptions, some of the more important metals and their ores are briefly discussed in the following list.

Aluminum: Aluminum, the second most abundant metallic element in the Earth's crust after silicon, is one of the most important metals of industry and is primarily derived from bauxite. It weighs about one-third as much as steel or copper; is malleable, ductile, and easily machined and cast; and has excellent corrosion resistance and durability.

Copper: Copper, one of the most useful of all metals, is usually found in nature in association with sulfur. Copper is one of the oldest metals ever used and has been one of the important materials in the development of civilization. Because of its properties of high ductility, malleability, and thermal and electrical conductivity, and its resistance to corrosion, copper has become a major industrial metal, ranking third after iron and aluminum in terms of quantities consumed.

Gold: Gold has been prized since the dawn of history for its great beauty and the fact that it is soft enough to be easily fashioned into coins, jewelry, and other valuable objects. Gold occurs as native gold and is typically found in quartz veins and in association with the mineral pyrite.

Lead: Galena, the most important source of lead, occurs in a wide variety of rocks, including igneous, sedimentary, and metamorphic. Lead is a very corrosion-resistant, dense, ductile, and malleable blue-tray metal that has been used for at least 5,000 years. Lead is used in the manufacture of paints, as a type of metal, pipe, solder, metal alloys, shot, and as shielding materials to protect against radioactivity and X rays.

Mercury: The most abundant ore of mercury is cinnabar (also known as *quicksilver*), or mercuric sulfide. Mercury is the only common metal that is liquid at room temperature. Mercury has uniform volumetric thermal expansion, good electrical conductivity, and easily forms amalgams with almost all common metals except iron. Most mercury is used for the manufacture of industrial chemicals and for electrical and electronic applications.

Silver: Silver, a metal used for thousands of years as ornaments and utensils, for trade, and as the basis of many monetary systems, may occur as native silver or in one of several silver ore minerals. Silver also has many industrial applications such as in mirrors, electrical and electronic products, and photography, which is the largest single end use of silver.

Tin: Although widely distributed in small amounts, tin occurs in commercial quantities in igneous rocks, where it is commonly associated with quartz, topaz, galena, and tourmaline. Tin is one of the earliest metals known and used. Most tin is used as a protective coating or as an alloy with other metals such as lead or zinc.

Iron: Iron ore is a mineral substance that, when heated in the presence of a reductant (coke or charcoal), will yield metallic iron. Iron ore is the source of primary iron for the world's iron and steel industries. It is therefore essential for the production of steel, which in turn is essential to maintain a strong industrial base.

Zinc: Zinc, a rather common mineral, is a metal of considerable economic importance. It is found in veins in igneous, sedimentary, and metamorphic rocks, and as replacement deposits in limestone. Zinc is used in galvanizing steel, and in the manufacture of paint, cosmetics, types metal, dry cell batteries, and for a multitude of other purposes.

Radioactive Minerals

Radioactive minerals have come to play an ever-increasing part in modern technology. Certain of these minerals are widely used as sources of energy for nuclear power plants, in nuclear medicine, and in modern weapon systems. Examples of common radioactive minerals include autunite, brannerite, carnotite, monazite, thorianite, and uraninite. The vast majority of the radioactive content in minerals or ores is either unranium-238 or thorium-232.

Nonmetallic Minerals

Nonmetallic minerals refer to a vast category that includes coal, petroleum, sulfur, fertilizer, building stones, and gemstones. Industrial minerals include, for example, flake graphite, silicon carbide, garnet, talc, marble, slate, granite, salt, sulfur, asbestos, and others.

References and Recommended Reading

Asimov, I., Back, D. F. (Illustrator). 1992. *Atom: Journey across the subatomic cosmos.* New York: Dutton/Plume.

Levi, R., & Rosenthal, R. (Translator). 1986. *The periodic table (American).* New York: Knopf Publishing Group.

Mebane, R. C., & Rybolt, T. R. 1998. *Adventures with atoms and molecules: Chemistry experiments for young people,* vol 1. New Jersey: Enslow.

Nardo, D. 2001. *Atoms.* New York: Gale Group.

Senese, F. 2004. What is chemistry? Accessed 05/25/08 from http://antoine.frostburg.edu/chem/101/intro/faq/what-is-chemsitry.shtml.

Spellman, F. R. 2006. *Chemistry for non-chemists.* Lanham, MD: Government Institutes Press.

Stwertka, A. 2002. *Guide to the elements.* New York: Oxford University Press.

USGS. 2008. Minerals information. Accessed 05/26/08 @ minerals.usgs.gov/minerals/pubs/commodity/mica/ mecamybo/.pdf.

CHAPTER 14

Igneous Rocks and Magma Eruption

> I wanted the gold, and I sought it,
> I scrabbled and mucked like a slave.
> Was it famine or scurvy—I fought it;
> I hurled my youth into a grave.
> I wanted the gold, and I got it—
> Came out with a fortune last fall,—
> Yet somehow life's not what I thought it,
> And somehow the gold isn't all.
>
> —Robert W. Service, *The Spell of the Yukon*

Igneous (Latin for *fire*) rocks are those rocks that have solidified from an original molten silicate state. The occurrence and distribution of igneous rocks and igneous rock types (Figure 14.1) can be related to the operation of plate tectonics. The molten rock material from which igneous rocks form is called *magma*. Magma, characterized by a wide range of chemical compositions and with high temperature, is a mixture of liquid rock, crystals, and gas. Magmas are large bodies of molten rock deeply buried within the earth. These magmas are less dense than surrounding rocks, and will therefore move upward. In the upward movement, sometimes magmatic materials are poured out on the surface of the Earth as, for example, when lava flows from a volcano. These igneous rocks are volcanic or **extrusive rocks**; they form when the magma cools and crystallizes on the surface of the earth. Under certain other conditions, magma does not make it to the Earth's surface and cools and crystallizes within Earth's crust. These intruding rock materials harden and form **intrusive** or **plutonic rocks**.

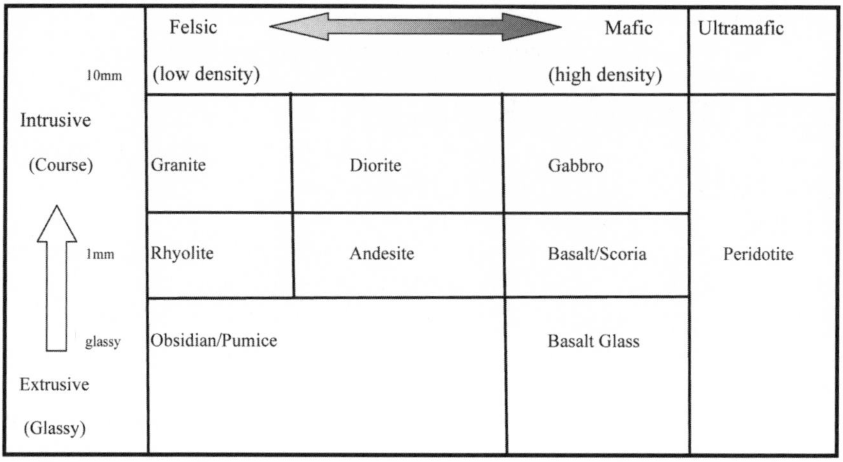

	Felsic	⟵⟶	Mafic	Ultramafic
10mm	(low density)		(high density)	
Intrusive (Course)	Granite	Diorite	Gabbro	
1mm	Rhyolite	Andesite	Basalt/Scoria	Peridotite
glassy	Obsidian/Pumice		Basalt Glass	
Extrusive (Glassy)				

Figure 14.1. Igneous rock chart.

Magma

Magma is molten silicate material and may include already formed crystals and dissolved gases. The term *magma* applies to silicate melts within the Earth's crust. When magma reaches the surface it is referred to as *lava*. The chemical composition of magma is controlled by the abundance of elements in the Earth. These include oxygen, silicon, aluminum, hydrogen, sodium, calcium, iron, potassium, and manganese, which make up 99% of magma. Because oxygen is so abundant, chemical analyses are usually given in terms of oxides. Silicon dioxide (SiO_2, also known as *silica*) is the most abundant oxide. Because magma gas expands as pressure is reduced, magmas have an explosive character. The flow (or viscosity) of magma depends on temperature, composition, and gas content. Higher silicon dioxide content and lower temperature magmas have higher viscosity.

Magma consists of three types: basaltic, andesitic, and rhyolitic. Table 14.1 summarizes the characteristics of each type.

Table 14.1. Characteristics of Magma Types

Magma Type	Solidified Volcanic	Solidified Plutonic	Chemical Composition	Temperature
Basaltic	Basalt	Gabbro	45%–55% silicon dioxide	1,000–1,200° C
Andesitic	Andesite	Diorite	55%–65% silicon dioxide	800–1,000° C
Rhyolitic	Rhyolite	Granite	65%–75% silicon dioxide	650–800° C

Intrusive Rocks

Intrusive (or plutonic) rocks are rocks that have solidified from molten mineral mixtures beneath the surface of the Earth. Intrusive rocks that are deeply buried tend to cool slowly and develop a coarse texture. On the other hand, those intrusive rocks near the surface that cool more quickly are more finely textured. The shape, size, and arrangement of the grains composing it determine the texture of igneous rocks. Because of crowded conditions under which mineral particles are formed, they are usually angular and irregular in outline. Typical intrusive rocks include:

- **Gabbro**—a heavy, dark-colored igneous rock consisting of coarse grains of feldspar and augite
- **Peridotite**—a rock in which the dark minerals are predominant
- **Granite**—the most common and best-known of the coarse-textured intrusive rocks
- **Syenite**—resembles granite, but is less common in its occurrence and contains little or no quartz

Extrusive Rocks

Extrusive (or volcanic) rocks are those that pour out of craters of volcanoes or from great fissures or cracks in the Earth's crust and make it to the surface of the Earth in a molten state (liquid lava). Extrusive rocks tend to cool quickly, and typically have small crystals (because fast cooling does not allow large crystals to grow). Some cool so rapidly that no crystallization occurs and volcanic glass is produced instead.

Some of the more common extrusive rocks are felsite, pumice, basalt, and obsidian.

- **Felsite**—very fine-textured igneous rocks
- **Pumice**—frothy lava that solidifies while steam and other gases bubble out of it
- **Basalt**—world's most abundant fine-grained extrusive rock
- **Obsidian**—volcanic glass; cools so fast that there is no formation of separate mineral crystals

Bowen's Reaction Series

The geologist, Norman L. Bowen, back in the early 1900s was able to explain why certain types of minerals tend to be found together while others are almost never associated with one another. Bowen found that minerals tend to form in specific sequences in igneous rocks, and these sequences could be assembled into a composite sequence. The idealized progression that he determined is still accepted as the general model (see Figure 14.2) for the evolution of magmas during the cooling process.

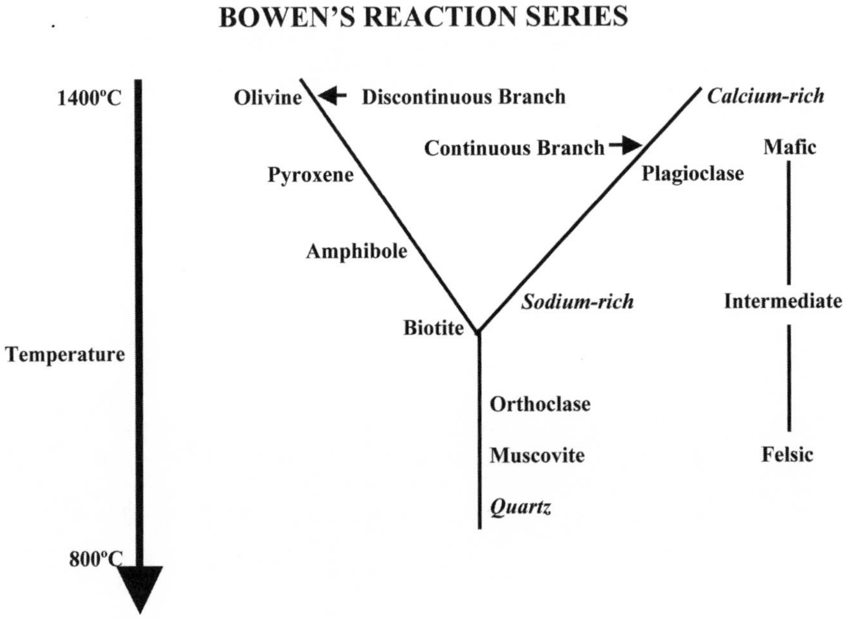

BOWEN'S REACTION SERIES

Figure 14.2.   Bowen's reaction series.

To better understand Bowen's reaction series, it is important to define key terms:

- **Magma**—molten igneous rock
- **Felsic**—white pumice
- **Pumice**—textured form of volcanic rock; a solidified frothy lava
- **Extrusion**—magma intruded or emplaced beneath the surface of the earth
- **Feldspar**—the family of minerals including microcline, orthoclase, and plagioclase
- **Mafic**—a mineral containing iron and magnesium
- **Aphanitic**—mineral grains too small to be seen without a magnifying glass
- **Phaneritic**—mineral grains large enough to be seen without a magnifying glass

- **Reaction series**—a series of minerals in which a mineral reacts to change to another
- **Rock-forming mineral**—the minerals commonly found in rocks; Bowen's reaction series lists all of the common minerals found in igneous rocks
- **Specific gravity**—the relative mass or weight of a material compared with the mass or weight of an equal volume of water

Some igneous rocks are named according to textural criteria:

- Scoria—porous
- Pumice
- Obsidian—glass
- Tuff—cemented ash
- Breccia—cemented fragments
- Permatite—extremely large crystals
- Aplite—sugary texture, quartz & feldspar
- Porphyry—fine matrix, large crystals

THE DISCONTINUOUS REACTION SERIES

The left side of Figure 14.2 shows a group of mafic or iron-magnesium–bearing minerals—olivine, pyroxene, amphibole, and biotite. If the chemistry of the melt is just right, these minerals react discontinuously to form the next mineral in the series. If there is enough silica in the igneous magma melt, each mineral will change to the next mineral lower in the series as the temperature drops. Descending down Bowen's reaction series, the minerals increase in the proportions of silica in their composition. In basaltic melt, as shown in Figure 14.2, olivine will be the first mafic mineral (silicate mineral rich in magnesium and iron) to form. When the temperature is low enough to form pyroxene, all of the olivine will react with the melt to form pyroxene, and pyroxene will crystallize out of the melt. At the crystallization temperature of amphibole, all the pyroxene will react with the melt to form amphibole, and amphibole will crystallize. At the crystallization temperature of biotite, all of the amphibole will react to form biotite, and biotite will crystallize. Thus all igneous rocks should only have biotite; however, this is not the case. In crystallizing olivine, if there is not enough silica to form pyroxene, then the reaction does not occur and olivine remains. Additionally, if the temperature drops too fast for the reaction to take place (volcanic magma eruption), then the reaction does not have time to occur, the rock solidifies quickly, and the mineral remains olivine.

THE CONTINUOUS REACTION SERIES

The right side of Figure 14.2 shows the plagioclases. Plagioclase minerals have the formula $(Ca, Na)(Al, Si)_3O_8$. The highest temperature plagioclase has only calcium (Ca). The lowest temperature plagioclase has only sodium (Na). In between, these ions mix in a continuous series from 100% Ca and 0% Na at the highest temperature to 50% Ca and 50% Na at the middle temperature to 0% Ca and 100% Na at the lowest temperature. In a basaltic melt, for example, the first plagioclase to form could be 100% Ca and 0% Na plagioclase. As the temperature drops, the crystal reacts with the melt to form 99% Ca and 1% Na plagioclase and 99% Ca and 1% Na plagioclase crystallizes. Then those react to form 98% Ca and 2% Na and the same composition crystallizes and so forth. All of this happens continuously, provided there is enough time for the reactions to take place and enough sodium, aluminum, and silica in the melt to form each new mineral. The end result is a rock with plagioclases with the same ratio of Ca to Na as the starting magma.

✓ **Key Point:** In regard to the Bowen reaction series, on both sides of the reaction series shown in Figure 14.2, the silica content of the minerals increases as the crystallization trend heads downward. Biotite has more silica than olivine. Sodium plagioclase has more silica than calcium plagioclase.

✓ **Important Point:** The magma temperature and the chemical composition of the magma determine what minerals crystallize and thus what kind of igneous rock we get.

Eruption of Magma

The volcanic processes that lead to the deposition of extrusive igneous rocks can be studied in action today, and help us to explain the textures of ancient rocks with respect to depositional processes. Some of the major features of volcanic processes and landforms are discussed in this section. The following information is from the U.S. Geological Survey (2008) *Principal Types of Volcanoes*.

TYPES OF VOLCANOES

Geologists generally group volcanoes into four main kinds: cinder cones, composite volcanoes, shield volcanoes, and lava domes.

- **Cinder cones** are the simplest type of volcano. They are built from particles and blobs of congealed lava ejected from a single vent. As the gas-charged lava is blown violently into the air, it breaks into small fragments that solidify and fall as cinders around the vent to form a circular or oval cone. Most cinder cones have a bowl-shaped crater at the summit and rarely rise more than a thousand feet or so above their surroundings. Cinder cones are numerous in western North America as well as throughout other volcanic terrains of the world.
- **Composite volcanoes**, sometimes called *stratovolcanoes*, have created some of Earth's grandest mountains. They are typically steep-sided, symmetrical cones of large dimension built of alternating layers of lava flows, volcanic ash, cinders, blocks, and bombs, and might rise as much as 8,000 feet above their bases. Most composite volcanoes have a crater at the summit that contains a central vent or a clustered group of vents. Lava either flows through breaks in the crater wall or issue from fissures on the flanks of the cone. Lava, solidified within the fissures, forms dikes that act as ribs that greatly strengthen the cone. The essential feature of a composite volcano is a conduit system through which magma from a reservoir deep in the Earth's crust rises to the surface. The volcano is built up by the accumulation of material erupted through the conduit and increases in size as lava, cinders, ash, and other materials, are added to its slopes.

Probably the best-known active composite volcano at the present time is Mount Saint Helens (Figure 14.3a, b, and c) in Washington State. Mount Saint Helens is most notorious for its catastrophic eruption on May 18, 1980, which was the deadliest and most economically destructive volcanic event in the history of the United States. Fifty-seven people were killed; 250 homes, 47 bridges, 15 miles of railways, and 185 miles of highway were destroyed.

- **Shield volcanoes** are built almost entirely of fluid lava flows. Flow after flow pours out in all directions from a central summit vent, or group of vents, building a broad, gently sloping cone of flat, domical shape, with a profile much like that of a warrior's shield. They are built up slowly by the accretion of thousands of highly fluid lava flows called *basalt lava* that spread widely over great distances, and then cool as thin, gently dipping sheets. Lava also commonly erupts from vents along fractures (rift zones) that develop on the flanks of the cone. Some of the largest volcanoes in the world are shield volcanoes (Figure 14.4).
- **Lava Domes** are formed by relatively small bulbous masses of lava too viscous to flow any great distance; consequently, on extrusion, the lava piles over and

Figure 14.3a. Mt. St. Helens north face (blast face) taken from start of Truman Ridge.

Figure 14.3b. Mt. St. Helens north face Lehar (mudslide) runoff area Toutle River basin.

Figure 14.3c.   Mt. St. Helens inside the crater. Notice lava dome building with steam rising off upper side.

Photo by Frank R. Spellman

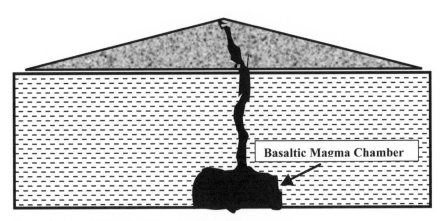

Basaltic Magma Chamber

Figure 14.4.   Cross-Section of a Shield Volcano.

around its vent. A dome grows largely by expansion from within. As it grows, its outer surface cools and hardens, then shatters, spilling loose fragments down its sides. Some domes form craggy knobs or spines over the volcanic vent, whereas others form short, steep-sided lava flows known as *coulees* (French "to flow"). Volcanic domes commonly occur within the craters or on the flanks of large composite volcanoes.

TYPES OF VOLCANIC ERUPTIONS

The type of volcanic eruption is often labeled with the name of a well-known volcano where characteristic behavior is similar; hence, the use of such terms as *Strombolian, Vulcanian, Vesuvian, Pelean, Hawaiian,* and others.

- A **strombolian-type eruption** is in constant action with huge clots of molten lava bursting from the summit crater to form luminous arcs through the sky. Collecting on the flanks of the cone, lava clots combine to steam down the slopes in fiery rivulets.
- A **Vulcanian-type eruption** is characterized by very viscous lavas; a dense cloud of ash-laden gas explodes from the crater and rises high above the peak. Steaming ash forms a whitish cloud near the upper level of the cone.
- A **Pelean-type eruption (or *nuee ardente*—"glowing cloud")** is characterized by its explosiveness. It erupts from a central crater with violent explosions that eject great quantities of gas, volcanic ash, dust, incandescent lava fragments, and large rock fragments.
- An **Hawaiian-type (or quiet) eruption** is characterized by less viscous lava that permits the escape of gas with a minimum of explosive violence. In fissure-type eruptions, a fountain of fiery lava erupts to a height of several hundred feet or more. Such lava might collect in old pit craters to form lava lakes, to form cones, or to feed radiating flows.
- A **Vesuvian eruption** is characterized by great quantities of ash-laden gas that violently discharge to form a cauliflower-shaped cloud high above the volcano.

Types of Lava

The two Hawaiian words, *pahoehoe* and *aa*, are used to describe how lava flows. *Pahoehoe* (Pa-hoy-hoy) is the name for smooth or ropy lava. Cooler lava hardens on the surface; hotter, more fluid lava flows under it, often leaving caves or tubes behind. *Aa* (ah-ah) is the name of rough, jagged lava. The molten lava is much less fluid and usually moves more slowly. A crust never hardens on the surface, but chunks of cooler rock tumble along the top and sides instead. *Aa* can be impassable.

- A **phreatic (or steam-blast) eruption** is driven by explosive expanding steam resulting from cold ground or surface water coming into contact with hot rock or magma. The distinguishing feature of phreatic explosions is that they only blast out fragments of preexisting solid rock from the volcanic conduit; no new magma is created.
- A **Plinian eruption** is a large explosive event that forms enormous dark columns of tephra (solid material ejected) and gas high into the stratosphere. Such eruptions are named for Pliny the Younger, who carefully described the disastrous eruption of Vesuvius in 79 CE. This eruption generated a huge column of tephra into the fall. Many thousands of people evacuated areas around the volcano, but about 2,000 were killed, including Pliny the Older.

Lava Flow Terminology (U.S. Geological Survey, 2000)

- Some **lava flows** are associated with volcanoes and others are the result of fissure flow. These masses of molten rock pour onto the Earth's surface during an effusive eruption. Both moving lava and the resulting solidified deposit are referred to as *lava flows*. Because of the wide range in (1) viscosity of the different lava types (basalt, andesite, dacite, and rhyolite), (2) lava discharge during eruptions, and (3) characteristics of the erupting vent and topography over which lava travels, lava flows come in a great variety of shapes and sizes.
- A **lava cascade** is similar to a cascade of water that creates a small waterfall as water descends over rocks. In similar fashion, a *lava cascade* refers to the rush or descent of lava over a cliff. In Hawaii, lava cascades typically occur when lava spills over the edge of a crater, a fault scarp, or a sea cliff into the ocean.
- **Lava drapery** is the cooled, congealed rock on the face of a cliff, crater, or fissure formed by lava pouring or cascading over their edges.
- **Lava channels** are narrow, curved or straight open pathways through which lava moves on the surface of a volcano. The volume of lava moving down a channel fluctuates so that the channel may be full or overflowing at times and nearly empty at other times. During overflow, some of the lava congeals and cools along the banks to form natural levees that may eventually enable the lava channel to build a few meters above the surround ground.
- **Standing waves** in a fast-moving lava flow appear to be stationary relative to the lava that moves over the land through them, similar to the standing waves in a water stream. In Hawaii, standing waves as high as 3 m have been observed.
- **Lava spillways** are confined lava channels on the sides of a volcanic cone or shield that form when lava overflows the rim of the vent.
- **Lava surges** are intermittent surges or accelerations in the forward advance of lava that can occur when the supply of lava to a flow suddenly increases

or a flow front gives way. The supply of lava might increase as consequence of a higher discharge of lava from the vent, a sudden change in the vent geometry so that a great volume of lava escapes (e.g., the collapse of a vent wall), or by the escape of ponded lava from along a channel. Lava surges may be accompanied by thin, short-lived breakouts of fluid lava form the main channel and flow front.

- **Methane explosions** are sudden explosions of methane gas that occur frequently near the edges of active lava flows. Methane gas is generated when vegetation is covered and heated by molten lava. The explosive gas travels beneath the ground through cracks and fills abandoned lava tubes for long distances around the margins of the flow. Methane gas explosions have occurred at least 100 m from the leading edge of a flow, blasting rocks and debris in all directs.
- **Volcanic domes** are rounded, steep-sided mounds built by very viscous magma, usually either dacite or rhyolite. Such magmas are typically too viscous (resistant to flow) to move far from the vent before cooling and crystallizing. Domes may consist of one or more individual lava flows. Volcanic domes are also referred to as *lava domes.*

Did You Know?

The longest historical dome-building eruption is still occurring at Santiaguito Dome, which is erupting on the southeast flank of Santa Maria volcano in Guatemala. The dome began erupting in 1922.

Intrusions

Intrusive (or plutonic igneous) rocks have been intruded or injected into the surrounding rocks. Some of these intrusions are invisible because they are imbedded at great depth; consequently igneous intrusive bodies may be seen only after the underlying rocks have been removed by erosion.

Intrusions are of two types: Concordant intrusions, which are parallel to layers of rocks, and discordant intrusions, which cut across layers. Some of the more common intrusive bodies (plutons) are discussed in the following sections.

CONCORDANT INTRUSIONS

- **Sills** are tabular bodies of igneous rocks that spread out as essentially thin, horizontal sheets between beds or layers of rocks.

- **Laccoliths** are lens-like, mushroom-shaped, or blister-like intrusive bodies, usually near the surface, that have relatively flat under surfaces and arched or domed upper surfaces. They differ from sills in that they are thicker in the center and become thinner near their margins.
- **Lopoliths** are mega-sills, usually of gabbro or diorite that might cover hundreds of square kilometers and be kilometers thick. They often have a concave structure and are differentiated. That is, they take so long to harden that heavy minerals have a chance to sink and light minerals can rise.

DISCORDANT INTRUSIONS

- **Dikes** are thin, wall-like sheets of magma intruded into fractures in the crust.
- **Stocks** or **plutons** are small, irregular intrusions.
- **Batholiths** are the largest of igneous intrusions and are usually granitic and cover hundreds or thousands of square kilometers.

Did You Know?

Most obsidian is black, but red, green, and brown obsidian is known. Obsidian forms when magma is cooled so quickly that individual minerals cannot crystallize.

Volcanic Landforms

Volcanic landforms (or volcanic edifices) are controlled by the geological processes that form them and act on them after they have formed. Four principal types of volcanic landforms are plateau basalts or lava plains, volcanic mountains, craters, and calderas.

Plateau basalts and lava plains are formed when great floods of lava are released by fissure eruptions instead of central vents and spread in sheet-like layers over the Earth's surface, forming broad plateaus. Some of these plateaus are quite extensive. For example, the Columbia River Plateau of Oregon, Washington, Nevada, and Idaho is covered by 200,000 square miles of basaltic lava.

Volcanic mountains are mountains composed of the volcanic products of central eruptions, and are classified as cinder cones (conical hills), composite cones (stratovolcanoes), and lava domes (shield volcanoes).

Volcanic craters are circular, funnel-shaped depressions, usually less than 1 km in diameter, that form as a result of explosions that emit gases and tephra.

Calderas are much larger depressions, circular to elliptical in shape, with diameters ranging from 1 km to 50 km. Calderas form as a result of collapse of a volcanic structure. The collapse results from evacuation of the underlying magma chamber.

Thermal Areas

Thermal areas are locations where volcanic or other igneous activity takes place as is evidenced by the presence or action of volcanic gases, steam, or hot water escaping from the earth.

Fumaroles are vents where gases, either from a magma body at depth, or steam from heated groundwater, emerge at the surface of the earth.

Hot springs or thermal springs are areas where hot water comes to the surface of the Earth. Cool groundwater moves downward and is heated by a body of magma or hot rock. A hot spring results if this hot water can find its way back to the surface, usually along fault zones.

Geysers result if the hot spring has a plumbing system that allows for the accumulation of steam from the boiling water. When the steam pressure builds so that it is higher than the pressure of the overlying water in the system, the steam will move rapidly toward the surface, causing the eruption of the overlying water. Some geysers, like Old Faithful in Yellowstone Park, erupt at regular intervals, but most geysers are quite erratic in their performance. The time between eruptions is controlled by the time it takes for the steam pressure to build in the underlying plumbing system.

References and Recommended Reading

Abbott, P. L. 1996. *Natural disasters*. New York: Wm. C. Brown Publishing Co.

Anderson, J. G., Bodin, P., Brune, H. N., Prince, J., Singh, S. K., Quaas, R., & Onate, M. 1986. Strong ground motion from the Michoacan, Mexico earthquake. *Science,* vol. 233, pp. 1043–1049.

Browning, J. M. 1973. Catastrophic rock slides. Mount Huascaran, north-central, Peru, May 32, 1970. *Bulletin American Association of Petroleum Geologists,* vol. 57, pp. 1335–1341.

Coch, N. K. 1995. *Geohazards, natural and human*. New York: Prentice Hall.

Eagleman, J. 1983. *Severe and unusual weather*. New York: Van Nostrand Reinhold.

Francis, P. 1993. *Volcanoes: A planetary perspective*. New York: Oxford University Press.

GeoMan. 2008. *Bowen's reaction series.* Accessed from http://jersey.uorgeon.edu/ ~mstrick/AskGeoMan/geoQuerry32.html.

Keller, E. A. 1985. *Environmental geology,* 4th ed. New York: Merrill Publishing Co.

Kiersh, G. A. 1964. Vaiont reservoir disaster. *Civil Engineering,* vol. 34.

Murck, B. W., Skinner, B. J., & Porter, S. C. 1997. *Dangerous Earth: An introduction to geologic hazards.* New York: John Wiley & Sons.

Skinner, B. J., & Porter, S. C. 1995. *The dynamic earth: An introduction to physical geology,* 3rd ed. New York: John Wiley & Sons.

Spellman, F. R., & Whiting, N. E. 2006. *Environmental science and technology,* 2nd ed. Rockville, MD: Government Institutes Press.

Stephens, J. C., et al. 1984. Organic soils subsidence. *Geological Society of American Reviews in Engineering Geology,* vol. 6, p. 3.

Swanson, D. A., Wright, T. H., & Helz, R. T. 1975. Linear vent systems and estimated rates of magma production and eruption of the Yakima basalt on the Columbia Plateau. *American Journal of Science,* vol. 275, pp. 877–905.

Tilling, R. I. 1984. *Eruptions of Mount St. Helens: Past, present, and future.* Washington, DC: Department of the Interior, U.S. Geological Survey.

U.S. Geological Survey. 1989. Lessons learned from the Loma Prieta, California, earthquake of October 17, 1989. Circular 1045. Washington, DC: U.S. Geological Survey.

U.S. Geological Survey. 2000. *Photo glossary of volcano terms.* Accessed 06/01/08 from http://volcanoes.usgs.gov/Products/Pglossary/LavaCascade.html.

U.S. Geological Survey. 2008. *Principal types of volcanoes.* Accessed 05/31/08 from http://pubs.usgs.gov/gip/volc/types.html.

Williams, H., & McKinney, A. R. 1979. *Volcanology.* New York: Freeman & Copper Co.

CHAPTER 15

Sedimentary Rocks

A sedimentary rock is like a history book.

—U.S. Geological Survey (2006)

The rocks where the exposure to storms is greatest, and where only ruin seems to be the object, are all the more lavishly clothed upon with beauty. . . . In like manner do men find themselves enriched by storms that seem only big with ruin.

—John Muir, 1881

When rocks are exposed on the Earth's surface, they are especially vulnerable to the surface agents of erosion (weathering, erosion, rain, streamflow, wind, wave action, ocean circulation). When eroded, these rock fragments (called *detritus*) are commonly picked up and transported by wind, water, and ice, and when the transporting agent has dropped them because the transporting energy is not strong enough to carry them, they are generally referred to as *sediments*. Sediments are composed of components including small fragments (gravel, sand, or silt size), new minerals (mainly clays), and dissolved portions the source rock (dissolved salts in river and ocean water). Sediments on the Earth's surface may form by mere mechanical accumulation (wind or water) such as gravel and sand deposits in a river or sand dunes in a desert; by chemical precipitation, such as salt and calcite precipitation in shallow seas and lakes; and by activity of organisms, such as carbonate accumulation in coral reefs, or accumulation of organic matter in swamps (coal precursor). Sediments are typically deposited in layers or beds called *strata*. When sediments become compacted and cemented together (a process known as *lithification*), they form sedimentary rocks.

This compaction or lithification of sedimentary materials into stratified layers is probably the most significant feature of sedimentary rocks. These stratified layers are like pages in the ultimate history book—Earth's history—with each page dedicated to a specific or particular timeframe, earliest to present. Sediments of any particular period form a distinct layer that is underlain and overlain by equally distinct layers of respectively older and younger times. These layers, composed of such common rock types as sandstone, shale, and limestone, make up about 75% of the rocks exposed on the Earth's surface. Geologists can study sedimentary rocks in the making; therefore, they probably know more details about the origin of these rocks than that of igneous and metamorphic rocks combined.

Types of Sedimentary Rocks

Several different types of sedimentary rocks can be distinguished according to the source of rock materials that form them. The main groupings are:

- **Clastic (or detrital) sedimentary rocks**, subdivided into:
  - Conglomerates
  - Sandstones
  - Mudstones and shales
- **Chemical and biochemical sedimentary rocks**, subdivided into:
  - Limestone and dolostone
  - Evaporites
  - Carbonaceous rocks

CLASTIC SEDIMENTARY ROCKS

Clastic sedimentary rocks are the group of rocks most people think of when they think of sedimentary rocks. *Clasts* are fragmented material from other rocks and clastic sedimentary rocks are made up of pieces of preexisting rocks. Pieces of rock are loosened by weathering, and then transported by water, wind, gravity, or glacial action to some basin or depression where sediment is trapped. If the sediment is buried deeply, it becomes compacted and cemented, forming sedimentary rock. Depending on grain size, they are subdivided into conglomerate, sandstone, siltstone, and shale (see Table 15.1).

The formation of a clastic sedimentary rock involves the following processes:

- **Transportation:** Sediments move to their final destination by sliding down slopes; being picked up by the wind; or by being carried by running water

Table 15.1.   Classification of Clastic Sedimentary

Name of Particle	Size Range	Loose Sediment	Consolidated Rock
Boulder	>256 mm	Gravel	Conglomerate
Cobble	64–256 mm	Gravel	Conglomerate
Pebble	2–64 mm	Gravel	Conglomerate
Sand	1/16–2 mm	Sand	Sandstone
Silt	1/256–1/16 mm	Silt	Siltstone
Clay	<1/256 mm	Clay	Claystone, mudstone, and shale

Source: Adapted from James Madison University, 2000.

in streams, rivers, or ocean currents. During transport the sediment particles will be sorted according to size and density and will be rounded by abrasion. The distance the sediment is transported and the energy of the transporting medium all leave clues in the final sediment that tell us something about the mode of transportation.

- **Deposition:** Sediment is deposited when the energy of the transporting medium becomes too low to continue the transport process. In the deposition process, when the velocity of the transporting medium becomes too low to transport sediment, the sediment will fall out and become deposited.

- **Diagenesis:** Diagenesis is the process of chemical and physical change that turns sediment into rock. The first step in the process is compaction, which occurs when the weight of the overlying material increases. As the grains of the material are compacted together, pore space is reduced and water is eliminated from the substance. The free water usually carries mineral components in solution, and these constituents may later precipitate as new minerals in the pore spaces. This causes cementation, which then starts to bind the individual particles together and can be seen in quartz, calcite, iron oxide, clay, glauconite, and feldspar. The next stage of diagenesis involves alteration. Consider limestone and plagioclase; for example, both of these primary rocks can be converted or altered, respectively, to dolomite and to albite. This alteration occurs through pressure solution, which occurs when carbonate rock begins to dissolve under pressure, either of deep burial or tectonic squeezing. Finally, limestone is precipitated and formed in the recrystallization process. In addition, an absence of oxygen during the compaction process may cause other alterations to the original sediment.

Some of the more common types of clastic sedimentary rocks are described in the following list.

- **Sandstone:** Sandstones, composed essentially of cemented sand, comprise about 30% of all sedimentary rocks. The most abundant mineral in sandstone is quartz, along with lesser amounts of calcite, gypsum, and various iron compounds. Sandstone is used as an abrasive (for sandpaper) and as a building stone.

- **Conglomerate:** These are consolidated gravel deposits with variable amounts of sand and mud between the pebbles, and are the least abundant sediment type. Conglomerates accumulate in stream channels, along the margins of mountain ranges, and may also accumulate on beaches. Conglomerates composed largely of angular pebbles are called *breccias;* those formed in glacial deposits are called *tillites.*

- **Shale:** Shale consists of consolidated clay and other fine particles (mud) that have hardened into rock. It is the most abundant of all sedimentary rocks, comprising about 60% to 70% of the sedimentary rocks on Earth. Characteristically fine-grained and thinly bedded, shale is split easily along dividing (bedding) planes. Shale is classified or typed by composition. For example, shale containing large amounts of clay is called *argillaceous* shale. Shale containing appreciable amounts of sand is called *arenaceous* shale. Shale high in organic matter is typically black in color and known as *carbonaceous* shale. Shale that contains large amounts of lime is known as *calcareous shale,* and is used in the manufacture of Portland cement. Another type of shale, *oil* shale, is currently of great interest worldwide because of (at the time of this writing) the increasing cost of crude oil. Oil shale contains a fossilized insoluble organic materials called *kerogen,* which is converted into petroleum products and may be a short-term solution to crude oil shortage problems, as discussed in the following section.

United States: Coal-Shale Deposits

J. R. Dyni (2005) of the Unites States Giological Survey (USGS) points out that Numerous deposits of oil shale, ranging from the Precambrian to the Tertiary age, are present in the United States. The two most important deposits are in the Eocene Green River Formation in Colorado, Wyoming, and Utah, and in the Devonian–Mississippian black shales in the eastern United States. Oil shale associated with coal deposits of Pennsylvanian age also appears in the eastern United States. Other deposits are known to be in Nevada, Montana, Alaska, Kansas, and elsewhere, but these are either too small, too low grade, or have not yet been well enough explored (Russell, 1990). Because of their size and grade, most investigations have focused on the Green River and the Devonian Mississippian deposits.

GREEN RIVER FORMATION

Geology

Lacustrine sediments of the Green River Formation were deposited in two large lakes that occupied 65,000 km^2 in several sedimentary-structural basins in Colorado, Wyoming, and Utah during the early Eocene period through the middle Eocene period. The Uinta Mountain uplift and its eastward extension, the Axial Basin anticline, separate these basins. The Green River lake system was in existence for more than 10 million years during a time of a warm temperate to subtropic climate. During parts of their history, the lake basins were closed, and the waters became highly saline.

Fluctuations in the amount of inflowing stream waters caused large expansions and contractions of the lakes as evidenced by widespread intertonguing of marly lacustrine strata with beds of land-derived sandstone and siltstone. During arid times, the lakes contracted, and the waters became increasingly saline and alkaline. The lake-water content of soluble sodium carbonates and chloride increased, whereas the less soluble divalent Ca + Mg + Fe carbonates were precipitated with organic-rich sediments. During the driest periods, the lake waters reached salinities sufficient to precipitate beds of nahcolite, halite, and trona. The sediment pore waters were sufficiently saline to precipitate disseminated crystals of nahcolite, shorite, and dawsonite along with a host of other authigenic (generated where it was found or observed) carbonate and silicate minerals (Milton, 1977).

A noteworthy aspect of the mineralogy is the complete lack of authigenic sulfate minerals. Although sulfate was probably a major anion in the stream waters entering the lakes, the sulfate ion was presumably totally consumed by sulfate-reducing bacteria in the lake and sediment waters according to the following generalized oxidation-reduction reaction:

$$2CH_2O + SO_4^{-2} \rightarrow 2HCO_3^{-1} + H_2S \text{ (hydrogen sulfide)}$$

Note that two moles of bicarbonate are formed for each mole of sulfate that is reduced. The resulting hydrogen sulfide could either react with available Fe^{++} to precipitate as ion sulfide minerals or escape from ht sediments as a gas (Dyni, 1998). Other major sources of carbonate include calcium carbonate–secreting algae, hydrolysis of silicate minerals, and direct input from inflowing streams.

The warm alkaline lake waters of the Eocene Green River lakes provided excellent conditions for the abundant growth of blue-green algae (cyanobacteria) that are thought to be the major precursor of the organic matter in the oil shale. During times of freshening waters, the lakes hosted a variety of fishes, rays, bivalves, gastropods, ostracodes, and other aquatic fauna. Areas peripheral to the lakes supported a large and varied assemblage of land plants, insects, amphibians, turtles, lizards, snakes, crocodiles, birds, and numerous mammalian animals (McKenna, 1960; MacGinitie, 1969; Grande, 1984).

Historical Developments

The occurrence of oil shale in the Green River Formation in Colorado, Utah, and Wyoming has been known for many years. During the early 1900s, it was clearly established that the Green River deposits were a major resource of shale oil (Woodruff and Day, 1914; Winchester, 1916; Gavin, 1924). During this early period, the Green River and other deposits were investigated, including oil shale of the marine Phosphoria Formation of Permian age in Montana (Bowen, 1917; Condit, 1919) and oil shale in Tertiary lake beds near Elko, Nevada (Winchester, 1923).

In 1967, the U.S. Department of Interior began an extensive program to investigate the commercialization of the Green River oil-shale deposits. The dramatic increases in petroleum prices resulting from the OPEC oil embargo of 1973–1974 triggered another resurgence of oil-shale activities during the 1970s and into the early 1980s. In 1974, several parcels of public oil-shale lands in Colorado, Utah, and Wyoming were put up for competitive bid under the Federal Prototype Oil Shale Leasing Program. Two tracts were leased in Colorado (C-a and C-b) and two in Utah (U-a and U-b) to oil companies.

Large underground mining facilities, including vertical shafts, room-and-pillar entries, and modified in-situ retorts, were constructed on Tracts C-a and C-b, but little or no shale oil was produced. During this time, Unocal Oil Company was developing its oil-shale facilities on privately owned land on the south side of the Piceance Creek Basin. The facilities included a room-and-pillar mine with a surface entry, a 10,000 barrel/day (1,460 ton/day) retort, and an upgrading plant. A few miles north of the Unocal property, Exxon Corporation opened a room-and-pillar mine with a surface entry, haulage roads, waste-rock dumpsite, and a water-storage reservoir and dam.

In 1977–1978, the U.S. Bureau of Mines opened an experimental mine that included a 723-meter-deep shaft with several room-and-pillar entries in the northern part of the Piceance Creek Basin to conduct research on the deeper deposits of oil shale, which are commingled with nahcolite and dawsonite. The site was closed in the later 1980s.

About $80 million were spent on the U-a/U-b tracts in Utah by three energy companies to sink a 313-m-deep vertical shaft and inclined haulage way to a high-grade zone of oil shale and to open several small entries. Other facilities included a mine services building, water and sewage-treatment plants, and a water-retention dam.

The Seep Ridge project sited south of the U-a/U-b tracts, funded by Geokinetics, Inc. and the U.S. Department of Energy, produced shale oil by a shallow in-situ retorting method. Several thousand barrels of shale oil were produced.

The Unocal oil-shale plant was the last major project to produce shale oil from the Green River Formation. Plant construction began in 1980, and capital investment for constructing the mine, retort, upgrading plant, and other facilities was $650 million. Unocal produced 657,000 tons (about 4.4 million bbls) of shale oil, which were shipped to Chicago for refining into transportation fuels and other products under a program partly subsidized by the U.S. Government.

The average rate of production in the last months of operation was about 875 tons (about 5,900 barrels) of shale oil per day; the facility was closed in 1991.

In the past few years, Shell Oil Company began an experimental field project to recover shale oil by a proprietary in-situ technique. Some details about the project have been publicly announced, and the results to date (2006) appear to favor continued research.

Shale-Oil Resources

As the Green River oil-shale deposits in Colorado became better known, estimates of the resource increased from about 20 billion barrels in 1916 to 900 billion barrels in 1961 and to 1 trillion barrels (approximately 147 billion tons) in 1989 (Winchester, 1916; Donnell, 1961; Pitman et al., 1989).

The Green River oil-shale resources in Utah and Wyoming are not as well known as those in Colorado. Trudell et al. (1983) calculated the measured and estimated resources of shale oil in an area of about 5,200 km² in eastern Uinta Basin, Utah, to be 214 billion barrels (31 billion tons) of which about one-third is in the rich Mahogany oil-shale zone. Culbertson et al. (1980) estimated the oil-shale resources in the Green River Formation in the Green River Basin in southwest Wyoming to be 244 billion barrels (35 billon tons) of shale oil.

Additional resources are also in the Washakie Basin east of the Green River Basin in southwest Wyoming. Trudell et al. (1973) reported that several members of the Green River Formation on Kinney Rim on the west side of the Washakie Bain contain sequences of low to moderate grades of oil shale in three core holes. Two sequences of oil shale in the Laney Member, 11- and 42-meters thick, average 63 l/t and represent as much as 8.7 million tons of in-situ shale oil per square kilometer.

Other Mineral Resources

In addition to fossil energy, the Green River oil-shale deposits in Colorado contain valuable resources of sodium carbonate minerals including nahcolite ($NaHCO_3$) and dawsonite [$NaAl(OH)_2CO_3$]. Both minerals are commingled with high-grade oil shale in the deep northern part of the basin. Dyni (1974) estimated the total nahcolite resource at 29 billion tons. Beard et al. (1974) estimated nearly the same amount of nahcolite and 17 billion tons of dawsonite. Both minerals have value for soda as (Na_2CO_3) and dawsonite also has potential value for its alumina (Al_2O_3) content. The latter mineral is most likely to be recovered as a byproduct of an oil-shale operation. One company is solution mining nahcolite for the manufacture of sodium bicarbonate in the northern part of the Piceance Creek Basin at depths of about 600 m (Day, 1998). Another company stopped solution mining nahcolite in 2004, but now processes soda ash obtained from the Wyoming trona (hydrated sodium bicarbonate carbonate) deposits to manufacture sodium bicarbonate.

The Wilkins Peak Member of the Green River Formation in the Green River Basin in southwestern Wyoming contains not only oils but also the world's largest known source of natural sodium carbonate as trona ($Na_2CO_3 \cdot NaHCO_3 \cdot 2H_2O$). The trona resource is estimated at more than 115 billion tons in 22 beds ranging from 1.2 to 9.8 m in thickness (Wiig et al., 1995). In 1997, trona production from five mines was 16.5 million tons (Harris, 1997). Trona is refined into soda ash (Na_2CO_3) used in the manufacture of bottle and flat glass, baking soda, soap and detergents, waste treatment chemicals, and many other industrial chemicals. One ton of soda ash is obtained from about two tons of trona ore. Wyoming trona supplies about 90% of U.S. soda ash needs; in addition, about one-third of the total Wyoming soda ash produced is exported.

In the deeper part of the Piceance Creek Basin, the Green River oil shale contains a potential resource of natural gas, but its economic recovery is questionable (Cole and Daub, 1991). Natural gas is also present in the Green River oil-shale deposits in southwest Wyoming, and probably in the oil shale in Utah, but in unknown quantities.

EASTERN DEVONIAN–MISSISSIPPIAN OIL SHALE

Depositional Environment

Black organic-rich marine shale and associated sediments of Late Devonian and Early Mississippian age underlie about 725,000 km^2 in the eastern United States. These shales have been exploited for many years as a resource of natural gas, but have also been considered as a potential low-grade resource of shale oil and uranium (Roen & Kepferle, 1993; Conant & Swanson, 1961).

Over the years, geologists have applied many local names to these shales and associated rocks, including the Chattanooga, New Albany, Ohio, Sunbury, Antrim, and others. A group of papers detailing the stratigraphy, structure, and gas potential of these rocks in eastern United States have been published by the U.S. Geological Survey (Roen and Kepferle, 1993).

The black shales were deposited during Late Devonian and Early Mississippian time in a large epeiric sea that covered much of middle and eastern United States east of the Mississippi River. The area includes the broad, shallow, Interior Platform on the west that grade eastward into the Appalachian Basin. The depth to the base of the Devonian–Mississippian black shales ranges from surface exposures on the Interior Platform to more than 2,700 meters along the depositional axis of the Appalachian Basin (De Witt et al., 1993).

The Late Devonian sea (Paleozoic era; ~416 to 359.2 million years ago) was relatively shallow with minimal current and wave action much like the environment in which the Alum Shale of Sweden was deposited in Europe. A large part of the organic matter in the black shale is amorphous bituminite, although a few structured fossil organisms such as Tasmanites, Botryococcus, Foerstia, and others have been recognized. Conodonts (extinct primitive fishlike chordates with cone-like teeth) and linguloid (small, oval, and inarticulate) brachiopods are sparingly distributed through some beds. Although much of the organic matter is amorphous and of uncertain origin, it is generally believed that much of it was derived from planktonic algae.

In the distal (farthest away) parts of the Devonian sea, the organic matter accumulated very slowly along with very fine-grained clayey sediments in poorly oxygenated waters free of burrowing organisms. Conant and Swanson (1961) estimated that 30 cm of the upper part of the Chattanooga Shale deposited on the Interior Platform in Tennessee could represent as much as 150,000 years of sedimentation.

The black shales thicken eastward into the Appalachian Basin owning to increasing amounts of clastic sediments that were shed into the Devonian sea from the Appalachian highland lying to the east of the basin. Pyrite and marcasite are abundant authigenic materials, but carbonate minerals are only a minor fraction of the mineral matter.

Resources

The oil-shale resource is in that part of the Interior Platform where the black shales are the richest and closest to the surface. Although long known to produce oil upon retorting, the organic matter in the Devonian–Mississippian black shale yields only about half as much as the organic matter of the Green River oil shale, which is thought to be attributable to differences in the type of organic matter (or type of organic carbon) in each of the oil shales. The Devonian–Mississippian oils shale has a higher ratio of aromatic to aliphatic organic carbon than Green River oil shale, and is shown by material balance Fischer assays to yield much less shale oil and a higher percentage of carbon residue (Miknis, 1990).

Hydroretorting Devonian–Mississippian oil shale can increase the oil yield by more than 200 percent of the value determined by Fischer assay. In contrast, the conversion of organic matter to oil by hydroretorting is much less for Green River oils shale, about 130 to 140% of the Fischer assay value. Other marine oil shales also respond favorably to hydroretorting, with yields as much as, or more than, 300% of the Fischer assay (Dyni et al. 1990).

Mathews et al. (1980) evaluated the Devonian–Mississippian oils shales in areas of the Interior Platform where the shales are rich enough in organic matter and close enough to the surface to be mineable by open pit. Results or investigations in Alabama, Illinois, Indiana, Kentucky, Ohio, Michigan, eastern Missouri, Tennessee, and West Virginia indicated that 98% of the near-surface mineable resources are in Kentucky, Ohio, Indiana, and Tennessee (Matthews, 1983).

The criteria for the evaluation of the Devonian–Mississippian oil-shale resource used by Matthews et al. (1980) were:

1. Organic carbon content: ≥ 10 weight percent
2. Overburden: ≤ 200 m
3. Stripping ratio: $\leq 2.5{:}1$
4. Thickness of shale bed: ≥ 3 m
5. Open-pit mining and hydroretorting

On the basis of these criteria, the total Devonian–Mississippian shale oil resources were estimated to be 423 billion barrels (61 billion tons) (Dyni, 2005).

Chemical Sedimentary Rocks

Chemical and organic sedimentary rocks are the other main group of sediments besides clastic (lithified) sediments. They are formed by weathered material in solution precipitating from water or as biochemical rocks made of dead marine organisms. Usually special conditions are required for these rocks to form, such as high temperature, high evaporation, and high organic activity. Some chemical sediment is deposited directly from the water in which the material is dissolved; for example, solution upon evaporation of seawater. Such deposits are generally referred to as *inorganic chemical sediments*. Chemical sediments that have been deposited by or with the assistance of plants or animals are said to be *organic* or *biochemical sediments*. Accumulated carbon-rich plant material may form coal. Deposits made mostly of animal shells may form limestone, chert, or coquina.

CHEMICAL SEDIMENTARY ROCKS

Sedimentary rocks formed from sediments created by inorganic processes are discussed in the following list.

Limestone: Calcite ($CaCO_3$) is precipitated by organisms usually to form a shell or other skeletal structure. Accumulation of these skeletal remains results in the most common type of chemical sediment, limestone. Limestone might form by inorganic precipitation as well as by organic activity.

Dolomite: Dolomite consists of carbonate mineral (known as *magnesium limestone*) [$CaMg(CO_3)_2$], and occurs in more or less the same settings as limestone. Dolomite is formed when some of the calcium in limestone is replaced by magnesium.

Evaporites: Evaporites are sedimentary rocks (true chemical sediments) that are derived from minerals precipitated from seawater. Thus they consist mostly of halite (salt [$NaCl$]) and gypsum ($CaSO_4$) by chemical precipitation—high evaporation rates cause concentration of solids to increase as the result of water loss by evaporation.

CHEMICAL SEDIMENTARY ROCKS: INORGANIC PROCESSES

These rocks consist of sediments formed from the remains or secretions of organisms including *fossiliferous limestone, coquina* (type of limestone composed of shells and coarse shell fragments), *chalk* (a porous, fine-textured variety of limestone composed of calcareous shells), *lignite* (brown coal), and *bituminous* (soft) *coal.*

Physical Characteristics of Sedimentary Rocks

Sedimentary rocks possess definite physical characteristics and display certain features that make them readily distinguishable from igneous or metamorphic rocks. Some of the most important sedimentary characteristics are:

- **Stratification:** Probably the most characteristic feature of sedimentary rocks is their tendency to occur in strata or beds. These strata are formed as geological agents such as wind, water, or ice gradually deposit sediment.
- **Cross-bedding:** Sets of beds that are inclined relative to one another. The beds are inclined in the direction that the wind or water was moving at the time of deposition. Boundaries between sets of cross beds usually represent an erosional surface. They are very common in beach deposits, sand dunes, and river deposited sediment.
- **Texture:** The size, shape, and arrangements of materials derived by processes of weathering, transportation, deposition, and diagenesis determine the texture of sedimentary rocks. Again, the textures we find in sediment and sedimentary rocks are dependent on processes that occur during each stage of formation. These include:
  - Source materials, the nature of wind and water currents present, distance materials transported or time in the transportation process, biological activity, and exposure to various chemical environments.

- **Graded Bedding:** In a stream, as current velocity wanes, first the larger or denser particles are deposited, followed by smaller particles. This results in bedding showing a decrease in grain size from the bottom of the bed to the top of the bed (fine sediment on top and coarse at bottom).
- **Ripple Marks:** Characteristic of shallow water deposition. Caused by small waves or winds that often leave ripples of sand on the surface of a beach or bottom of a stream or sand. Ripples of this type have also been preserved in certain sedimentary rocks and may provide the geologist with information about the conditions of deposition when the sediment was originally deposited.
- **Mud Cracks:** It is not uncommon to find mud cracks that result from the drying out of wet sediment on the bottom of dried-up lakes, ponds, or stream beds. These many-sided (polygonal) shapes give a honeycomb appearance on the surface. If preserved in sedimentary rocks, such shapes suggest that the rock was subjected to alternate periods of flooding and drying.
- **Concretions:** Spherical or flattened masses of rock enclosed in some shales or limestones that are generally harder than the rock enclosing them are called *concretions.* Because concretions are usually harder than the enclosing rock, they are often left behind after the surrounding rock has been eroded away.
- **Fossil:** *Fossils* are remains or evidence of once-living organisms that have been preserved in the Earth's crust. Because life has evolved, fossils give clues to relative age of the sediment; they can be important indicators of past climates.
- **Color:** Hematites (iron oxides) produce a pink or red color in such areas as the Grand Canyon and Painted Desert.

Sedimentary Rock Facies

A sedimentary facies is a group of characteristics that describe an accumulation of deposits that have distinctive characteristics and grade laterally into other sedimentary deposits as a result of changing environments and original deposits.

References and Recommended Reading

American Society for Testing and Materials (ASTM). (1966). Designated D 388-66—Specifications for classification of coals by rank. In *Annual book of American Society for Testing and Materials Standards,* pp. 66–71. West Conshohocken, PA: ASTM.
American Society for Testing and Materials (ASTM). (1984). Designation D 3904-80—Standard test method for oil from oil shale. In *Annual book of American Society for Testing and Materials Standards,* pp. 513–525. West Conshohocken, PA: ASTM.

Beard, T. M., Tait, D. B., & Smith, J. W. (1974). *Nahcolite and dawsonite resources in the Green River Formation, Piceance resources of the Piceance Creek Basin, 25th field conference*, pp. 101–109. Denver, CO: Rocky Mountain Association of Geologists.

Bowen, F. F. (1917). Phosphatic oil shales near Dell and Dillon, Beaverhead Country, Montana. *U.S. Geological Survey Bulletin, 661*, 315–320.

Cole, R. D., & Daub, G. T. (1991). Methane occurrences and potential resources in the lower Parachute Creek Meander of Green River Formation, Piceance Creek Basin, Colorado in 24th Oil Shale Symposium Proceedings. *Colorado School of Mines Quarterly, 83*(4), 1–7.

Conant, L. C., & Swanson, V. E. (1961). Chattanooga shale and related rocks of central Tennessee and nearby areas. *U.S. Geological Survey Professional Paper 357*, 91 p.

Condit, D. D. (1919). *Oil shale to western Montana, southeastern Idaho, and adjacent parts of Wyoming and Utah. U.S. Geological Survey Bulletin 711*, 15–40.

Cross, T. A., & Homewood, P. W. (1997). Amanz Gressly's role in founding modern stratigraphy. *Geological Society of America Bulletin 109*(12), 1617–1630.

Culbertson, W. C., Smith, J. W., & Trudell, L. G. (1980). *Oil shale resources and geology of the Green River Formation in the Green River Basin*, Wyoming: U.S Department of Energy Laramie Energy Technology Center LETC/RI-80/6.

Day, R. L. (1998). Solution mining of Colorado nahcolite. In *Proceedings of the First International Soda Ash Conference*, Rocks Springs, Wyoming, June 10–12, 1997. *Wyoming State Geological Survey Public Information Circular, 40*, pp. 121–130.

De Witt, W., Jr., Roen J. B., & Wallace, L. G. (1993). Stratigraphy of Devonian black shales and associated rocks in the Appalachian Basin. In *Petroleum Geology of the Devonian and Mississippian black shale of eastern North America: U.S. Geological Survey Bulletin 1909*, Chapter B, pp. B1–B57.

Donnell, J. R. (1961). Tertiary geology and oil-shale resources of the Piceance Creek Basin between the Colorado and White Rivers, northwestern Colorado. *U.S. Geological Survey Bulletin 1082-L*, pp. 835–891.

Dyni, J. R. (1974). *Stratigraphy and nahcolite resources of the saline facies of the Green River Formation in northwest Colorado, in Guidebook to the Energy resources of the Piceance Creek Basin, 25th field conference*, pp. 111–122. Denver, CO: Rocky Mountain Association of Geologists.

Dyni, J. R., Anders, D. E., & Rex, R. C., Jr. (1990). Comparison of hydro-retorting, Fischer assay, and Rock-Eval analyses of some world oil shales. In *Proceedings 1989 Eastern Oil Shale Symposium*, pp. 270–286. Lexington: University of Kentucky Institute of Mining and Minerals Research.

Dyni, J. R. (1998). Prospecting for Green River-type sodium carbonate deposits. *Proceedings of the First International Soda Ash Conference*, vol. II. *Wyoming State Geological Survey Information Circular 40*, pp. 37–47.

Dyni, J.R. (2005). USGS. *Geology and Resources of Some World Oil-Shale Deposits*. In USGFS Pubs, pubs.usgs.gov. Reston, VA: United States Geological Survey.

Gavin, M. J. (1924). Oil shale, an historical, technical, and economic study. *U.S. Bureau of Mines Bulletin 210*, pp. 1–215.

Grande, L. (1984). Paleontology of the Green River Formation with a review of the fish fauna. *Geological Survey of Wyoming Bulletin, 63*, p. 333.

Harris, R. E. (1997). Fifty years of Wyoming trona mining. In *Prospect to pipeline: Casper, Wyoming geological associates: 48th guidebook.* p. 177–182. Casper: Wyoming Geological Associates.

Hutton, A. C. (1987). Petrographic classification of oil shales. *International Journal of Coal Geology, 8,* 203–231.

Hutton, A. C. (1988). *Organic petrography of oil shales: U.S. Geological Survey short course,* January 25–29, Denver, CO, unpublished.

Hutton, A. C. (1991). Classification, organic petrography and geochemistry of oil shale. In *Proceedings 1990 Eastern Oil Shale Symposium,* pp. 16–172. Lexington: University of Kentucky Institute for Mining and Minerals Research.

James Madison University. (2000). *A basic sedimentary rock classification.* Accessed 05/31/08 from http://csmres.jmu.edu/geolab/fichter/sedrx/basicclass.html. URL no longer exists.

MacGinitie, H. D. (1969). *The Eocene Green River flora of northwestern Colorado and northeastern Utah.* Berkeley: University of California Press.

Matthews, R. D. (1983). The Devonian-Mississippian oil shale resource of the United States. In Gary, H. H. (Ed.), *Sixteenth Oil Shale Symposium Proceedings,* pp. 14–25. Golden: Colorado School of Mines Press.

Matthews, R. D., Janka, J. C., & Dennison, J. M. (1980). *Devonian oil shale of the eastern United States, a major American energy resource [preprint]:* Evansville, IN: American Association of Petroleum Geologists Meeting, Oct. 1–3, 1980.

McKenna, M. C. (1960). *Fossil mammalia from the early Wasatchian Four mile fauna: Eocene of northwest Colorado.* Berkeley: University of California Press.

Miknis, F. P. (1990). Conversion characteristics of selected foreign and domestic oil shales. In *Twenty-third Oil Shale Symposium Proceedings.* Golden: Colorado School of Mines Press.

Milton, C. (1977). Mineralogy of the Green River Formation. *The Mineralogy Record, 8,* 368–379.

Pitman, J. K., Pierce, F. W., & Grundy, W. D. (1989). *Thickness, oil-yield, and kriged resource estimates for the Eocene Green River Formation.* Piceance Creek Basin, CO: U.S. Geological Survey Oil and Gas Investigations Cart OC-123.

Reading, H. G. (Ed.). (1996). *Sedimentary environments and facies.* New York: Blackwell Scientific.

Roen, J. B., & Kepferle, R. C. (Eds.). (1993). *Petroleum geology of the Devonian and Mississippian black shale of eastern North America: U.S. Geological Survey Bulletin 1909,* Chapters A–N, pp. 117–131.

Russell, P. L. (1990). *Oil shales of the world, their origin, occurrence and exploitation.* New York: Pergamon Press.

Schora, F. C., Janka, J. C., Lynch, P. A., & Feldkirchner, H. (1983). Progress in the commercialization of the Hytort Process. In *Proceedings 1982 Eastern Oils Shale Symposium,* pp. 183–190. Lexington: University of Kentucky, Institute for Mining and Minerals Research.

Stach, E., Taylor, G. H., Machowsky, M.-Th., Chandra, D., Teichmuller, M., & Teichmuller, R. (1975). *Stach's textbook of coal petrology,* Berlin: Gebruder Borntradger.

Stanfield, K. E., & Frost, I. C. (1949). Method of assaying oil shale by a modified Fischer retort. *U.S. Bureau of Mines Report of Investigations, 4477*, pp. 18–20.

Trudell, L. G., Roehler, H. W., & Smith, J. W. (1973). Geology of Eocene rocks and oil yields of Green River oil shales on part of Kinney Rim, Washakie Basin. *U.S. Bureau of Mines Report of Investigations 7775*, p. 22–26.

Trudell, L. G., Smith, J. W., Beard, T. N., & Mason, G. M. (1983). *Primary oil-shale resources of the Green River Energy Laramie Energy Technology Center*, DOE/LC/RI-82-4.

U.S. Geological Survey. (2004). *Sedimentary rocks.* Accessed 06/02/08 from http://geomaps.wer.usgs.gov/par/rxmin/ rock2.html. URL no longer exists.

U.S. Geological Survey. (2006). *The making of sedimentary rocks.* Accessed 06/03/08 from http://education.usgs.gov/schoolyard/rocks sedimentary.html. URL no longer exists.

Wiig, S. V., Grundy, W. D., & Dyni, J. R. (1995). *Trona resources in the Green River Formation, southwest Wyoming: U.S. Geological Survey Open-File Report* 95-476.

Winchester, D. E. (1916). Oil shale in northwestern Colorado and adjacent areas. *U.S. Geological Survey Bulletin 641-F,* 139–198.

Winchester, D. E. (1923). Oil shale of the Rocky Mountain region. *U.S. Geological Survey Bulletin 729,* p. 117.

Woodruff, E. G., & Day, D. T. (1914). Oil shales of northwestern Colorado and northeastern Utah. *U.S. Geological Survey Bulletin 581,* 1.

CHAPTER 16

Metamorphism and Deformation

The land ethic simply enlarges the boundaries of the community to include soils, waters, plants, and animals, or collectively: the land. This sounds simple: do we not already sing our love for and obligation to the land of the free and the home of the brave? Yes, but just what and whom do we love? Certainly not the soil, which we are sending helter-skelter downriver. Certainly not the waters, which we assume have no function except to turn turbines, float barges, and carry off sewage. Certainly not the plants, of which we exterminate whole communities without batting an eye. Certainly not to animals, of which we have already extirpated many of the largest and most beautiful species. A land ethic of course cannot prevent the alteration, management, and use of these "resources," but it does affirm their right to continued existence, and, at least in spots, their continued existence in a natural state.

—Aldo Leopold, 1948

In geology, *metamorphism* (Greek: *meta,* "change"; *morph,* "form": *metamorphism* means "to change form") is the process of change that rocks (mineral assemblage and texture) within the earth undergo when exposed to increasing temperatures and pressures at which their mineral components are no longer stable. Metamorphism may be *local*—contact metamorphism is due to igneous intrusion—or *regional*—as takes place in mountain building when slate, schist, and gneiss are formed. Contrary to the popular view that metamorphism can't occur unless tremendous heat is generated, metamorphism may take place in a solid state, without melting.

Recall that our earlier discussion of sedimentary rocks pointed out that sedimentary rocks also go through a process of changing form known as *diagenesis*. In geology, however, we restrict these sedimentary processes to those that occur at temperatures below 200° C and pressures below about 300 mega pascal (MPa). This is equivalent to approximately 3,000 atmospheres (atm) of pressure. On the other hand, metamorphism occurs at temperatures and pressure higher than 200° C and 300 MPa. The upper limit of metamorphism occurs at the pressure and temperature of wet partial melting of the affected rock. Keep in mind, however, that once melting begins, the process changes to an igneous process.

Metamorphism: Source of Heat and Pressure

Metamorphism occurs because some minerals are stable only under certain conditions of pressure and temperature. The heat involved with metamorphic processes is derived from uranium and thorium and other elements with lead and radiation added. Potassium-40 to calcium-40 or argon plus radiation provides another internal source of heat. In regard to pressure, recall that air pressure at sea level is measured at 14 psi (1 atm or 1 Bar = 100,000 pascals). In comparison, the pressure beneath 33 feet of water is equal to 1 atm (1 Bar). It is interesting to note that this same level of pressure is measured beneath 10 feet of rock. Pressure measured in the deepest part of the ocean is equivalent to 1,000 Bar. The pressure measured under 1 mile of rock is 500 Bars (1,000 Bars or 1 kilobar [kb] is the measured pressure beneath 2 miles of rock). The point is that rocks can be subjected to higher temperatures and pressure as they become buried deeper in the Earth. Such burial usually takes place as a result of tectonic processes such as continental collisions or subduction (discussed in detail later). These same processes, especially tectonic uplift and erosion, eventually are involved with projecting metamorphic rocks to the surface.

Types of Metamorphism

As mentioned, metamorphism consists of two types: contact metamorphism and regional metamorphism.

- **Contact metamorphism** occurs around or adjacent to igneous intrusions under low pressure but high temperatures in local, shallow areas of 0–6 km. Because only a small area surrounding the intrusion is heated by the magma, metamorphism is restricted to a zone surrounding the intrusion.

- **Regional metamorphism** occurs over large areas that were subjected to high pressure, causing deformation. Strongly foliated metamorphic rocks such as schists, slates, and gneisses are formed (Figure 16.1).

Did You Know?

Limestone is formed by the mineral calcite. Calcite is very stable over a wide range of temperature and pressure. Consequently, when metamorphism of limestone occurs, the original calcite crystals grow larger. The resultant rock produced is marble.

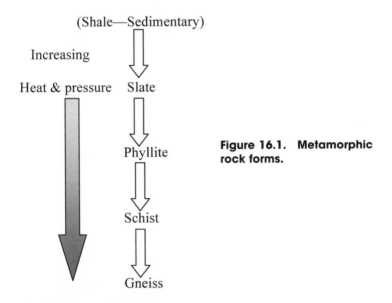

Figure 16.1. Metamorphic rock forms.

Major Metamorphic Rock Types

The three major metamorphic rock types include slate, schist, and gneiss.

- **Slate** is fine-grained chlorite and clay minerals; it is generally a foliated (banded), homogenous, metamorphic rock derived from an original sedimentary shale form through low-grade regional metamorphism. Figure 16.2 shows the response of shale rock to increasing metamorphic processes.
- **Schist** is a crystalline rock formed primarily from basalt, an igneous rock; shale, a sedimentary rock; or slate, a metamorphic rock. Tremendous heat

Original Shale

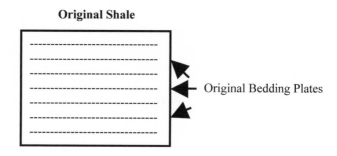

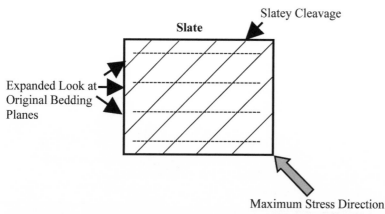

Figure 16.2. Metamorphism of clay and quartz minerals (shale) where maximum stress is applied at an angle to the original bedding planes, so that the slatey cleavage has developed at an angle to the original bedding.

and pressure caused these rocks to be transformed into schist rock. Schist rock has the tendency to split into irregular planar layers (schistosity). Most schist is composed largely of platy minerals such as muscovite, chlorite, talc, biotite, and graphite. Figure 16.3 shows schistosity of metamorphic schist rock.

- **Gneiss** (sounds like *niece*) is a metamorphic rock—metamorphosed primarily from granite and diorite—characterized by banding caused by segregation of different types of rocks, typically light and dark silicates. In the formation of gneiss, metamorphism continues and the sheet silicates become unstable, and dark-colored minerals like hornblende and pyroxene start to grow. As mentioned, these dark-colored minerals tend to become segregated in distinct bands through the rock, giving the rock a gneissic banding. The name *gneissic* actually is more suitable for its texture. Again, *gneissic texture* refers to the segregation of light and dark minerals.

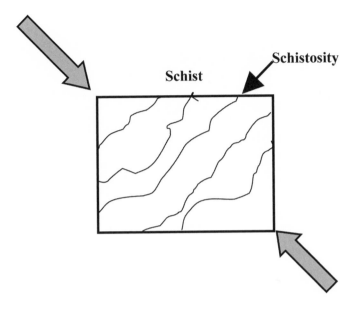

Maximum Stress Direction

Figure 16.3.   Shows the irregular planar foliation (Schistosity) of metamorphic schist rock.

Metamorphic Facies

As mentioned earlier in our discussion of sedimentary facies, a *facies* is a body of rock with specified characteristics. Metamorphic facies is similar to the concept of sedimentary facies in that a sedimentary facies is also a set of environmental conditions present during deposition. Keep in mind, however, that in general, metamorphic rocks do not undergo significant changes in chemical composition during metamorphism. In metamorphism—changes in rock assemblages—it is all about pressure and temperature, and variations from location to location, from rock type to rock type, and assemblage to assemblage. Consequently, the mineral assemblages that are observed are an indication of the temperature and pressure environment that they were subjected to. Moreover, it is important to keep in mind that, though we are speaking of changes in rocks because of metamorphism, not all changes in rocks are due to metamorphism—changes also occur as the result of surface and agents such as weathering (discussed later) and sedimentary processes such as diagenesis (discussed earlier).

Deformation of Rock

Although rock beds, especially sedimentary rocks, are generally deposited in almost horizontal beds, they are generally found to be tilted or distorted if followed over any considerable distance. Such structures are the result of forces within the earth that continually bend them, twist them, or fracture them. When rocks bend, twist, or fracture, we say they *deform* (change shape or size). When rocks deform (change in size, shape, or volume) they are said to *strain*. The intensity of the changes is proportional to the intensity of deformation and depth of burial. The forces that cause deformation of rock are referred to as *stresses*. The most intense zones of rock deformation are associated with mountain building processes.

To better understand rock deformation we must first understand the forces of stress.

- **Stress** is a force applied over an area (force/unit area).
- **Uniform stress** is a stress wherein the forces act equally from all directions (also called *pressure*). The pressure from the weight of overlying rocks within the earth is uniform stress but is usually referred to as *confining stress*.
- **Differential stress** is a stress in which the stress is not equal from all directions. Three kinds of differential stress occur.
  - Tensional or extensional stress stretches rock.
  - Compressional stress squeezes rock.
  - Shear stress results in slippage and translation.

TYPES OF DEFORMATION

Increasing stress causes rock to deform in three successive stages.

- **Elastic deformation:** Strain is reversible.
- **Ductile deformation:** Strain is irreversible.
- **Fracture:** Strain is irreversible; the material breaks.

Did You Know?

Brittle materials have a small or large region of elastic behavior but only a small region of ductile behavior before they fracture. On the other hand, ductile materials have a small region of elastic behavior and a large region of ductile behavior before they fracture.

How a particular material behaves depends on several factors, including:

- **Temperature:** When temperature is high materials are ductile. At low temperature, materials are brittle.
- **Confining pressure:** High surrounding pressure tends to hinder fracture. At low confining pressure, material fractures sooner.
- **Strain rate:** High strain rates cause materials to fracture. Low strain rates favor ductile behavior.
- **Composition:** Deals with the chemical bond types that hold rocks together. Quartz, olivine, and feldspars are very brittle while others such as micas, clay, and calcite minerals are more ductile. The presence of water is another compositional factor. Water weakens chemical bonds and allows rock to act in a ductile manner; dry rocks, however, tend to behave in a brittle manner.

Did You Know?

Geologists study rock deformation for many reasons. For example, it is important to know how beds and rocks are deformed to determine the location of coal seams, water aquifers, and other geological phenomena. Also, the location of ore deposits and petroleum traps is of economic importance.

DEFORMATION IN PROGRESS

Deformation of rocks is ongoing process. However, to the casual observer, this ongoing deformation process usually is not evident because it is slow and gradual. In many cases, it is so slow that it can only be observed and documented by using sensitive measuring instruments over periods of months, years, and decades, or even longer. Again, unless deformation is abrupt along fault lines caused by the fracture of rocks on a time scale of seconds or minutes, as in an earthquake, we are unlikely to notice deformation in progress.

GEOLOGICAL EVIDENCE OF FORMER DEFORMATION

When we observe crustal rock formations, evidence of deformation that has occurred in the past is very evident. For example, lava flows and sedimentary strata generally follow the law of original horizontality (sediments pile up sandwich-fashion and lave simply piles up). Thus, when we observe strata that is no longer horizontal but instead is inclined, it is clear that deformation has occurred sometime in past. Geologists are able to uniquely define and communicate to others the orientation of planar features by using two special terms—*strike* and *dip*.

Dip of a bed is a measure of its slope or tilt in relation to the horizontal. For an inclined plane the *strike* is the compass direction of any horizontal line on the plane. Again, stated in relation to strike, the *dip* is the angle between a horizontal plan and the inclined plane, measured perpendicular to the direction of strike. The *direction of dip* is the direction of maximum slope (or the direction a ball would run over the bed if its surface were perfectly flat). The *angle of dip* is the acute angle this direction makes with a horizontal plane.

Dip-strike symbols are used in most geological maps to record strike and dip measurements. The generally used symbol has a long line oriented parallel to the compass direction of the strike. A short tick line is placed in the center of the line on the side to which the inclined plane dips, and the angle of dip is recorded next to the strike and dip symbols as shown in Figure 16.4. For beds with a 90-degree vertical dip, the short line crosses the strike line, and for beds with no horizontal dip a circle with a cross inside is used, as shown in Figure 16.4.

Strike and dip symbol for vertical beds

Strike and dip symbol for horizontal beds

Figure 16.4. Strike and dip symbols for geo-logical maps.

Faults—the cause of earthquakes—involve brittle rock fractures and relative movement of rock units. In nearly all cases, the rocks involved were originally in a horizontal position. The amount of movement might vary from less than a few inches to many thousands of feet vertically and to more than 100 miles horizontally. Different types of faults are produced by different compressional and tensional stresses, and they also depend on the rock type and geological setting.

TYPES OF FAULTS

Depending on the direction of relative displacement, faults can be divided into several different types. Because faults are planar features, the concepts of *strike* and *dip* also apply, and thus the strike and dip of a fault plane can be measured. Types of faults are discussed in the following list.

- **Dip slip faults** have an inclined fault plane along which the relative displacement or offset has occurred along the dip direction.

✓ **Important Point:** In looking at the displacement of any fault, we don't know which side actually moved or if both sides moved. All we can determine is the relative sense of motion.

The block above the fault, for any inclined plane, is known as the *hanging wall block* and the block below the fault is the *footwall block*.

- **Normal faults** are faults in which relative downward movement has taken place down the upper face or hanging wall of the fault plane (Figure 16.5).

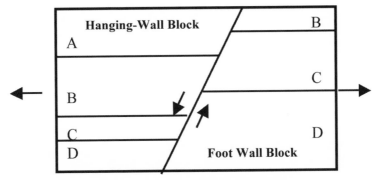

Figure 16.5. Normal fault with extensional stress.

- **Horsts and grabens** are normal faults that often occur in a series because of the tensional stress, which causes adjacent faults to dip in opposite directions. In this situation, the down-dropped fault blocks form *grabens* and the uplifted fault blocks form *horsts* (Figure 16.6). Rift valleys are graben structures hundreds of miles in length. The most prominent is that along the Red Sea, but the basin and range province of Nevada, Utah, and Idaho is also an example. In the basin and range, the basins are elongated grabens than now form valleys, and the ranges are uplifted horst blocks. They also occur below the oceans along the crests of the mid-oceanic ridges.

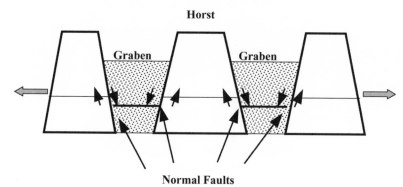

Figure 16.6.   Horst and Grabens.

- **Half-grabens** are bounded by only one fault instead of the two that form a normal graben (Figure 16.7).

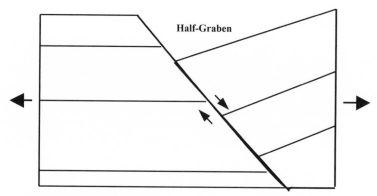

Figure 16.7.   Half-Graben.

- **Reverse faults** have relative upward movement of the hanging wall of the fault plane (Figure 16.8). They occur in areas of horizontal compressional stresses and folding such as mountain belts.

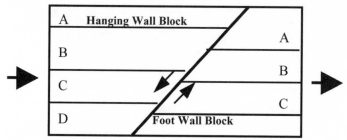

Figure 16.8.   Reverse fault compressional stress.

• **Thrust faults** are exemplified in Chief Mountain in Montana, which is an example of a special case of a reverse fault where the dip of the fault is less than 15 degrees (low dip). Thrust faults can have considerable displacement, measuring hundreds of miles, and can result in older strata overlying younger strata (Figure 16.9).

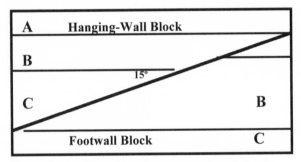

Figure 16.9.   Thrust fault compressional stress.

• **Strike-slip faults (tear or transform faults)** are those where shearing stress has produced horizontal movement. The San Andreas Fault (of 1906 San Francisco earthquake fame) is an example of a strike-slip fault (Figure 16.10), or, more precisely, is an example of a transform fault (two horizontal plates that slide past one another in a horizontal manner).

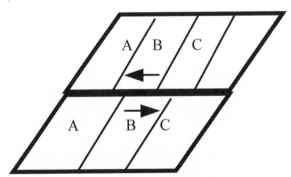

Figure 16.10.   Strike-slip fault (left-lateral).

MOVEMENTS ON FAULTS

Except in desert areas, quarries, and cliff faces, faults are rarely seen at the surface, but their presence is indicated by one or more of the following features:

• **Fault Breccias**—occur where the ground up rocks of the fault zone are scattered into angular, irregular-sized, crumbled up rock fragments.

- **Slikensides**—are polished striations or flutings (scratch marks) that are left on the fault plane as one block moves relative to the other. They are often found in the fault zone.

Folding of Rocks

Folds are wrinkles or flexures in stratified, ductile rocks. Folds result from compressional stresses acting over considerable time. They sometimes occur in isolation, but especially in mountain ranges they are more often packed together. Upfolds (where the originally horizontal strata have been folded upward) are called *anticlines* (Figure 16.11), and downfolds (where the two limbs [sides] of the fold dip inward toward the hinge of the fold) are called *synclines* (Figure 16.12).

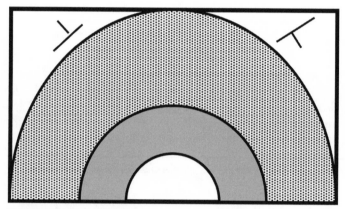

Figure 16.11.　Anticline.

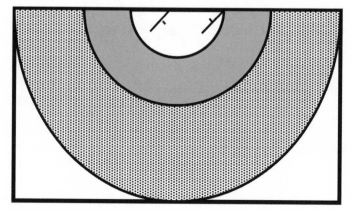

Figure 16.12.　Syncline.

CLASSIFICATION OF FOLDS

Folds can be classified based on their appearance.

- **Symmetrical folds** are those in which the axial plane is vertical with limbs dipping in opposite directions at the same inclinations (angles).
- **Asymmetrical folds** are those having an inclined axial plane and, like symmetrical folds, have limbs dipping in opposite directions, but at different inclinations.
- **Overturned folds** occur when the folding is so intense that the strata on one limb of the fold becomes nearly upside down.
- **Isoclinal folds** (*iso* means "name," and *cline* means "angle"; so *isoclinal* means the limbs have the same angle) occur when compressional stresses that cause the folding are intense, causing the fold to close up and have limbs that are parallel to each other.
- **Recumbant folds** occur when an overturned fold has an axial plant that is nearly horizontal.
- **Chevron folds** have no curvature in their hinge and straight-side limbs that form a zigzag pattern.

Did You Know?

All rocks do not behave the same under stress. When subjected to the same stress, some rocks will fracture or fault, while others will fold. When such contrasting rocks occur in the same area, such as brittle rocks overlaid by ductile rocks, the ductile rocks may bend or fold over the fault while the brittle rocks may fault.

Deformation of the Crust: Mountain Building

One of the most conspicuous and spectacular results of deformation acting within the crust of the Earth is the formation of mountain ranges. Any isolated, upstanding mass can be called a mountain. There is no minimum height or shape involved. There are three types of mountains originated by three processes (two of which are related to deformation):

1. **Fault block mountains:** As the name implies, fault block mountains originate by faulting—normal and reverse faulting—in any type of rock. The Sierra Nevada represents the uplifted block of granite 400 miles long and 100 miles wide.

2. **Fold and thrust mountains:** These mountains all have similar characteristics. They are the result of large compressional forces that cause continental crustal areas to collide, causing folding and thrusting of crustal blocks. The Alps, Himalayan, and Appalachian Mountains were formed by such processes.

3. **Volcanic mountains:** These include some of the most spectacular and beautiful mountains. Mount Rainer and Mount Saint Helens in Washington State, Mount Vesuvius in Italy, Mount Hood in Oregon, and the mountains of the Hawaiian Islands and Iceland are all characteristically steep and symmetrical volcanic mountains.

References and Recommended Reading

Dinwiddie, R., et al. (2005). *Universe*. New York: DK Publishing/Penguin Books.
Earth. (2003). New York: DK Publishing/Penguin Books.

CHAPTER 17

Weathering

In regard to surface rock formations, weathering and erosion are both creators and executioners! As soon as rock is lifted above sea level, weather starts to break it up. Water, ice, and chemicals split, dissolve, or rot the rocky surface until it crumbles. Mixed with water, air, and plant and animal remains, crumbled rock forms soil.

One can't obtain even the slimmest edge of understanding of geology without understanding the process of weathering, the first step in the erosion process that causes the breakdown of rocks, either to form new minerals that are stable on the surface of Earth or to break the rocks down to smaller particles. Simply, *weathering,* which projects itself on all surface material above the water table, is the general term used for all the ways in which a rock may be broken down.

Factors That Influence Weathering

The factors that influence weathering include:

- **Rock type and structure:** Each mineral contained in rocks has a different susceptibility to weathering. A rock with bedding planes, joints, and fractures provides pathways for the entry of water, leading to more rapid weathering. Differential weathering (rocks erode at differing rates) can occur when rock combinations consist of rocks that weather faster than more resistant rocks.
- **Slope:** On steep slopes weathering products might be quickly washed away by rains. Wherever the force of gravity is greater than the force of friction holding particles on a slope, they tend to slide downhill.
- **Climate:** Higher temperatures and high amounts of water generally cause chemical reactions to run faster. Rates of weathering are higher in warmer climates than in colder dry climates.

- **Animals:** Rodents, earthworms, and ants that burrow into soil bring material to the surface where it can be exposed to the agents of weathering.
- **Time:** The role time plays depends on slope, animals, and climate.

Categories of Weathering Processes

Although weathering processes are separated, it is important to recognize that these processes work in tandem to break down rocks and minerals to smaller fragments. Geologists recognize two categories of weathering processes

1. **Physical (or mechanical) weathering** is the disintegration of rocks and minerals by a physical or mechanical process.
2. **Chemical weathering** involves the decomposition of rock by chemical changes or solution.

PHYSICAL WEATHERING

Physical weathering involves the disintegration of a rock by physical processes. These include freezing and thawing of water in rock crevices, disruption by plant roots or burrowing animals, and the changes in volume that result from chemical weathering with the rock. These and other physical weathering processes are discussed in the following list.

- **Development of joints:** Joints are another way that rocks yield to stress. Joints are fractures or cracks in which the rocks on either side of the fracture have not undergone relative movement. Joints form as a result of expansion caused by cooling or relief of pressure as overlying rocks are removed by erosion. They form free space in rock by which other agents of chemical or physical weathering can enter (unlike faults that show offset across the fracture). They play an important part in rock weathering as zones of weakness and water movement.
- **Crystal growth:** The water that percolates through fractures and pore spaces might contain ions that precipitate to form crystals. When crystals grow they can cause the necessary stresses needed for mechanical rupturing of rocks and minerals.
- **Heat:** It was once thought that daily heating and cooling of rocks was a major contributor to the weathering process. This view is no longer shared by most practicing geologists. However, it should be pointed out that sudden heating of rocks from forest fires may cause expansion and eventual breakage of rock.

- **Biological activities:** Plant and animal activities are important contributors to rock weathering. Plants contribute to the weathering process by extending their root systems into fractures and growing, causing expansion of the fracture. Growth of plants and their effects are evident in many places where they are planted near cement work (streets, brickwork, and sidewalks). Animal burrowing in rock cracks can break rock.
- **Frost wedging:** Frost wedging is often produced by alternate freezing and thawing of water in rock pores and fissures. Expansion of water during freezing causes the rock to fracture. Frost wedging is more prevalent at high altitudes where there may be many freeze–thaw cycles. One classic and striking example of weathering of Earth's surface rocks by frost wedging is illustrated by the formation of Hoodoos in Bryce Canyon National Park, Utah (see Figure 18.1). "Although Bryce Canyon receives a meager 18 inches of precipitation annually, it's amazing what this little bit of water can do under the right circumstances!" (National Park Service [NPS], 2008).

Approximately 200 freeze–thaw cycles occur annually in Bryce. During these periods, snow and ice melt in the afternoon and water seeps into the joints of the Bryce or Claron Formation. When the sun sets, temperatures plummet and the water re-freezes, expanding up to 9% as it becomes ice. This frost wedging process exerts tremendous pressure or force on the adjacent rock and shatters and pries the weak rock apart. The assault from frost wedging is a powerful force but, at the same time, rain water (the universal solvent), which is naturally acidic, slowly dissolves away the limestone, rounding off the edges of these fractured rocks and washing away the debris. Small rivulets of water round down Bryce's rime, forming gullies. As gullies are cut deeper, narrow walls of rock, known as *fins*, being to emerge. Fins eventually develop holes known as *windows.* Windows grow larger until their roofs collapse, creating hoodoos (Figure 17.1). As old hoodoos age and collapse, new ones are born (NPS, 2008).

Did You Know?

Bryce Canyon National Park lies along the high eastern escarpment of the Paunsaugunt Plateau in the Colorado Plateau region of southern Utah. Its extraordinary geological character is expressed by thousands of rock chimneys (hoodoos) that occupy amphitheater-like alcoves in the Pink Cliffs, whose bedrock host is the Claron Formation of the Eocene age (Davis & Pollock, 2003).

Figure 17.1.   Frost-wedged-formed Hoodoos. Bryce Canyon National Park, Utah.

Photo by Frank R. Spellman.

Did You Know?

Hoodoo pronunciation: hu-du
Noun.
Etymology: West African; from voodoo
A natural column of rock in western North American often in fantastic form.

—Merriam-Webster Online

CHEMICAL WEATHERING

Chemical weathering involves the decomposition of rock by chemical changes or solution. Rocks formed under conditions present deep within the Earth are exposed to conditions quite different (i.e., surface temperatures and pressures are lower on the surface and copious amounts of free water and oxygen are available) when uplifted onto the surface. The chief processes are oxidation, carbonation and hydration, and solution in water above and below the surface.

The Persistent Hand of Water

Because of its unprecedented impact on shaping and reshaping Earth, it is important to point out that, given time, nothing, absolutely nothing on Earth is safe from the heavy hand of water. The effects of water sculpting by virtue of movement and accompanying friction are covered later in the text. For now, in regard to water exposure and chemical weathering, it is sufficient to note that the main agent responsible for chemical weathering reactions is not water movement but instead is water and weak acids formed in water.

The acids formed in water are solutions that have abundant free hydrogen$^+$ ions. The most common weak acid that occurs in surface waters is carbonic acid. Carbonic acid (H_2CO_3) is produced when atmospheric carbon dioxide dissolves in water; it exists only in solution. Hydrogen ions are quite small and can easily enter crystal structures, releasing other ions into the water.

$$H_2O + CO_2 \rightarrow H_2CO_3 \rightarrow H^+ + HCO_3^-$$

Where H_2O is water, CO_2 is carbon dioxide, H_2CO_3 is carbonic acid, H^+ is hydrogen ion, and HCO_3^- is bicarbonate ion.

Types of Chemical Weathering Reactions

As mentioned, chemical weathering breaks rocks down chemically by adding or removing chemical elements, and changes them into other materials. Again, as stated, chemical weathering consists of chemical reactions, most of which involves water. Types of chemical weathering include:

- **Hydrolysis** is a water–rock reaction that occurs when an ion in the mineral is replaced by H^+ or OH^-.
- **Leaching** causes ions to be removed by dissolution into water.
- **Oxidation** occurs because oxygen is plentiful near Earth's surface; thus, it may react with minerals to change the oxidation state of an ion.
- **Dehydration** occurs when water or a hydroxide ion is removed from a mineral.
- **Complete dissolution** solution weathering because of the low pH rainfall mixing with various chemicals present in the ground layer.

References and Recommended Reading

American Society for Testing and Materials (ASTM). (1969). *Manual on water.* Philadelphia: American Society for Testing and Materials.

Bailey, J. (1992). *The way nature works.* New York: Macmillan.

Brady, N. C., & Weil, R. R. (1996). *The nature and properties of soils,* (11th ed.). Upper Saddle River, NJ: Prentice-Hall.

Carson, R. (1962). *Silent spring.* Boston: Houghton Mifflin Company.

Ciardi, J. (1997). Stoneworks. In E. M. Cifelli (Ed.). *The collected poems of John Ciardi.* Fayettville: University of Arkansas Press.

Davis, G. H., & Pollock, G. L. (2003). Geology of Bryce Canyon National Park, Utah. In Sprinkel, D. A., et al. (Eds.). *Geology of Utah's parks and monuments* (2nd ed.). Salt Lake City: Utah Geological Association.

Eswaran, H. (1993). Assessment of global resources: Current status and future needs. *Pedologie 40*(3), 19–39.

Foth, H. D. (1992). *Fundamentals of soil science* (6th ed.). New York: John Wiley and Sons.

Franck, I., & Brownstone, D. (1992). *The green encyclopedia.* New York: Prentice-Hall.

Hoodoo. (2012). *Merriam-Webster Online.* Accessed from www.merriam-webster.com.

Kemmer, F. N. *Water: The Universal Solvent.* Oak Ridge, IL: NALCO Chemical Company.

Konigsburg, E. M. (1996). *The view from Saturday.* New York: Scholastic.

Mowet, F. (1957). *The dog who wouldn't be.* New York: Willow Books.

National Park Service (NPS). (2008). *The hoodoo.* Washington, DC: National Park Service.

Spellman, F. R. (1998). *The science of environmental pollution.* Boca Raton, FL: CRC Press.

Tomera, A. N. (1989). *Understanding basic ecological concepts.* Portland, ME: J. Weston Walch.

U.S. Department of Agriculture (USDA) (1975). *Soil taxonomy: A basic system of soil classification for making and interpreting soil surveys.* Washington, DC: USDA Natural Resources Conservation Service.

U.S. Department of Agriculture (USDA) Soil Survey Staff (1975). *Soil classification: A comprehensive system.* Washington, DC: USDA Natural Resources Conservation Service.

U.S. Department of Agriculture (USDA) Soil Survey Staff (1994). *Keys to soil taxonomy.* Washington, DC: USDA Natural Resources Conservation Service.

CHAPTER 18

Wind Erosion, Mass Wasting, and Desertification

The winds wander, the snow and rain and dew fall, the earth whirls—all but to prosper a poor lush violet.

—John Muir, 1913

Wind Erosion

During a recent research outing to several national parks in the western United States, we stopped at several locations and photographed various natural wonders. One of the focal points of study was the weathering processes discussed in this chapter. The natural bridges, such as the one shown in Figure 18.1 and the natural arches or windows (which eventually will become hoodoos) shown in Figures 18.2 and 18.3 all are a result of some form of weathering; thus, they are highlighted in this chapter.

There was a time not that long ago when many believed that the main difference between natural bridges (see Figure 18.1) and the natural windows (performing hoodoos) shown in Figures 18.2 and 18.3 was that the natural bridges were formed by water erosion and natural arches were formed by wind erosion. Contrary to popular belief or myth, however, wind is not a significant factor in the formation of natural arches or other natural formations. Substantial studies have shown that natural arches and natural bridges are formed by many different processes of erosion that contribute to the natural, selective removal of rock. Every process relevant to natural arch formation involves the action of water, gravity, temperature variation, or tectonic pressure on rock.

Again, wind is not a significant agent in natural arch formation. Wind does act to disperse the loose grains that result from microscopic erosion.

Figure 18.1.   Rainbow Bridge—the world's largest known natural bridge, Lake Powell/Colorado River region, Utah.
Photo by Frank R. Spellman

Figure 18.2.   Window forming in future hoodoos. Bryce Canyon, Utah.
Photo by Frank R. Spellman

Figure 18.3.   Weathered-window forming in hoodoo formation. Bryce Canyon, Utah.
Photo by Frank R. Spellman

Moreover, sandstorms can scour and polish already existing arches. The bottom line (and the point to remember) is that wind never alone creates arches (Barnes, 1987; Vreeland, 1994).

As prefaced earlier, wind action or erosion is very limited in extent and effect. It is largely confined to desert regions, but even there it is limited to a height of about 18 inches above ground level. Wind does have the power, however, to transport, to deposit, and to erode sediment. In this chapter we discuss each of the aspects of the wind because they are important in any study of geology.

Did You Know?

Wind is common in deserts because the air near the surface is heated and rises and cooler air comes in to replace the hot, rising air. This movement of air results in winds. Also arid desert regions have little or no soil moisture to hold rock and mineral fragments.

Wind Sediment Transport

Sediment near the ground surface is transported by wind in a process called *saltation* (Latin, *saltus,* "leap"). *Wind saltation,* which is similar to what occurs in the bed load of streams, refers to short jumps (leaps) of grains that are dislodged from the surface and leap a short distance. As the grains fall back to the surface, they dislodge other grains that then get carried by wind until they collide with ground to dislodge other particles. Above ground level, wind can swoop down to the surface and lift smaller particles, suspending them in the wind and making them airborne; they may travel long distances.

Did You Know?

Sand ripples occur as a result of large grains accumulating as smaller grains are transported away. Ripples form in lines perpendicular to wind direction. Sand-sized particles generally do not travel very far in the wind, but windblown dust made up of smaller fragments can be suspended in the wind for much larger distances.

Wind-Driven Erosion

As mentioned, wind by itself has little if any effect on solid rock. But in arid and semiarid regions, wind can be an effective geologic agent anywhere that it possesses a velocity high enough to pick up a load of rock fragments, which may become effective tools of erosion in the land-forming process. Wind can erode by *deflation* and *abrasion.*

DEFLATION

The process of deflation (or blowing away) is the lowering of the land surface resulting from removal of fine-grained particles by the wind. Deflation concentrates the coarser-grained particles at the surface, eventually resulting in a relatively smooth surface composed only of the coarser-grained fragments that cannot be transported by the wind. Such a coarse-grained surface is called *desert pavement* (Figure 18.4). Some of these coarser-grained fragments may exhibit a dark, enamel-like coat of iron or manganese called *desert varnish.*

Deflation may create several types of distinctive features. For example, *lag gravels* are formed when the wind blows away finer rock particles, leaving behind

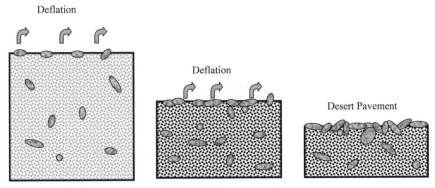

Figure 18.4.   Wind-driven deflation processes.

a residue of coarse gravel and stones. *Blowouts* may be developed where wind has scooped out soft unconsolidated rocks and soil.

ABRASION

The wind abrades (sand blasts) by picking up sand and dust particles, which are transported as part of its load. Abrasion is restricted to a distance of about 1 or 2 m above the surface because sand grains are lifted a short distance. The destructive action of these windblown abrasives might wear away wooden telephone poles and fence posts, and abrade, scour, or groove solid rock surfaces.

Wind abrasion also plays a part in the development of such landforms as isolated rocks (pedestals and table rocks) that have had their bases undercut by windblown sand and grit (Figure 18.5). *Ventifacts* are another interesting and relatively common product of wind erosion. These are any bedrock surface, stone, or pebble that has been abraded or shaped by windblown sediment in a process similar to sand blasting. Ventifacts are formed when the wind blows sand against the side of the stone, shaping it into a flat, polished surface. At a much larger scale, elongate ridges called *yardangs* form by the abrasion and streamlining rock structures oriented parallel to the prevailing wind direction.

Wind Deposition

The velocity of the wind and the size, shape, and weight of the rock particles determines the manner in which wind carries its load. Wind-transported materials are most commonly derived from flood plains, beach sands, glacial deposits, volcanic explosions, and dried lake bottoms—places containing light ash and loose, weathered rock fragments.

Figure 18.5. Seeming to defy gravity, Balanced Rock (Arches National Park, Moab, Utah) has a harder cap rock that somewhat protects the more easily eroded base; eventually the double hammering of water and wind erosion will cause it to disintegrate, leaving a pile of rocky debris as a reminder of the power of erosion.
Photo by Frank R. Spellman

The wind is capable of transporting large quantities of material for very great distances. The wind deposits sediment when its velocity decreases to the point where the particles can no longer be transported. Initially (in a strong wind), part of the sediment load rolls or slides along the ground (bed load). Some sand particles move by a series of leaping or bounding movements (saltation). And lighter dust may be transported upward (suspension) into higher, faster moving wind currents, traveling many thousands of miles.

As mentioned, the wind will begin to deposit its load when its velocity is decreased, or when the air is washed clean by falling rain or snow. A decrease in wind velocity may also be brought about when the wind strikes some barrier-type obstacle (fences, trees, rocks, human-made structures) in its path. As the air moves over the top of the obstacle, streamlines converge and the velocity increases. After passing over the obstacle, the streamlines diverge and the velocity decreases. As the velocity decreases, some of the load in suspension can no longer be held in suspension, and thus drops out to form a deposit. The major types of windblown or eolian deposits are dunes and loess.

DUNES

Sand dunes are asymmetrical mounds with a gentle slope in the upwind direction and steep slope on the downwind side (see Figure 19.6). Dunes vary greatly in size and shape and form when there is a ready supply of sand, a steady wind, and some kind of obstacle or barrier such as rocks, fences, or vegetation to trap some of the sand. Sand dunes form when moving air slows down on the downwind side of an obstacle (Figure 18.6). Dunes may reach heights up to 500 m and cover large areas. Types of sand dunes include Barchan, transverse, longitudinal, and parabolic.

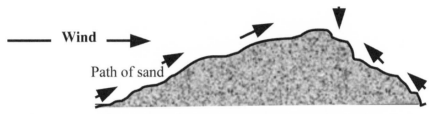

Figure 18.6.   Profile of typical sand dune. Arrows denote paths of wind currents.

- **Barchan dunes** are crescent-shaped dunes characterized by two long, curved extensions pointing in the direction of the wind, and a curved slip face on the downwind side of the dune (Figure 18.7[a]). These dunes are formed in areas where winds blow steadily and from a single direction.
- **Transverse dunes** form along sea coasts and lake shores and might be 15 feet high and half a mile in length. Transverse dunes develop with their long axis at right angles to the wind (Figure 18.7[b]).
- **Longitudinal dunes** are long, ridge-like dunes that develop parallel to the wind (Figure 18.7[c]).
- **Parabolic dunes** are U-shaped dunes with an open end facing upwind. They occur where there is a constant wind direction, an abundant sand supply, and abundant vegetation, which usually stabilizes them (Figure 18.7[d]).

In the United States, the most significant sand dune formations can be found in Great Sand Dunes National Park and Preserve located in southwest Colorado (Figure 18.8[a–c]). The Great Sand Dunes dunefield is actually just one of four primary components of the Great Sand Dunes geological system.

The *mountain watershed* of Great Sand Dunes receives heavy snow and rain each year. Creeks flow from alpine tundra and lakes, down through subalpine and montane woodlands, and finally around the main dunefield. Sand that has blown from the valley floor is captured and carried back toward the valley. When

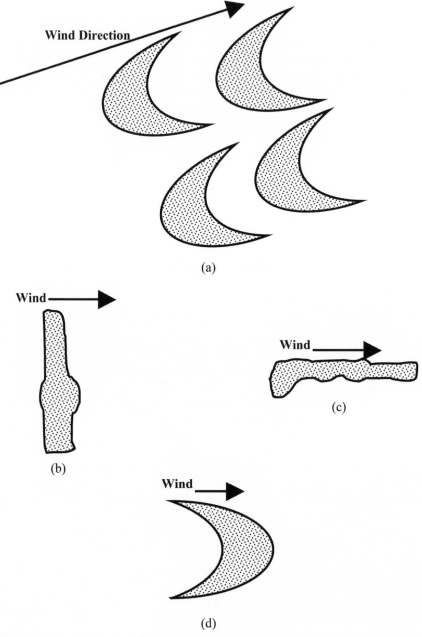

Figure 18.7.   a. Barchan dune; b. Transverse dune; c. Longitudinal dune; d. Parabolic dune.

Figure 18.8a. Medeno Creek. Great Sand Dunes National Park and Pre-serve, Colorado.
Photo by Frank R. Spellman

Figure 18.8b. Active dunefield. Great Sand Dunes National Park and Pre-serve, Colorado.
Photo by Frank R. Spellman

Figure 18.8c.   Active dunefield. Great Sand Dunes National Park and Preserve, Colorado.
Photo by Frank R. Spellman

creeks disappear into the valley floor, sand is again picked up and carried into the main dunefield. This recycling action of water and wind contributes to the great height of this dunefield.

The 30-square-mile (78 sq km) active dunefield is where the tallest (~750 ft) dunes reside. It is stabilized by opposing wind directions (southwesterly and northeasterly), creeks that recycle sand back into it, and 7% moisture content below the dry surface. The dunefield is composed of reversing dunes, transverse dunes, star dunes, and a few barchan dunes. It is estimated to contain over 5 billion cubic meters of sand.

The *sand sheet* is the largest components of the Great Sand Dunes geological system, made up of sandy grasslands that extend around three sides of the main dunefield. Almost 90% of the sand deposit is found here, while only about 10% is found in the main dunefield. The sand sheet is the primary source of sand for the Great Sand Dunes. Small parabolic dunes form here, then migrate into the main dunefield. Nebkha (coppice) dunes form around vegetation.

The sabkha forms where sand is seasonally saturated by rising ground water. When the water evaporates away in late summer, minerals similar to baking soda cement sand grains together into a hard, white crust. Areas of sabkha can be found throughout western portions of the sand sheet, wherever the water table meets the surface. Some wetlands in the sabkha are deeper with plentiful plants and animals, while others are shallow and salty.

Did You Know?

The Great Sand Dunes tiger beetle is found nowhere else on Earth. Its specially adapted long legs and fine hairs on its underside help it survive sand temperatures of 140° F (60° C).

LOESS

Loess is a yellowish, fine-grained, nonstratified material carried by the wind and accumulated in deposits of dust. The materials forming loess are derived from surface dust originating primarily in deserts, river flood plains, deltas, and glacial outwash deposits. Loess is cohesive and possesses the property of forming steep bluffs with vertical faces such as the deposits found in the pampas of Argentina and the lower Mississippi River Valley.

Mass Wasting

Mass wasting, or mass movement, takes place as Earth materials (loose, unce-mented mixture of soil and rock particles known as *regolith*) move downslope in response to gravity without the aid of a transporting medium, such as water, ice, or wind—though these factors play a role in regolith movement. This type of erosion is apt to occur in any area with slopes steep enough to allow downward movement of rock debris. Some of the factors that help gravity overcome this resistance are discussed in the following sections.

GRAVITY

The heavy hand of gravity constantly pulls everything, everywhere toward Earth's surface. On a flat surface, parallel to Earth's surface, the constant force of gravity acts downward. This downward force prevents gravitational movement of any material that remains on or parallel to a flat surface.

On a slope, the force of gravity can be resolved into two components: a component acting perpendicular to the slope and a component acting tangential to the slope. Thus, material on a slope is pulled inward in a direction that is perpendicular (the glue) to the slope (Figure 18.9[a]). This helps prevent material from sliding downward. However, as stated previously, on a slope, another component of gravity exerts a force (a constant tug) that acts to pull material down a slope parallel to the surface of the slope. Known as *shear stress*, this force of gravity exerts stress in direct relationship to the steepness of the slope. That is, shear stress increases as the slope steepens. In response to increased shear stress, the perpendicular force (the glue) of gravity decreases (Figure 18.9[b]).

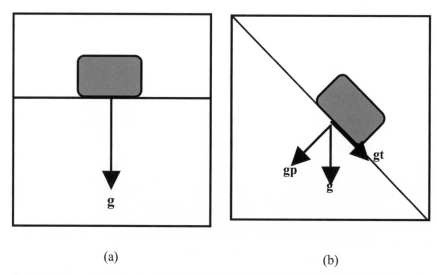

(a)   (b)

Figure 18.9. (a) Gravity acting perpendicular to the surface. (b) The perpendicular component (the glue) of gravity, *gp*, helps to hold the material in place on the slope.

Did You Know?

When shear on a slope decreases, material may still be stuck to the slope and prevented from moving downward by the force of friction. It may be held in place by the frictional contact between the particles making up that material. Contact between the surfaces of the particles creates a certain amount of tension that holds the particles in place at an angle. The steepest angle at which loose material on a slope remains motionless is called the *angle of repose* (generally about 35 degrees). Particles with angled edges that catch on each other also tend to have a higher angle of repose than those that have become rounded through weathering and that simply roll over each other.

WATER

Even though mass wasting may occur in either wet or dry materials, water greatly facilitates downslope movements; it is an important agent in the process of mass wasting. Water will either help hold material together (act as glue—demonstrated in building beach sandcastles with slightly dampened sand), increasing its angle of repose, or cause it to slide downward like a liquid (acting like a lubricant). Water may soften clays, making them slippery, add weight to the rock mass, and, in large amounts, may actually force rock particles apart, thus reducing soil cohesion.

FREEZING AND THAWING

Earlier, the erosive power of frost wedging (water contained in rock and soil expands when frozen) was discussed. Mass wasting in cold climates is governed by the fact that water is frozen as ice during long periods of the year, especially in high-altitude regions. Ice, although it is solid, does have the ability to flow (glacial-movement effect), and alternate periods of freezing and thawing can also contribute to movement; in some instances ice expansion may be great enough to force rocks downhill.

UNDERCUTTING

Undercutting occurs when streams erode their banks or surf action along a coast, making it unstable. Undercutting can also occur when human-made excavations remove support and allow overlying material to fall.

ORGANIC ACTIVITIES

Whenever animals burrow into the ground, they disturb soil materials, casting rocks out of their holes as they dig; these are commonly piled up downslope. Eventually, weather conditions and the constant force applied by gravity can put these piles into motion. Animals also contribute to mass wasting whenever they walk on soil surfaces; their motions can knock materials downhill.

SHOCK WAVES OR VIBRATIONS

A sudden strong shock or vibration, such as an earthquake, faulting, blasting, and heavy traffic can trigger slope instability. Minor shocks like heavy vehicles rambling down the road, trees blowing in the wind, or human-made explosions can also trigger mass-wasting events such as landslides.

KINDS OF MASS MOVEMENTS

A landslide is a mass movement that occurs suddenly and violently. In contrast, soil creep is mass movement that is almost imperceptible. These processes can be divided into two broad categories: rapid and slow movements. *Rapid movements* include landslides, slumps, mudflows, and earthflows. *Slow movements* include soil creep and solifluction.

Rapid Movements

- **Landslides** are by far the most spectacular and violent of all mass movements. Landslides are characterized by the sudden movement of great quantities of rock and soil downslope. Such movements typically occur on steep slopes that have large accumulations of weathered material. Precipitation in the form of rain or snow may seep into the mass of steeply sloping rock debris, adding sufficient weight to start the entire mass sliding.
- **Slumps** are special landslides that occur along a curved surfaces. The upper surface of each slump block remains relatively undisturbed, as do the individual blocks. Slumps leave arcuate (Latin, "curved like a bow") scars or depressions on the hill slope. Heavy rains or earthquakes usually trigger slumps. Slump is a common occurrence along the banks of streams or the walls of steep valleys.
- **Mudflows** are highly fluid, high-velocity mixtures of sediment and water that have a consistency of wet concrete. Mass wasting of this type typically occurs when certain arid or semiarid mountainous regions are subjected to unusually heavy rains.
- **Earthflows** are usually associated with heavy rains and move at velocities between several centimeters and hundreds of meters per year. They usually remain active for long periods. They generally tend to be narrow, tongue-like features that begin at a scarp or small cliff.

Slow Movements

- **Soil creep** is a continuous movement, usually so slow as to be imperceptible, that normally occurs on almost all slopes that are moist but not steep enough for landslides. Soil creep is usually accelerated by frost wedging, by alternate thawing and freezing, and by certain plant and animal activities. Evidence for creep is often seen in bent trees, offsets in roads and fences, and inclined utility poles.

• **Solifluction** is a downslope movement typical of areas where the ground is normally frozen to considerable depth—arctic, subarctic, and high mountain regions. The actual soil flowage occurs when the upper portion of the mantle rock thaws and becomes water saturated. The underlying, still frozen subsoil acts as a slide for the sodden mantle rock, which will move down even the gentlest slope.

Did You Know?

Landslides constitute a major geologic hazard because they are widespread, occur in all 50 states and U.S. territories, and caused $1–2 billion in damages and more than 25 fatalities on average each year. Expansion of urban and recreational developments into hillside areas leads to more people being threatened by landslides each year. Landslides commonly occur in connection with other major natural disasters such as earthquakes, volcanoes, wildfires, and floods (U.S. Geological Survey [USGS], 2008).

DESERTIFICATION

Deserts are areas where the amount of precipitation received is less than the potential evaporation (<10 in/year); they cover roughly 30% of the Earth's land surface—areas we think of as arid. *Desertification* occurs in hot areas far from sources of moisture, areas isolated from moisture by high mountains, in coastal areas along which there are onshore winds and cold-water currents, and high-pressure areas, where descending air masses produce warm, dry air.

According to the U.S. Geological Survey (1997), the world's great deserts were formed by natural processes interacting over long intervals of time. During most of these times, deserts have grown and shrunk independent of human activities. Desertification does not occur in linear, easily mapable patterns. Deserts advance erratically, forming patches on their borders. Scientists question whether desertification, as a process of global change, is permanent or how and when it can be halted or reversed.

References and Recommended Reading

Barnes, F. A. (1987). *Canyon country arches and bridges*. Author.
Goodwin, P. H. (1998). *Landslides, slumps, and creep*. New York: Franklin Watts.
Jennings, T. (1999). *Landslides and avalanches*. North Mankato, MN: Thame-side Press.

U.S. Geological Survey (USGS) (1997). Desertification. Accessed 7/08/08 from http://pubs.usgs.gov/gip/deserts/desertificaiton/.

U.S. Geological Survey (USGS) (2008). *Landslide hazards program.* Accessed 7/22/08 from http://landslides.usgs.gov/.

Vreeland, R. H. (1994). *Nature's bridges and arches: Vol. 1. General Information* (2nd ed.). Author.

Walker, J. (1992). *Avalanches and landslides.* New York: Gloucester Press.

CHAPTER 19

Glaciation

Those who dwell among the beauties and mysteries of the
earth are never alone or weary of life.

—Rachel Carson

Approximately 10,000 to 12,000 years ago, many parts of Earth were covered
with massive sheets of ice. Moreover, the geologic record shows that this most
recent ice-sheet covering of large portions of Earth's surface is not a one-time
phenomenon; instead, Earth has experienced several glaciation periods as well as
interglacial periods like the one we are presently experiencing. Although the ice
that we discussed earlier in the story of Yurk and the cat has now retreated from
most of Europe, Asia, and North America, it has left traces of its influence across
the whole face of the landscape in jagged mountain peaks; gouged-out upland
valleys; swamps; changed river courses; and boulder-strewn, table-flat prairies in
the lowlands.

Ice covers about 10% of all land and approximately 12% of the oceans.
Most of this ice is contained in the polar ice sea, polar sheets and ice caps, valley
glaciers, and piedmont glaciers formed by valley glaciers merging on a plain. In
the grand scheme of geology of the present time, the glaciers of today are not
that significant. It is the glaciation of the past with its accompanying geologic
evidence left behind by ancient glaciers that is important. This geologic record
indicates that the Earth's climate has undergone fluctuations in the past, and that
the amount of the Earth's surface covered by glaciers has been much larger in the
past than in the present. In regard to the effects of past glaciation, one need only
look at the topography of the western mountain ranges in the northern part of
North America to view the significant depositional processes of glaciers.

Glaciers

A *glacier* is a thick mass of slowing moving ice, consisting largely of recrystallized snow that shows evidence of downslope or outward movement caused by the pull of gravity. Glaciers can only form at latitudes or elevations above the snowline (the elevation at which snow forms and remains present year round). Glaciers form in these areas if the snow becomes compacted, forcing out the air between the snowflakes. The weight of the overlying snow causes the snow to recrystallize and increase its grain-size, until it increases its density and becomes a solid block of ice.

TYPES OF GLACIERS

There are various types of glaciers, including the following:

- **Mountain glaciers** are relatively small glaciers that occur at higher elevations in mountainous regions. A good example of mountain glaciers can be seen in the few remaining glaciers of Glacier National Park, Montana (Figure 19.1).

Figure 19.1.   Remnants of a glacier. Glacier National Park, Montana.
Photo by Frank R. Spellman

Note: The low snow and ice content of the glacier shown in Figure 19.1 is due to Earth's recent warming trend and the time of the year when the photos were taken (July 2008).

- **Continental glaciers (ice sheets)** are the largest glaciers. They cover Greenland and Antarctica and contain about 95% of all glacial ice on Earth.
- **Ice shelves** are sheets of ice floating on water and attached to land. They may extend hundreds of miles from land and reach thicknesses of several thousand feet.
- **Polar glaciers** are always below the melting point at the surface and do not produce any melt water.
- **Temperate glaciers** are at a temperature and pressure level near the melting point throughout the body except for a few feet of ice. This layer is subjected to annual temperature fluctuations.

GLACIER CHARACTERISTICS

The primary characteristics displayed by glaciers are changes in size and movement. A glacier changes in size by the addition of snowfall, compaction, and recrystallization. This process is known as *accumulation.* Glaciers also shrink in size (as a result of temperature increases). This process is known as *ablation.*

Earth's gravity, pushing, pulling, and tugging almost everything toward Earth's surface is involved with the movement of glaciers. Gravity moves glaciers to lower elevations by two different processes:

- **Basal sliding** is a type of glacier movement that occurs when a film of water at the base of the glacier reduces friction by lubricating the surface and allowing the whole glacier to slide across its underlying bed.
- **Internal flow**, called *creep,* forms fold structures and results from deformation of the ice crystal structure; the crystals slide over each other like a deck of cards. This type of flow is conducive to the formation of crevasses in the upper portions of the glacier. Generally, crevasses form when the lower portion of a glacier flows over sudden changes in topography.

Did You Know?

Within a glacier, the velocity constantly changes. The velocity is low next to the base of the glacier and where it is in contact with valley walls. The velocity increases toward the center and upper parts of the glacier.

Glaciation

Glaciation is a geological process that modifies land surface by the action of glaciers. For those who study glaciation and glaciers, the fact that glaciations have occurred so recently in North America and Europe provides the opportunity to study the undeniable results of glacial erosion and deposition. This is the case, of course, because the forces involved with erosion—weathering, mass wasting, and stream erosion—have not had enough time to remove the traces of glaciation from Earth's surface. Glaciated landscapes are the result of both glacial erosion (glaciers transport rocks and erode surfaces) and glacial deposition (glaciers transport material that melts and deposits material).

GLACIAL EROSION

Glacial erosion has a powerful effect on land that has been buried by ice and has done much to shape our present world. Both valley and continental glaciers acquire tens of thousands of boulders and rock fragments, which, frozen into the sole of the glacier, act like thousands of files, gouging and rasping the rocks (and everything else) over which the glaciers pass. The rock surfaces display fluting, striation, and polishing effects of glacial erosion. The form and direction of these grooves can be used to show the direction in which the glaciers move.

Glacial erosion manifests itself in small-scale erosional features, landform production by mountain glaciers, and landforms produced by ice caps and ice sheets. These are described in the following.

- **Small scale erosional features** include glacial striations and polish. *Glacial striations* are long, parallel scratches and glacial grooves that are produced at the bottom of temperate glaciers by rocks embedded in the ice scraping against the rock underlying the glacier. *Glacial polish* is characteristic of rock that has a smooth surface produced as a result of fined-grained material embedded in the glacier acting like sandpaper on the underlying surface.
- **Landforms produced by mountain glaciers** are erosion-produced features that include:
  - **Cirques** are bowl-shaped valleys formed at the heads of glaciers and below arêtes and horned mountains; they often contain a small lake called a *tarn*.

    As cirque glaciers grow larger, they might spread into valleys and flow down the valleys as *valley glaciers*. Valley glaciers are tongues of ice that spill down a valley as snow and ice accumulate, filling it with ice, perhaps for scores of miles. When a valley glacier extends down to sea level, it might carve a narrow valley into the coastline. These are called *fjord glaciers,* and the narrow valleys they carve, which later become filled with seawater after the ice has melted, are *fjords.* When a valley glacier extends down a valley and then covers a gentle slope beyond the mountain range, it is called a *piedmont glacier.* If valley glaciers cover a mountain range, they are called *ice caps.*

○ **Glacial valleys** are valleys that once contained glacial ice and become eroded into a "U" shape in cross section. "V"-shaped valleys are the result of stream erosion.

• **Aretes** are sharp ridges formed by headward glacial erosion.

○ **Horns** are sharp, pyramidal mountain peaks formed when headward erosion of several glaciers intersect.

○ **Hanging valleys** are exemplified by Yosemite's Bridalveil Falls, a waterfall that plunges over a hanging valley. Generally, hanging valleys result in tributary streams that are not able to erode to the base level of the main stream; therefore, the tributary stream is left at higher elevation than the main stream, creating a hanging valley and sometimes spectacular waterfalls.

○ **Fjords** are submerged, glacially deepened, narrow inlets with sheer, high sides, a U-shaped cross profile, and a submerged seaward sill largely formed of end moraine.

• **Landforms produced by ice caps and ice sheets** include the following:

○ **Abrasional features** are small-scale abrasional features in the form of glacial polish and striations that occur in temperate environments beneath ice caps and ice sheets.

○ **Streamlined forms**, sometimes called "basket of eggs" topography, occur when the land beneath a moving continental ice sheet is molded into smooth, elongated forms called *drumlins.* Drumlins are aligned in the direction of ice flow; their steeper, blunter ends point toward the direction from which the ice came.

GLACIAL DEPOSITS

All sediment deposited as a result of glacial erosion is called *glacial drift.* The sediment deposited, glacial drift, consists of rock fragments that are carried by the glacier on its surface, within the ice and at its base.

• **Ice land deposits** are the results of glacial ice deposited on land.

• **Till (or rock flour)** is nonsorted glacial drift deposited directly from ice. Consisting of a random mixture of different-sized fragments of angular rocks in a matrix of fine grained, sand- to clay-sized fragments, till was produced by abrasion within the glacier. After undergoing diagenesis and turning to rock, till is called *tillite*.

- **Erratics** are a glacially deposited rock, fragment, or boulder that rests on a surface made of different rock. Erratics are often found miles from their source and by mapping the distribution pattern of erratics, geologists can often determine the flow directions of the ice that carried them to their present locations.
- **Moraines** are mounds, ridges, or ground coverings of unsorted debris, deposited by the melting away of a glacier. Depending on where it formed in relation to the glacier, moraines can be:
  - **Ground moraines** are till-covered areas deposited beneath the glacier that result in a hummocky topography with lots of enclosed small basins.
  - **End moraines and terminal moraines** are ridges of unconsolidated debris deposited at the low elevation end of a glacier as the ice retreats as the result of ablation (melting). They usually reflect the shape of the glacier's terminus.
  - **Lateral moraines** are till deposits that were deposited along the sides of mountain glaciers.
  - **Medial moraines** occur when two valley glaciers meet to form a larger glacier, and the rock debris along the sides of both glaciers merge to form a medial moraine (runs down the center of a valley floor).
- **Glacial marine drift (icebergs)** are glaciers that reach lake shores or oceans and calve off into large icebergs, which then float on the water surface until they melt. The rock debris that the icebergs contain is deposited on the lakebed or ocean floor when the iceberg melts.
- **Stratified drift** is glacial drift that can be picked up and moved by melt-water streams, which can then deposit that material as stratified drift.
- **Outwash plains** is melt runoff at the end of a glacier that is usually choked with sediment and forms braided streams, which deposit poorly sorted stratified sediment in an outwash plain—they usually are flat, interlocking alluvial fans.
- **Outwash terraces** form if the outwash streams cut down into their outwash deposits, forming river terraces.
- **Kettle holes** are depressions caused by melting of large blocks of stagnant ice, found in any typical glacial deposit. They are sometimes filled by lakes; Minnesota, the "land of a thousand lakes," is an example.
- **Kames** are isolated hills of stratified material formed from debris that fell into openings in retreating or stagnant ice.
- **Eskers** are long, narrow, and often branching sinuous ridges of poorly sorted gravel and sand formed by deposition from former glacier streams.

References and Recommended Reading

Associated Press (1997, December 7). Does warming feed El Niño? *Virginian Pilot* (Norfolk, VA), p. A-15.

Associated Press (1998, September 25). Tougher air pollution standards too costly, Midwestern states say. *Lancaster New Era* (Lancaster, PA).

Associated Press (1998, September 28). Ozone hole over Antarctica at record size. *Lancaster New Era* (Lancaster, PA).

Chernicoff, S. (1999). *Geology.* Boston: Houghton Mifflin Company.

Dolan, E. F. (1991). *Our poisoned sky.* New York: Cobblehill Book.

Global warming: It's here . . . and almost certain to get worse (1998, August 24). *Time Magazine.*

Global warming: politics and economics further complicate the issue (1997, December 1). *USA Today,* p. A-1, 2.

Hansen, J. E., et al. (1986). Climate sensitivity to increasing greenhouse gases. In Barth, M. C., & Titus, J. G. (Eds.). *Greenhouse effect and sea level rise: A challenge for this generation.* New York: Van Nostrand Reinhold.

Hansen, J. E., et al. (1989). Greenhouse effect of chlorofluorocarbons and other trace gases. *Journal of Geophysical Research 94*(November), 16,417 16,421.

National Oceanic and Atmospheric Association (NOAA). (2008). *Global warming: Frequently asked questions.* Accessed 11/21/08 from http://lwf.ncdc.noaa.gov/oa/climate/globalwarming.html.

Tarbuck, E. J. & Lutgens, F. K. (2000). *Earth science.* Upper Saddle River, NJ: Prentice Hall.

CHAPTER 20

Earthquakes

It's been raining a lot, or very hot—it must be earthquake weather!

FICTION: Many people believe that earthquakes are more common in certain kinds of weather. In fact, no correlation with weather has been found. Earthquakes begin many kilometers (miles) below the region affected by surface weather. People tend to notice earthquakes that fit the pattern and forget the ones that don't. Also, every region of the world has a story about earthquake weather, but the type of weather is whatever they had for their most memorable earthquake.

—U.S. Geological Survey (2008)

What Causes Earthquakes?

Anyone who has witnessed (been exposed to) or studied one of over a million or so earthquakes that occur each year on Earth is unlikely to forget such occurrences. Even though most earthquakes are insignificant, a few thousand of these produce noticeable effects such as tremors and ground shaking. The passage of time has shown that about 20 earthquakes each year cause major damage and destruction. It is estimated that about 10,000 people die each year because of earthquakes.

Over the millennia, the effect of damaging earthquakes has been obvious to those who witnessed the results. However, the cause of earthquakes has not been as obvious. For example, earthquakes have been blamed on everything from super-incantations of mythical beasts to the wrath of Gods to unexplainable magical occurrences to normal, natural phenomena occasionally required to retain Earth's structural integrity; that is, providing Earth with a periodic form of feedback to keep the planet in balance. We can say, overall, that an earthquake on Earth provides our planet with a sort of a geological homeostasis needed to maintain life as we know it.

Through the ages, earthquakes have also come under the attention and eventually the pen of the world's greatest writers. Consider, for example, Voltaire's classic satirical novel, *Candide,* published in 1759, in which he mercilessly satirizes science and, in particular, earthquakes. Voltaire based the following comments on the 1755 Great Lisbon, Portugal, Earthquake, which was blamed for the deaths of more than 60,000 people. On viewing the total devastation of Lisbon, Dr. Pangloss says to Candide:

> [T]he heirs of the dead will benefit financially; the building trade will enjoy a boom. Private misfortune must not be overrated. These poor people in their death agonies, and the worms about to devour them, are playing their proper and appointed part in God's master plan.

Although we still do not know what we do not know about earthquakes and their causes, we have evolved from using witchcraft or magic to explain their origins to the scientific methods employed today. In the first place, we do know that earthquakes are caused by the sudden release of energy along a fault. Earthquakes are usually followed by a series of smaller earthquakes that we called *aftershocks.* Aftershocks represent further adjustments of rock along the fault. There are currently no reliable methods for predicting when earthquakes will occur.

In regard to the causes or origins of earthquakes, we have developed a couple of theories. One of these theories explains how earthquakes occur via *elastic rebound.* That is, according to elastic rebound theory, subsurface rock masses subjected to prolonged pressures from different directions will slowly bend and change shape. Continued pressure sets up strains so great that the rocks will eventually reach their elastic limit and rupture (break) and suddenly snap back into their original, unstrained state. It is the snapping back (elastic rebound) that generates the seismic waves radiating outward from the break. The greater the stored energy (strain), the greater the release of energy.

Many active volcanic belts coincide with major belts of earthquake activity (*seismic* and *volcanic activity*), which indicates that volcanoes and earthquakes may have a common cause. Plate interactions commonly cause both earthquakes (tectonic earthquakes) and volcanoes.

Seismology

Even though *seismology* is the study of earthquakes, it is actually the study of how seismic waves behave in the Earth. The source of an earthquake is called the *hypocenter* or *focus* (i.e., the exact location within the Earth were seismic waves are generated). The *epicenter* is the point on the Earth's surface directly above the focus. Seismologists want to know where the focus and epicenter are located so a comparative study of the behavior of the earthquake event can be made with previous events in an effort to further understanding.

Seismologists use instruments to detect, measure, and record seismic waves. Generally, the instrument used is the *seismograph,* which has been around for a long time. Modern updates have upgraded these instruments from the paper or magnetic tape strip to electronically recorded data that is input into a computer. A study of the relative arrival times of the various types of waves at a single location can be used to determine the distance to the epicenter. To determine the exact epicenter location, records from at least three widely separated seismograph stations are required.

Seismic Waves

As mentioned, some of the energy released by an earthquake travels through the Earth. The speed of a seismic wave depends on the density and elasticity of the materials through which they travel. Seismic waves come in several types, as described in the following list:

- **P-waves:** Primary, pressure, or push-pull waves (arrive first—first detected by seismograph) are compressional waves (expand and contract) that travel through the earth (solids, liquids, or gases) at speeds of from 3.4 to 8.6 miles per second. P-waves move faster at depth, depending on the elastic properties of the rock through which they travel. P-waves are the same thing as sound waves.
- **S-waves:** Secondary or shear waves travel with a velocity (between 2.2 and 4.5 miles per second) that depends only on the rigidity and density of the material through which they travel. They are the second set of waves to arrive at the seismograph and will not travel through gases or liquids; thus the velocity of S-waves through gas or liquids is zero.
- **Surface waves:** Several types of surface waves travel along the Earth's outer layer or surface or on layer boundaries in the Earth. These are rolling, shaking waves that are the slowest waves and that do the damage in large earthquakes.

Earthquake Magnitude and Intensity

The size of an earthquake is measured using two parameters: energy released (magnitude) and damage caused (intensity).

EARTHQUAKE MAGNITUDE

The size of an earthquake is usually given in terms of its Richter magnitude. Richter magnitude is a scale devised by Charles Richter that measures the amplitude (height) of the largest recorded wave at a specific distance from the earthquake. A better measure is the Richter scale, which measures the total amount of energy released by an earthquake as recorded by seismographs. The amount of energy released is related to the Richter scale by the equation:

$$\text{Log } E = 11.8 + 1.5\ M$$

where

Log = the logarithm to the based 10
E = the energy released in ergs
M = the Richter magnitude

In using the equation to calculate Richter magnitude, it quickly becomes apparent that we see that each increase of 1 in Richter magnitude yields a 31-fold increase in the amount of energy released. Thus, a magnitude 6 earthquake releases 31 times more energy than a magnitude 5 earthquake. A magnitude 9 earthquake releases 31×31 or 961 times more energy than a magnitude 7 earthquake.

Did You Know?

While it is correct to say that for each increase in 1 in the Richter magnitude, there is a 10-fold increase in amplitude of the wave, it is incorrect to say that each increase of 1 in Richter magnitude represents a 10-fold increase in the size of the earthquake.

EARTHQUAKE INTENSITY

Earthquake intensity is a rough measure of an earthquake's destructive power (i.e., size and strength—how much the earth shook at a given place near the source of an earthquake). To measure earthquake intensity, Mercalli in 1902 devised an intensity scale of earthquakes based on the impressions of people involved, movement of furniture and other objects, and damage to buildings. The shock is most intense at the epicenter, which, as noted earlier, is located on the surface directly above the focus.

Mercalli's intensity scale uses a series of numbers based on a scale of 1 to 12 to indicate different degrees of intensity (Table 20.1). Keep in mind that this

Table 20.1.   Modified Mercalli Intensity Scale

Intensity	Description
I	Not felt except under unusual conditions.
II	Felt by only a few on upper floors.
III	Felt by people lying down or seated.
IV	Felt indoors by many, by few outside.
V	Felt by everyone, people awakened.
VI	Trees sway, bells ring, some objects fall.
VII	Causes alarm, walls and plaster crack.
VIII	Chimneys collapse, poorly constructed buildings seriously damaged.
IX	Some houses collapse, pipes break.
X	Ground cracks, most buildings collapse.
XI	Few buildings survive, bridges collapse.
XII	Total destruction occurs.

scale is somewhat subjective, but it provides a qualitative, but systematic, evaluation of earthquake damage.

Internal Structure of Earth

Information obtained from seismographs and other instruments indicate that the lithosphere might be divided into three zones: the crust, mantle, and core (Figure 20.1).

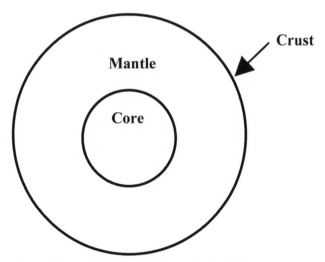

Figure 20.1.   Internal structure of the earth.

- **Earth's crust:** The outermost and thinnest layer of the lithosphere is called the *crust.* There are two different types of crust: thin (as little as 4 miles in places) oceanic crust (compose primarily of basalt) that underlies the ocean basins and thicker continental crust (primarily granite 20 to 30 miles thick) that underlies the continents.
- **Earth's mantle:** Beneath the crust is an 1,800-mile thick intermediate, dense, hot zone of semisolid rock called the *mantle.* It is thought to be composed mainly of olivine-rich rock.
- **Earth's core:** Earth's core is about 4,300 miles in diameter. It is thought to be composed of a very hot, dense iron and nickel alloy. The core is divided into two different zones. The outer core is a liquid because the temperatures there are adequate to melt the iron-nickel alloy. The highly pressurized inner is core solid because the atoms are tightly crowded together.

References and Recommended Reading

U.S. Geological Survey. (2008). *Earthquake hazards program.* Accessed 09/14/08 from http://earthquakes.usgs.gov/learning/topics/megaqk_facts_fantasy.php.
Voltaire. (1991). *Candide.* New York: Dover (Original work published in 1759).

CHAPTER 21

Plate Tectonics

[W]ater plays a critical role in lubricating the motion of plates—without it there would be no plate tectonics. So water quickens life and the Earth itself.

—David Singleton

Within the past 45 or 50 years, geologists have developed the theory of plate tectonics (Greek, "builder"). The theory of plate tectonics deals with the formation, destruction, and large-scale motions of great segments of Earth's surface (crust), called *plates.* This theory relies heavily on the older concepts of continental drift (developed during the first half of the 20th century) and seafloor spreading (understood during the 1960s), which help to explain the cause of earthquakes and volcanic eruptions, and the origin of fold mountain systems.

Crustal Plates

Earth's crustal plates are composed of great slabs of rock (lithosphere) about 100 km thick, and cover many thousands of square miles (they are thin in comparison to their length and width); they float on the ductile asthenosphere, carrying both continents and oceans. Many geologists recognize at least eight main plates and numerous smaller ones. These **main** plates include:

• African Plate covering Africa—Continental plate
• Antarctic Plate covering Australia—Continental plate
• Australian Plate covering Australia—Continental plate
• Eurasian Plate covering Asia and Europe—Continental plate

- Indian Plate covering Indian subcontinent and a part of Indian Ocean—Continental plate
- Pacific Plate covering the Pacific Ocean—Oceanic plate
- North American Plate covering North America and northeast Siberia—Continental plate
- South American Plate covering south America—Continental plate

The **minor** plates include:

- Arabian Plate
- Caribbean Plate
- Juan de Fuca Plate
- Cocos Plate
- Nazea Plate
- Philippine Plate
- Scotia Plate

Plate Boundaries

As mentioned, the asthenosphere is the ductile, soft, plastic-like zone in the upper mantle on which the crustal plates ride. Crustal plates move in relation to one another at one of three types of plate boundaries: convergent (collision boundaries), divergent (spreading boundaries), and transform boundaries. These boundaries between plates are typically associated with deep-sea trenches, large faults, fold mountain ranges, and midoceanic ridges.

CONVERGENT BOUNDARIES

Convergent boundaries (or active margins) develop where two plates slide toward each other commonly forming either a subduction zone (if one plate subducts or moves underneath the other) or a continental collision (if the two plates contain continental crust). To relieve the stress created by the colliding plates, one plate is deformed and slips below the other.

DIVERGENT BOUNDARIES

Divergent boundaries occur where two plates slide apart from each other. Oceanic ridges, which are examples of these divergent boundaries, are where new oceanic, melted lithosphere materials well up, resulting in basaltic magmas that intrude and erupt at the oceanic ridge, in turn creating new oceanic lithosphere and crust (new ocean floor). Along with volcanic activity, the midoceanic ridges are also areas of seismic activity.

TRANSFORM PLATE BOUNDARIES

Transform, or shear and constructive boundaries, do not separate or collide; rather, they slide past each other in a horizontal manner with a shearing motion. Most transform boundaries occur where oceanic ridges are offset on the sea floor. The San Andreas Fault in California is an example of a transform fault.

References & Recommended Reading

Atkinson, L., & Sancetta, C. (1993). Hail and farewell. *Oceanography 6*(34).

Holmes, A. (1978). *Principles of physical geology* (3rd ed.). New York: John Wiley & Sons.

Lyman, J., & Fleming, R. H. (1940). Composition of seawater. *Journal of Marine Research 3,* 134–146.

McKnight, T. (2004). *Geographica: The complete illustrated atlas of the world.* New York: Barnes and Noble Books.

Oreskes, N. (Ed.). (2003). *Plate tectonics: An insiders history of the modern theory of the earth.* New York: Westview.

Stanley, S. M. (1999). *Earth system history* (pp. 211–222). New York: W. H. Freeman.

Sverdrup, H. U., Johnson, M. W., & Fleming, R. H. (1942). *The oceans: Their physics, chemistry and general biology.* New York: Prentice-Hall.

Turcotte, D. L., & Schubert, G. (2002). *Geodynamics* (2nd ed.). New York: John Wiley & Sons.

CHAPTER 22

Historical Geology

Under heaven nothing is more soft and yielding than water.
Yet for attacking the solid and strong, nothing is better;
It has no equal.

—Lao Tzu, *Tao Te Ching*

The earth scorns our simplifications, and becomes much
more interesting in its derision. The history of life is not
a continuum of development, but a record punctuated by
brief, sometimes geologically instantaneous, episodes of
mass extinction and subsequent diversification. The geologic time scale maps this history, for fossils provide our
chief criterion in fixing the temporal order of rocks. . . .
Hence, the time scale is not the devil's ploy for torturing
students, but the chronicle of key moments in life's history. . . . I make no apologies for the central importance
of such knowledge.

—S. J. Gould, *Wonderful Life*

All things continue as they were from the beginning of the
creation.

—II Peter 3:3–6

In the study of geology, it is important to give some thought to historical geology, how geologists deal with time, and to the age of Earth. In this text, the focus
is on providing basic information on physical and not historical geology. However, it is important to note that historical geology is an important subset field
of geology. In historical geology, the principles of geology are used to reconstruct

and understand the history of Earth (Levin, 2003). Historical geology explains the sequence of geologic processes that change the Earth's surface and subsurface using stratigraphy, structural geology, and paleontology. In addition, the evolution of animals and plants during different periods in the geological timescale is also an area of focus. Radiometric dating techniques provide a means of deriving relative age versus absolute ages of geologic history.

- **Relative age (or relative time)** means that we can determine if something is younger or older than something else. This is accomplished by dating events in chronologic order of occurrence rather than in years. In other words, relative time gives us the sequence of events, but not how old something is. For example, World War I happened before World War II.
- **Absolute age (or absolute time)** means geologic time is measured in a more or less precise time (in years, minutes, seconds, or some other units of time) to the amount of time that has already passed. In other words, absolute time allows us to say how old something is. For example, the last major Ice Age occurred 11,000–12,000 years ago.

In geology, we use principles to determine relative ages, correlations, and absolute ages. For relative ages we use stratigraphy. For correlations we use physical criteria, fossils, and key beds. For absolute ages we use radiometric dating.

Stratigraphy

Stratigraphy is the geologic subscience dealing with the definition and interpretation of strata (layers) and stratified rocks in the Earth's crust, especially their lithology, sequence, distribution, and correlation.

Earth's layers have been characterized in different ways. One characterization commonly used is to compare the layers of Earth with the layers found in onions. When we cut an onion across, we find a series of layers surrounding a central core. A cross-section of Earth would show a similar structure. Earth's layers downward are more dense and heavier than the layers above.

Another way of characterizing the structure of earth is to compare it with a book. In fact, Saint Augustine characterized Earth as a book (in general and not geological terms) when he said, "The world is a book, and those who do not travel read only one page." For the purpose of this text, we can state Augustine's analogy in a different, "geological" way. We can say Earth's structural layers are like the pages of a book, and when we only study its outer layer, we read only one page. Earth's pages are not new, unread, or in pristine condition as you would expect in a brand-new library book. The pages that make up Earth's book are well-used, torn, cornered, crumpled, well-marked, faded in places, turned upside down, displaced, or lost. Earth's book binding (along with tremendous levels of pressure) is the force of gravity.

Sedimentary rock layers, or strata, were laid down on top of one another, again, like the pages in a book—Earth's book. Simply, when these pages are correctly read and interpreted, Earth's relative history is revealed. As mentioned, studying (reading) the layers of Earth is the study called *stratigraphy*.

Stratigraphic Laws

The reading of Earth's "pages" to determine its relative age is based on the stratigraphic laws. Developed in the 17th through 19th centuries by Niels Steno, James Hutton, and William Smith, among others, the modern-day geologist uses these principles to decipher the spatial and temporal relationships of rock layers. Basically, geologists studying Earth are concerned with the sequence of rocks and structures in time, and thus with the history of the Earth itself. It is important to point out, however, that these laws cannot determine the age of Earth's layers, but simply the relative order in which they were formed. Stratigraphic laws include the following:

- Original horizontality
- Lateral continuity
- Superposition
- Cross-cutting relations
- Laws of inclusions
- Law of faunal succession

✓ **Important Point:** To understand the stratigraphic laws, we must assume that the geologic processes of today were the same in the past. We call this *uniformitarianism*.

ORIGINAL HORIZONTALITY

The law of original horizontality states that all sedimentary rocks are originally deposited horizontally or nearly horizontal. If sedimentary rocks are no longer horizontal, an event occurred subsequent to the deposition that caused the layers to tilt or fold from their original position. In 1669, Niels Steno, a Danish geological pioneer, described this relationship: "strata either perpendicular to the horizon or inclined to the horizon were at one time parallel to the horizon" (Levin, 2005, p. 162). The law of original horizontality holds true in the deposition of most sedimentary material; however, one noted exception is sand deposited nonhorizontally at less than 15 degrees (or at the angle of repose) in sand dunes.

LATERAL CONTINUITY

Steno pointed out in 1669 that sedimentary rocks are laterally continuous over large geographic areas of Earth (deposits originally extended in all directions) unless some other solid bodies stood in the way, they thin out at edge, or grade into a different type of sediment (Levin, 2005, p. 162).

SUPERPOSITION

Another stratigraphic law states that because of Earth's gravity, the oldest layer of sediment is at the bottom of the sequence and the youngest at the top. Thus, in a sequence of layers that has not been disturbed or overturned by a later event, the oldest layers are on the bottom. Steno stated that "at the time when any given stratum was being formed, all the matter resting upon it was fluid, and, therefore, at the time when the lower stratum was being formed, none of the upper strata existed" (Levin, 2005, p. 162).

CROSS-CUTTING RELATIONS

In cross-cutting relations, Steno in 1669 pointed out that "if a body or discontinuity cuts across a stratum, it must have formed after that stratum" (Levin, 2005, p. 162). This law basically describes the relationship between existing rock and rock that intrudes by magma flow into exiting rock. This creates an intrusion, and the intrusion is always younger than the rock it invades.

LAW OF INCLUSIONS

The law of inclusions states that rock fragments in other rocks must be older than the rock containing or surrounding the fragments.

LAW OF FAUNAL SUCCESSION

The law of faunal succession, developed by William Smith (father of English geology, 1769–1839), recognized that fossil faunas (assemblages of animals that lived together at a given time and place) followed each other in a definite and determinable order. These faunas are distinctive for each portion of Earth history, and allowed geologists to develop a fossil stratigraphy, providing a means to correlate rocks. That is, by comparing these fossils the geologist is able to recognize deposits of the same ages.

STRUCTURAL BREAKS IN THE STRATIGRAPHIC RECORD

Because of uplift, subsidence, and deformation, the Earth's surface is continually changing, with surface erosion acting in some places and erosion of surface sediment occurring in other places. Whenever erosion is removing previous deposited sediment, or when sediment is not being deposited, there will be a break in the continuous record of sedimentation preserved in the rocks. This break in the stratigraphic record is called an *unconformity* (or *hiatus*). An unconformity is an uplifted surface with long periods of erosion or nondeposition. Geologists recognize three basic types of unconformities:

- **Angular unconformity** is an obvious, easily recognized type of unconformity because the beds above the nonconformity are not parallel to the beds below it. This type of nonconformity, commonly known as *Hutton's nonconformity* (1787) indicates that the lower series of rocks were tilted or folded prior to their erosion and the subsequent deposition of the overlying beds.
- **Disconformities** are known as *parallel unconformities* because the rock layers above and below the unconformity are parallel. Because there is often no angular relationship between sets of layers, disconformities are much harder to recognize in the field.
- **Nonconformity** is formed when overlying stratified, sedimentary rocks lie on an eroded surface of igneous or metamorphic rocks.

Relative versus Absolute Time

We cannot study geology without referring to geologic time. Geologic time is often discussed in two forms:

- **Relative time** is a chronostratic arrangement (in geologic column) of geologic events and periods in their proper order (displayed as the geologic time scale;

see Table 23.1). This is done by using the stratigraphic techniques (relative age relationships—vertical and stratigraphic positions) discussed earlier.

- **Absolute time** is a chronometric arrangement of numerical ages in millions of years or some other measurement; the time in years since the beginning or end of a period. These are commonly obtained via radiometric dating methods performed on appropriate rock types.

RELATIVE TIME: THE GEOLOGIC COLUMN

The *geologic column,* more commonly known as the *standard geologic column,* is the concept used to break relative geologic time into units of known relative age. In a sense, the column depicts snapshots of geologic time. The geologic column accomplishes this by scaling the oldest to the most recent rocks found in the entire Earth or in a given area.

Did You Know?

The largest unit of geologic time is an *era,* and each era is divided into smaller time units called *periods.* A period of geologic time is divided into *epochs,* which in turn may be subdivided into still smaller units. The geologic column provides a standard by which we can discuss the relative age of rock formations and the rocks and the fossils they contain. However, again, keep in mind that these time units are arbitrary and of unequal duration, and because we are dealing with relative time, we cannot be positive about the exact amount of time involved in each unit.

The *geologic time scale* (Table 22.1; based on present state of knowledge) is composed of named intervals of geologic time (relative) during which the rocks represented in the geologic column were deposited.

ABSOLUTE TIME: RADIOMETRIC TIME SCALE

Although it is easy to establish the relative ages of rocks based on the work of geologists who have applied the principles of stratigraphy, knowing how much time a geologic era, period, or epoch represents is much more difficult unless we have the absolute ages of rocks. Absolute time measurements are used to calibrate the relative time scale.

In the early years of geology, many attempts were made to establish absolute time measurements. Some of these give only a very rough approximation of the age of rocks; others (radiometric dating) are much more accurate.

Table 22.1. Geologic Time Scale (U.S. Geological Survey [USGS], 2007, 2008)

Erathem or Era	System, Subsytem or Period, Subperiod	Series or Epoch
Cenozoic 65 million years ago to present "Age of Recent Life"	Quaternary 1.8 million years ago to the present	Holocene 11,477 years ago (+/–85 years) to the present—Greek "holos" (entire) and "ceno" (new)
		Pleistocene 1.8 million to approx. 11,477 (+/–85 years) years ago—The Great Ice Age—Greek "pleistos" (most) and "ceno" (new)
	Tertiary	Pliocene 5.3 to 1.8 million years ago—Greek "pleion" (more) and "ceno" (new).
		Miocene 23 to 5.3 million years ago—Greek "meion" (less) and "ceno" (new).
		Oligocene 33.9 to 23 million years ago—Greek "oligos" (little, few) and "ceno" (new).
		Eocene 55.8 to 33.9 million years ago—Greek "eos" (dawn) and "ceno" (new).
		Paleocene 65.5 to 58.8 million years ago—Greek "palaois" (old) and "ceno" (new).
	Cretaceous 145.5 to 65.5 million years ago "The age of dinosaurs"	Late or Upper Early or Lower

(continued)

Table 22.1. (continued)

Erathem or Era	System, Subsytem or Period, Subperiod	Series or Epoch
Mesozoic 251 to 65.5 million years ago—Greek means "middle life"	Jurassic 199.6 to 145.5 million years ago	Late or Upper Middle Early or Lower
	Triassic 251 in 199.6 million years ago	Late or Upper Middle Early or Lower
	Permian 299 to 251 million years ago	Lopingian Guadalupian Cisuralian
	Pennsylvanian 381.1 to 299 million years ago "The coal age"	Late or Upper Middle Early or Lower
Paleozoic 542 to 251 million years ago "Age of Ancient Life"	Mississippian 359.2 to 318.1 million years ago	Late or Upper Middle Early or Lower
	Devonian 416 to 359.2 million years ago	Late or Upper Middle Early or Lower
	Silurian 443.7 to 416 million years ago	Pridoli Ludlow Wenlock Llandovery
	Ordovician 488.3 to 443.7 million years ago	Late or Upper Middle Early or Lower
	Cambrian 542 to 488.3 million years ago	Late or Upper Middle Early or Lower
Precambrian Approximately 4 billion to 542.0 million years ago		

- **Salinity of the sea:** The oceans were probably originally composed of fresh water; thus, the age of the Earth can be estimated on the basis of how long it would take the oceans to obtain their present salt content.
- **Rate of sedimentation:** If we knew how long it took to deposit all of the rock layers in the crust, we could get some idea as to the age of the Earth.
- **Radiometric time scale:** This is the most recent and accurate method yet devised for measuring absolute geologic time.

New vistas in science were opened by Henry Becquerel in 1896 when he discovered the natural radioactive decay of uranium. However, it wasn't until 1905 when the British physicist Lord Rutherford suggested using radioactivity as a tool for directly measuring geologic time. Two years later, in 1907, B. B. Boltwood, a radiochemist at Yale University, published a list of geologic ages based on radioactivity, which, during the current era, have been modified somewhat to reflect greater accuracy and proper application. Precise dating has been accomplished since 1950 (USGS, 2001).

To understand the methodology used to construct the radiometric time scale, it is important to know that a chemical element consists of atoms with a specific number of protons in their nuclei but different atomic weights owing to variations in the number of neutrons. Furthermore, atoms of the same element with differing atomic weights are called *isotopes*. Isotopes are formed spontaneously when the isotope (the parent) loses particles from its nucleus to form an isotope of a new element (the daughter). The rate of decay is expressed in terms of the time it takes for one-half (an isotope's half-life) of a particular radioactive isotope in a sample to decay. Most radioactive isotopes have short half-lives and lose their radioactivity within a few days or years. However, some isotopes decay slowly, and several of these are used as geologic clocks. The parent isotopes and corresponding daughter products most commonly used to determine the ages of ancient rocks are listed in Table 22.2.

Table 22.2. Parent Isotopes and Corresponding Daughter Products (USGS, 2001)

Parent Isotope	Stable Daughter Product	Half-Life Values
Uranium-238	Lead-206	4.5 billion years
Uranium-235	Lead-207	704 million years
Thorium-232	Lead-208	14 billion years
Rubidium-87	Strontium-87	48.8 billion years
Potassium-40	Argon-40	1.25 billion years
Samarium-147	Neodymium-143	106 billion years

The USGS (2001) points out that dating rocks by radioactive timekeepers is theoretically simple, but the laboratory procedures are complex. The principal difficulty lies in measuring precisely very small amounts of isotopes.

Because potassium is found in most rock-forming minerals, the potassium-argon method can be used on rocks as young as a few thousand years as well as on the oldest rocks known. The half-life of potassium's radioactive isotope potassium-40 is such that measurable quantities of argon (daughter) have accumulated in potassium-bearing minerals of nearly all ages, and the amounts of potassium and argon isotopes can be measured accurately, even in very small quantities.

Did You Know?

In practice and where feasible, two or more radioactive dating methods of analysis are used on the same specimen of rock to confirm the results.

Another important isotope used for dating purposes is based on the radioactive decay of the isotopearbon-14, which has a half-life of 5,730 years. As a result of the bombardment of nitrogen by neutrons from cosmic rays, the radiocarbon carbon-14 is produced continuously in the Earth's upper atmosphere. This radiocarbon becomes uniformly mixed with the nonradioactive carbon in the carbon dioxide of the air, and it eventually finds its way into all living plants and animals. In effect, all carbon in living organisms contains a constant proportion of radiocarbon to nonradioactive carbon. After the death of the organism, the amount of radiocarbon gradually decreases as it reverts to nitrogen-14 by radioactive decay. By measuring the amount of radioactivity remaining in organic materials, the amount of carbon-14 in the material can be calculated and the time of death can be determined. For example, if carbon from a sample of wood is found to contain only half as much carbon-14 as that from a living plant, the estimated age of the old wood is 5,730 years (USGS, 2001).

Did You Know?

The radiocarbon clock has become an extremely useful and efficient tool in dating the important episodes in the recent prehistory and history of man, but because of the relatively short half-life of carbon-14, the clock can be used for dating events that have taken place only within the past 50,000 years (USGS, 2001).

Table 22.3 lists a group of rocks and materials that have been dated by various atomic clock methods.

Table 22.3. Rock Groups dated by Atomic Clock Methods (USGS 2001)

Rock and Material Samples	Approximate Age in Years
Charcoal Sample, recovered from a bed near Crater Lake, Oregon, is from a tree burned in the violent eruption Mount Mazama, which created Crate Lake. This eruption blanketed several States with ash, providing geologists with an excellent time zone.	6,640
Charcoal Sample collected from the "Marmes Man" site in southeastern Washington. This rock shelter is believed to be among the oldest known inhabited sites in North America.	10,130
Spruce wood Sample from the Two Creeks forest bed near Milwaukee, Wisconsin, dates one of the last advances of the continental ice sheet into the United States.	11,640
Bishop Tuff Samples collected from volcanic ash and pumice that overlie glacial debris in Owens Valley, California. This volcanic episode provides an important reference datum in the glacial history of North America.	700,000
Volcanic ash Samples collected from strata in Olduvai Gorge, East Africa, which sandwich the fossil remains of Zinjanthropus and Homo habilis—possible precursors of modern man.	1,750,000
Monzonite Samples of copper-bearing rock from vast open-pit mine at Bingham Canyon, Utah.	37,500,000
Quartz monzonite Samples collected from Half Dome, Yosemite National Park, California.	80,000,000
Conway Granite Samples collected from Redstone Quarry in the White Mountains of New Hampshire.	180,000,000
Rhyolite Samples collected from Mount Rogers, the highest point in Virginia.	820,000,000

(continued)

Table 22.3. *(continued)*

Rock and Material Samples	Approximate Age in Years
Pikes Peak Granite	
Samples collected on top of Pikes Peak, Colorado.	1,030,000,000
Gneiss	
Samples from outcrops in the Karelian area of eastern Finland are believed to represent the oldest rocks in the Baltic region.	2,700,000,000
The Old Granite	
Samples from outcrops in the Transvaal, South Africa. These rocks intrude even older rocks that have been dated.	3,200,000,000
Morton Gneiss	
Samples from outcrops in southwestern Minnesota are believed to represent some of the oldest rocks in North America.	3,600,000,000

References and Recommended Reading

Blatt, H., Berry, W. B. N., & Brande, S. (1991). *Principles of stratigraphic analysis.* Boston: Blackwell Scientific Publications.

Gould, S. J. (1989). *Wonderful life.* New York: W.W. Norton and Co.

Harris, E. C. (1979). *Principles of archaeological stratigraphy.* New York: Academic Press.

Holmes, A. (1937). *The age of earth.* London: Nelson.

Levin, H. L. (2003). *The earth through time* (7th ed.). New Jersey: John Wiley & Sons.

Levin, H. L. (2005). *The earth through time* (8th ed.). New Jersey: John Wiley & Sons.

Palmer, D. (2005). *Earth time: Exploring the deep past from Victorian England to the Grand Canyon.* New York: John Wiley & Sons.

Tarbuck, E. J., Lutgens, F. K., & Tasa, D. (2007). *Earth—An introduction to physical geology* (9th ed). New Jersey: Prentice-Hall.

U.S. Geological Survey (USGS) (2001). *Radiometric time scale.* Washington, DC: Accessed 10/20/08 from http://pubs.usgs.gov/gip/geotime/radiometric.html.

U.S. Geological Survey (USGS) (2007). *U.S. Geological Survey fact sheet 2007–3015: U.S. Geological Survey Geologic Names Committee, 2007.* Washington, DC: USGS.

U.S. Geological Survey (USGS) (2008). *The geologic time scale.* Accessed 10/10/08 from http://vulcan.wr.usgs.gov/Glossary/geo_time_scale.html.

Part IV

HYDROSPHERE

CHAPTER 23

Earth's Blood

The watery environment in which single-cell organisms live provides them food and removes their wastes, a function that the human circulatory system provides for the 60–100 trillion cells in a human body. The circulatory system brings each cell its daily supply of nutritive amino acids and glucose, and carries away waste carbon dioxide and ammonia, to be filtered out of our systems and flushed away through micturition and excretory functions. The heart, the center of our circulatory system keeps blood moving on its predetermined circular path, so essential that if the pump fails, we quickly fail as well—we die.

As single-celled organisms no longer, humans sometimes assume they no longer need a watery environment in which to live—but they aren't paying close attention to the world around them. Actually, those of us who live on earth are as dependent upon the earth's circulatory system as we are on our own circulatory system. As our human hearts pump blood, circulating it through a series of vessels, and as our lives are dependent upon that flow of blood, so life on earth is, and our own lives are dependent on the earth's water cycle, and on water, in every aspect of our lives.

This cycle is so automatic that we generally ignore it until we are slapped in the face by it. Just as we don't control or pay attention to the beating of our hearts unless the beat skips or falters, unless we are confronted by flood or drought, unless our plans are disrupted by rain, we ignore the water cycle, preferring to believe that the water we drink comes out of the faucet, and not from deep within the belly of the earth, placed there by a process we only dimly comprehend. But water is as essential to us and to the earth as blood is in our bodies, and the constant cycle water travels makes our lives possible.

Earth's blood, water, is pumped, not by a heart, but by the hydrological cycle—the water cycle. A titanic force of nature, the water cycle is beyond our control—a fact that we ignore until weather patterns shift and suddenly inundated rivers flow where they will and not within human-engineered banks, floodwalls, dikes, and levees. In the water cycle, water evaporated from the oceans falls as rain, hail, sleet, or snow and it strikes the earth again; the cycle continues.

In cities, in summer, rain strikes hot cement and asphalt and swiftly evaporates, or runs into storm drains, swiftly rejoining the cycle. In fields, rain brings essential moisture to crops, and sinking deeper into the earth, ends as groundwater. If water strikes a forested area, the forest canopy breaks the force of the falling drops. The forest floor, carpeted in twigs, leaves, moss, dead and decaying vegetation, keeps the soil from splashing away in erosion as the water returns to the depths of the earth, or runs over the land to join a stream.

Whenever water strikes the earth, it flows along four pathways, which carry water through the cycle as our veins, arteries and capillaries carry our blood to our cells.

It may evaporate directly back into the air.
It may flow overland into a stream as runoff.
It may soak into the ground and be taken up by plants for evapotranspiration.
It may seep down to groundwater.

Whatever pathway water takes, one fact is certain: water is dynamic, vital, constantly on the move. And like human blood, which sustains our lives, earth's blood, to sustain us as well, must continue to flow.

—Frank R. Spellman (1996)

Setting the Stage

Whether we characterize it as ice, rainbow, steam, frost, dew, soft summer rain, fog, flood, or avalanche, or as stimulating as a stream or cascade, water is special. Water is strange. Water is different.

Water is the most abundant inorganic liquid in the world; moreover, it oc-curs naturally anywhere on earth. Literally awash with it, life depends on water, and yet water is so very different.

Water is scientifically different. With its rare and distinctive property of be-ing denser as a liquid than as a solid, it is different. Water is different in that it is the only chemical compound found naturally in solid, liquid, gaseous states. Water is sometimes called the *universal solvent*. This is a fitting name, especially when you consider that water is a powerful reagent, which is capable in time of dissolving everything on earth.

Water is different. It is usually associated with all the good things on earth. For example, water is associated with quenching thirst, with putting out fires, and with irrigating the earth. The question is, can we really say emphatically, definitively that water is associated with only those things that are good?

Not really!

Remember, water is different; nothing, absolutely nothing, is safe from it.

Water is different. This unique substance is odorless, colorless, and tasteless. Water covers 71% of the earth completely. Even the driest dust ball contains 10%–15% water.

Water and life—life and water—inseparable.

The prosaic becomes wondrous as we perceive the marvels of water.

The Earth is covered by 326 million cubic miles of water, but only 3% of this total is fresh with most locked up in polar ice caps, glaciers, in lakes; it flows through soil and in river and stream systems back to an ever increasingly saltier sea. Only 0.027% is available for human consumption. Water is different.

Salt water is different from fresh water. Moreover, this text deals with fresh-water and ignores salt water because salt water fails its most vital duty, which is to be pure, sweet, and serve to nourish us.

Standing at a dripping tap, water is so palpably wet, one can literally hear the drip-drop-plop.

Water is special—water is strange—water is different. More importantly, water is critical to our survival, yet we abuse it, discard it, foul it, curse it, dam it, and ignore it. At least this is the way we view the importance of water at this moment in time. However, because water is special, strange, and different, the dawn of tomorrow is pushing for quite a different view.

Along with being special, strange, and different, water is also a contradic-tion, a riddle.

How?

Consider the Chinese proverb that states, "Water can both float and sink a boat."

Water's presence everywhere feeds these contradictions. Lewis (1996) points out that "water is the key ingredient of mother's milk and snake venom, honey and tears" (p. 90).

Leonardo da Vinci gave us insight into more of water's apparent contradictions:

> Water is sometimes sharp and sometimes strong, sometimes acid and sometimes bitter;
> Water is sometimes sweet and sometimes thick or thin;
> Water sometimes brings hurt or pestilence, sometimes health-giving, sometimes poisonous.
> Water suffers changes into as many natures as are the different places through which it passes.
> Water, as with the mirror that changes with the color of its object, so it alters with the nature of the place, becoming: noisome, laxative, astringent, sulfurous, salt, incarnadined, mournful, raging, angry, red, yellow, green, black, blue, greasy, fat or slim.
> Water sometimes starts a conflagration, sometimes it extinguishes one.
> Water is warm and is cold.
> Water carries away or sets down.
> Water hollows out or builds up.
> Water tears down or establishes.
> Water empties or fills.
> Water raises itself or burrows down.
> Water spreads or is still.
> Water is the cause at times of life or death, or increase of privation, nourishes at times and at others does the contrary.
> Water, at times has a tang, at times it is without savor.
> Water sometimes submerges the valleys with great flood.
> In time and with water, everything changes.

Water's contradictions can be summed up by simply stating that, though the globe is awash in it, water is no single thing, but an elemental force that shapes our existence. Da Vinci's last contradiction, "In time and with water, everything changes" concerns us most in this text.

Many of da Vinci's water contradictions are apparent to most observers. The wide spectrum of the hydrosphere includes hydrology, limnology, hydrogeology, oceanography, marine biology, physical oceanography, and other specialized areas—again, many of these water specialty areas are apparent, whether we readily acknowledge them or not in the normal course of things. But with water there are other factors that do not necessarily stand out, that are not always so apparent. This is made clear by the following example—what you see on the surface is not necessarily what lies beneath.

STILL WATER

Consider a river pool, isolated by fluvial processes and time from the main stream flow. We are immediately struck by one overwhelming impression: It appears so still, so very still. Still enough to sooth us. The river pool provides a kind of poetic solemnity, if only at the pool's surface. No words of peace, no description of silence or motionless can convey the perfection of this place, in this moment stolen out of time.

We ask ourselves, "The water is still, but does the term *still* correctly describe what we are viewing? Is there any other term we can use besides *still*? Is there any other kind of still?"

Yes, of course, we know many ways to characterize *still*. For sound or noise, *still* can mean "inaudible," "noiseless," "quiet," or "silent." With lack of movement, *still* can mean "immobile," "inert," "motionless," or "stationary." At least this is how the pool appears on the surface to the casual visitor. The visitor sees no more than water and rocks.

The rest of the pool? We know very well that a river pool is more than just a surface. How does the rest of the pool (the subsurface, for example) fit the descriptors we tried to use to characterize its surface? Maybe they fit, maybe they don't. In time, we will go beneath the surface, through the liquid mass, to the very bottom of the pool to find out. For now, remember that images retained from first glances are almost always incorrectly perceived, incorrectly discerned, and never fully understood.

On second look, we see that the fundamental characterization of this particular pool's surface is correct enough. Wedged in a lonely riparian corridor—formed by river bank on one side and sand bar on the other—between a youthful, vigorous river system on its lower end and a glacier- and artesian-fed lake on its headwater end, almost entirely overhung by mossy old sitka spruce, the surface of the large pool, at least at this particular location, is indeed still. In the proverbial sense, the pool's surface is as still and as flat as a flawless sheet of glass.

The glass image is a good one because, like perfect glass, the pool's surface is clear, crystalline, unclouded, definitely transparent, yet perceptively deceptive as well. The water's clarity, accentuated by its bone-chilling coldness, is apparent at close range. Farther back, we see only the world reflected in the water—the depths are hidden and unknown. Quiet and reflective, the polished surface of the water perfectly reflects in mirror-image reversal the spring greens of the forest at the pond's edge, without the slightest ripple. Up close, looking straight into the bowels of the pool we are struck by the water's transparency. In the motionless depths, we do not see a deep, slow-moving reach with muddy bottom typical of a river or stream pool; instead, we clearly see the warm variegated tapestry of blues, greens, blacks stitched together with threads of fine, warm-colored sand that carpets the bottom, at least 12 feet below. Still waters can run deep.

No sounds emanate from the pool. The motionless, silent water doesn't, as we might expect, lap against its bank or bubble or gurgle over the gravel at its edge. Here, the river pool, held in temporary bondage, is patient, quiet, waiting, withholding all signs of life from its surface visitor.

Then the reality check: The present stillness, like all feelings of calm and serenity, could be fleeting, momentary, temporary, you think. And you would be correct, of course, because there is nothing still about a healthy river pool.

At this exact moment, true clarity is present; it just needs to be perceived. And it will be.

*   *   *

We toss a small stone into the river pool, and watch the concentric circles ripple outward as the stone drops through the clear depths to the pool bottom. For a brief instant, we are struck by the obvious: the stone sinks to the bottom, following the laws of gravity, just as the river flows according to those same inexorable laws—downhill in its search for the sea. As we watch, the ripples die away, leaving as little mark as the usual human lifespan creates in the waters of the world, then disappears as if it had never been. Now the river water is, as before, still. At the pool's edge, we look down through the massy depth to the very bottom—the substrate.

We determine that the pool bottom is not flat or smooth, but instead is pitted and mounded occasionally with discontinuities. Gravel mounds alongside small corresponding indentations—small, shallow pits—make it apparent to us that gravel was removed from the indentations and piled into slightly higher mounds. From our topside position, as we look down through the cool, quiescent liquid, the exact height of the mounds and the depth of the indentations is difficult for us to judge; our vision is distorted through several feet of water.

However, we can detect near the low gravel mounds (where female salmon buried their eggs, and where their young grow until they are old enough to fend for themselves), and actually through the gravel mounds, movement— water flow—an upwelling of groundwater. This water movement explains our ability to see the variegated color of pebbles. The mud and silt that would normally cover these pebbles has been washed away by the water's subtle, inescapable movement. Obviously, in the depths, our still water is not as still as it first appeared.

The slow, steady, inexorable flow of water in and out of the pool, along with the upflowing of groundwater through the pool's substrate and through the salmon redds (nests) is only a small part of the activities occurring within the pool, including the air above it, the vegetation surrounding it, and the damp bank and sandbar forming its sides.

Let's get back to the pool itself. If we could look at a cross-sectional slice of the pool, at the water column, the surface of the pool may carry those animals that can literally walk on water. The body of the pool may carry rotifers and protozoa and bacteria—tiny microscopic animals—as well as many fish. Fish will also inhabit hidden areas beneath large rocks and ledges, to escape predators. Going down farther in the water column, we come to the pool bed. This is called the *benthic zone,* and certainly the greatest number of creatures lives here, including larvae and nymphs of all sorts, worms, leeches, flatworms, clams, crayfish, dace, brook lampreys, sculpins, suckers, and water mites.

We need to go down even farther, down into the pool bed, to see the whole story. How far this goes and what lives here, beneath the water, depends on whether it is a gravelly bed or a silty or muddy one. Gravel will allow water, with its oxygen and food, to reach organisms that live underneath the pool. Many of the organisms that are found in the benthic zone may also be found underneath, in the hyporheic zone.

But to see the rest of the story we need to look at the pool's outlet, and where its flow enters the main river. This is the riffles—shallow places where water runs fast and is disturbed by rocks. Only organisms that cling very well, such as net-winged midges, caddisflies, stoneflies, some mayflies, dace, and sculpins can spend much time here, and the plant life is restricted to diatoms and small algae. Riffles are a good place for mayflies, stoneflies, and caddisflies to live because they offer plenty of gravel in which to hide.

* * *

At first, we struggled to find the "proper" words to describe the river pool. Eventually, we settled on "Still Waters." We did this because of our initial impression, and because of our lack of understanding—lack of knowledge. Even knowing what we know now, we might still describe the river pool as *still waters.* However, in reality, we must call the pool what it really is: a dynamic habitat. This is true, of course, because each river pool has its own biological community, all members interwoven with each other in complex fashion, all depending on each other. Thus, our river pool habitat is part of a complex, dynamic ecosystem. On reflection, we realize, moreover, that anything dynamic certainly can't be accurately characterized as "still"—including our river pool.

* * *

Maybe you have not had the opportunity to observe a river pool like the one described previously. Maybe such an opportunity does not interest you. However, the author's point can be made in a different manner.

Take a moment out of your hectic schedule and perform an action most people never think about doing. Hold a glass of water and think about the substance within the glass—about the substance you are getting ready to drink. You are aware that the water inside a drinking glass is not one of those items people usually spend much thought on, unless they are tasked with providing the drinking water—or dying of thirst.

As mentioned earlier, water is special, strange, and different. Some of us find water fascinating—a subject worthy of endless interest, because of its unique behavior, limitless utility, and ultimate and intimate connection with our existence. Perhaps you might agree with Tom Robbins, whose description of water follows.

> Stylishly composed in any situation—solid, gas or liquid—speaking in penetrating dialects understood by all things—animal, vegetable or mineral—water travels intrepidly through four dimensions, *sustaining* (Kick a lettuce in the field and it will yell "Water!") *destroying* (The Dutch boy's finger remembered the view from Ararat) and *creating* (It has even been said that human beings were invented by water as a device for transporting itself from one place to another, but that's another story). Always in motion, ever-flowing (whether at stream rate or glacier speed), rhythmic, dynamic, ubiquitous, changing and working its changes, a mathematics turned wrong side out, a philosophy in reverse, the ongoing odyssey of water is irresistible. (Robbins, 1976, pp. 1–2)

As Robbins said, water is always in motion. The one most essential characteristic of water is that it is dynamic: Water constantly evaporates from sea, lakes, and soil and transpires from foliage; is transported through the atmosphere; falls to earth; runs across the land; and filters down to flow along rock strata into aquifers. Eventually water finds its way to the sea again—indeed, water never stops moving.

A thought that might not have occurred to most people as they look at our glass of water is, "Who has tasted this same water before us?" Before us? Absolutely. Remember, water is almost a finite entity. What we have now is what we have had in the past. The same water consumed by Cleopatra, Aristotle, da Vinci, Napoleon, Joan of Arc (and several billion other folks who preceded us), we are drinking now—because water is dynamic (never at rest), and because water constantly cycles and recycles, as discussed in another section.

Water never goes away, disappears or vanishes; it always returns in one form or another. As Dove (1989) points out, "all water has a perfect memory and is forever trying to get back to where it was."

Earth's Blood is Life's Blood

The availability of a water supply adequate in terms of both quantity and quality is essential to our very existence. One thing is certain: History has shown that the provision of an adequate quantity of quality potable water has been a matter of major concern since the beginning of civilization.

Water—especially clean, safe water—we know we need it to survive—we know a lot about it—however, the more we know the more we discover we don't know.

Modern technology has allowed us to tap potable water supplies and to design and construct elaborate water distribution systems. Moreover, we have developed technology to treat used water (wastewater); that is, water we foul, soil, pollute, discard, and flush away.

Have you ever wondered where the water goes when you flush the toilet? Probably not.

An entire technology has developed around treating water and wastewater. Along with technology, of course, technological experts have been developed. These experts range from environmental/structural/civil engineers to environmental scientists, geologists, hydrologists, chemists, biologists, and others.

Along with those who design and construct water/wastewater treatment works, there is a large cadre of specialized technicians, spread worldwide who operate water and wastewater treatment plants. These operators are tasked, obviously, with either providing a water product that is both safe and palatable for consumption and/or with treating (cleaning) a waste stream before it is returned to its receiving body (usually a river or stream). It is important to point out that not only are water practitioners who treat potable and used water streams responsible for ensuring quality, quantity, and reuse of their product, they are also tasked with, because of the events of 9/11, protecting this essential resource from terrorist acts.

The fact that most water practitioners know more about water than the rest of us comes as no surprise. For the average person, knowledge of water usually extends to knowing no more than that water is good or bad; it is terrible tasting, just great, wonderful, clean and cool and sparkling, or full of scum/dirt/rust/chemicals, great for the skin or hair, very medicinal, and so on. Thus, to say the water "experts" know more about water than the average person is probably an accurate statement.

At this point, the reader is probably asking: What does all this have to do with anything? Good question.

What it has to do with water is quite simple. We need to accept the fact that we simply do not know what we do not know about water. We need to know more. To start with, let's talk a little about the way in which we view water.

Earlier brief mention was made about the water contents of a simple drinking water glass. Let's face it, drinking a glass of water is something that normally takes little effort and even less thought. The trouble is that our view of water and its importance is relative.

The situation could be different—even more relative, however. For example, consider the young woman who is an adventurer, an outdoors-person. She likes to jump into her four-wheel-drive vehicle and head out for new adventure. On this particular day, she decides to drive through Death Valley, California— one end to another and back on seldom used dirt road. She has done this a few times before. During her transit of this isolated region, she decides to take a side road that seems to lead to the mountains to her right.

She travels along this isolated, hardpan road for approximately 50 miles— then the motor in her four-wheel-drive vehicle quits. No matter what she does, the vehicle will not start. Eventually, the vehicle's battery dies; she had cranked on it too much.

Realizing that the vehicle is not going to start, she also realized she is alone and deep inside an inhospitable area. What she does not know is that the nearest human being is about 60 miles to the west.

She has another problem—a problem more pressing than any other. She does not have a canteen or container of water—an oversight on her part. Obviously, she tells herself, this is not a good situation.

What an understatement this turns out to be.

Just before noon, on foot, she starts back down the same road she had traveled. She reasons she does not know what is in any other direction other than the one she had just traversed. She also knows the end of this side road intersects the major highway that bisects Death Valley. She could flag down a car or truck or bus; she will get help, she reasons.

She walks—and walks—and walks some more. "Gee, if it wasn't so darn hot," she mutters to herself, to sagebrush, to scorpions, to rattlesnakes, and to cacti. The point is it is hot, about 107° F.

She continues on for hours, but now she is not really walking; instead, she is forcing her body to move along. Each step hurts. She is burning up. She is thirsty. How thirsty is she? Well, right about now just about anything liquid would do, thank you very much!

Later that night, after hours of walking through that hostile land, she can't go on. Deep down in her heat-stressed mind, she knows she is in serious trouble—trouble of the life-threatening variety.

Just before passing out, she uses her last ounce of energy to issue a dry, pathetic scream.

This scream of lost hope and imminent death is heard, but only by the sagebrush, the scorpions, the rattlesnakes, and the cacti—and by the vultures that are now circling above her parched, dead remains. The vultures are of no help, of course. They have heard these screams before. They are indifferent; they have all the water they need; their food supply isn't all that bad either.

The preceding case sheds light on a completely different view of water. Actually, it is a very basic view that holds: We cannot live without it. When dying of thirst, there is absolutely nothing we would not trade for a mouthful of it.

Historical Perspective

An early human, wandering alone from place to place, hunting and gathering to subsist, probably would have had little difficulty in obtaining drinking water because such a person would—and could—only survive in an area where drinking water was available with little travail.

The search for clean, fresh, and palatable water has been a human priority from the very beginning. The author takes no risk in stating that when humans first walked the Earth, many of the steps they took were in the direction of water.

When early humans were alone or in small numbers, finding drinking water was a constant priority, to be sure, but for us to imagine today just how big a priority finding drinking water became as the number of humans proliferated is difficult.

Eventually communities formed, and with their formation came the increasing need to find clean, fresh, and palatable drinking water, and also to find a means of delivering it from the source to the point of use.

Archeological digs are replete with the remains of ancient water systems (humanity's early attempts to satisfy that never-ending priority). Those digs (spanning the history of the last 20 or more centuries) testify to this. For well over 2,000 years, piped water supply systems have been in existence. Whether the pipes were fashioned from logs or clay or carved from stone or other materials is not the point—the point is they were fashioned to serve a vital purpose, one universal to the community needs of all humans: to deliver clean, fresh, and palatable water to where it was needed.

These early systems were not arcane. Today, we readily understand their intended purpose. As we might expect, they could be rather crude, but they were reasonably effective, though they lacked in two general areas we take for granted today.

First, of course, they were not pressurized, but instead relied on gravity flow, since the means to pressurize the mains was not known at the time—and even if such pressurized systems were known, they certainly would not have been used to pressurize water delivered via hollowed-out logs and clay pipe.

The second general area early civilizations lacked that we do not lack today in the industrialized world is sanitation. Remember, to know the need for something exists (in this case, the ability to sanitize, to disinfect water supplies), the nature of the problem must be defined. Not until the middle of the 1800s (after countless millions of deaths from waterborne disease over the centuries) did people realize that a direct connection between contaminated drinking water and disease existed. At that point, sanitation of water supply became an issue.

When the relationship between waterborne diseases and the consumption of drinking water was established, evolving scientific discoveries led the way toward the development of technology for processing and disinfection. Drinking water standards were developed by health authorities, scientists, and sanitary engineers.

With the current lofty state of effective technology that we in the United States and the rest of the developed world enjoy today, we could sit on our laurels, so to speak, and assume that because of the discoveries developed over time (and at the cost of countless people who died—and still die—from waterborne-diseases), that all is well with us—that problems related to providing us with clean, fresh, palatable drinking water are problems of the past.

Are they really problems of the past? Have we solved all the problems related to ensuring that our drinking water supply provides us with clean, fresh, and palatable drinking water? Is the water delivered to our tap as clean, fresh, and palatable as we think it is—as we hope it is? Does anyone really know?

What we do know is that we have made progress. We have come a long way from the days of gravity-flow water delivered via mains of logs and clay or stone. Many of us on this Earth have come a long way from the days of cholera epidemics.

However, to obtain a definitive answer to those questions, perhaps we should ask those who boiled their water for weeks on end in Sydney, Australia, in the fall of 1998. Or better yet, we should speak with those who drank the "city water" in Milwaukee in 1993, or in Las Vegas, Nevada—those who suffered and survived the onslaught of *Cryptosporidium,* from contaminated water out of their tap.

Or if we could, we should ask these same questions of a little boy named Robbie, who died of acute lymphatic leukemia, the probable cause of which is far less understandable to us: toxic industrial chemicals, unknowingly delivered to him via his local water supply.

If water is so precious, so necessary for sustaining life, then two questions arise: (1) Why do we ignore water? (2) Why do we abuse it (pollute or waste it)?

We ignore water because it is so common, so accessible, so available, so unexceptional (unless you are lost in the desert without a supply of it). Again, why do we pollute and waste water? There are several reasons; many will be discussed later in this text.

You might be asking yourself: Is water pollution really that big of a deal? Simply stated, yes, it is. Humanity has left its footprint (in the form of pollution) on the environment, including on our water sources. Humanity has a bad habit of doing this. What it really comes down to is "out of sight, out of mind" thinking. When we abuse our natural resources in any manner, maybe we think to ourselves, "Why worry about it? Let someone else sort it all out."

As this text proceeds, it will lead you down a path strewn with abuse and disregard for our water supply—then all (excepting the water) will become clear. One hopes that we will not have to wait until someone does sort it—and us—out. With time and everything else, there might be a whole lot of sorting out going on.

This text is designed to show how the obvious and unsatisfactory gap in knowledge dealing with the science of water is to be filled in. Having said this, now it is to welcome you the gap-filler: *The Science of Water: Concepts and Applications.*

Finally, before moving on with the rest of the text, it should be pointed out the view held throughout this work is that water is special, strange, and different— and very vital. This view is held for several reasons, but the most salient factor driving this view is the one that points to the fact that on this planet, *water is life.*

References and Recommended Reading

DeZuane, J. (1997). *Handbook of drinking water quality* (2nd ed.). New York: John Wiley and Sons.

Dove, R. (1989). *Grace notes.* New York: Norton.

Gerba, C. P. (1996). Risk assessment. In Pepper, Gerba, & Brusseau (Eds.). *Pollution science.* San Diego: Academic Press.

Hammer, M. J., & Hammer, M. J., Jr. (1996). *Water and wastewater technology* (3rd ed.). Englewood Cliffs, NJ: Prentice Hall.

Harr, J. (1995). *A civil action.* New York: Vintage Books.

Lewis, S. A. (1996). *The Sierra Club guide to safe drinking water.* San Francisco: Sierra Club Books.

Metcalf and Eddy, Inc. (1991). *Wastewater engineering: Treatment, disposal, reuse* (3rd ed.). New York: McGraw-Hill, Inc.

Meyer, W. B. (1996). *Human impact on Earth.* New York: Cambridge University Press.

Nathanson, J. A. (1997). *Basic environmental technology: Water supply, waste management, and pollution control.* Upper Saddle River, NJ: Prentice Hall.

Pielou, E. C. (1998). *Fresh water.* Chicago: University of Chicago.

Robbins, T. (1976). *Even cowgirls get the blues.* Boston: Houghton Mifflin.

CHAPTER 24

All about Water

Water can both float and sink a ship.

Unless you are thirsty, in real need of refreshment, when you look at a glass of water, you might ask, what could be more boring? The curious might wonder about the physical and chemical properties of water that make it so unique and necessary for living things. Pure water is virtually colorless and has no taste or smell. But the hidden qualities of water make it a most interesting subject.

When the uninitiated becomes initiated to the wonders of water, one of the first surprises is that the total quantity of water on Earth is much the same now as it was more than 3 or 4 billion years ago, when the 320 million cubic miles of it were first formed. Ever since then, the water reservoir has gone round and round, building up, breaking down, cooling, and then warming. Water is very durable, but remains difficult to explain because it has never been isolated in a completely undefiled state.

Remember, water is special, strange, and different.

How Special, Strange, and Different Is Water?

Have you ever wondered what the nutritive value of water is? Water has no nutritive value, yet it is the major ingredient of all living things. Consider yourself, for example. Think of what you need to survive. Food? Air? A Play Station 3? MTV? An iPod? Water? Naturally, the focus of this chapter is on water. Water is of major importance to all living things; up to 90% of some organisms' body weight comes from water. Up to 60% of the human body is water; the brain is composed of 70% water, and the lungs are nearly 90% water. About 83% of our blood is water, which helps digest our food, transport waste, and control body

313

temperature. Each day humans must replace 2.4 liters of water through drinking and from the foods eaten.

You and I would not exist without an ample liquid water supply on Earth. The unique qualities and properties of water are what make it so important and basic to life. The cells in our bodies are full of water. The excellent ability of water to dissolve so many substances allows our cells to use valuable nutrients, minerals, and chemicals in biological processes.

Water's "stickiness" (from surface tension) plays a part in the ability to transport these materials throughout our bodies. The carbohydrates and proteins that our bodies use as food are metabolized and transported by water in the bloodstream. No less important is the ability of water to transport waste material out of our bodies.

Water is used to fight forest fires. Yet we use water spray on coal in a furnace to make it burn better.

Chemically, water is hydrogen oxide. It turns out, however, on more advanced analysis to be a mixture of more than 30 possible compounds. In addition, all of its physical constants are abnormal (strange).

At a temperature of 2,900° C, some substances that contain water cannot be forced to part with it. And yet others that do not contain water will liberate it when even slightly heated.

When liquid, water is virtually incompressible; as it freezes, it expands by an eleventh of its volume.

For the these reasons, and for many others, we can truly say that water is special, strange, and different.

Characteristics of Water

To this point, many things have been said about water; however, it has not been said that water is plain. This is the case because nowhere in nature is plain water to be found. Here on Earth, with a geologic origin dating back over 3 to 5 billion years, water found in even its purest form is composed of many constituents. You probably know water's chemical description is H_2O—that is one atom of oxygen bound to two atoms of hydrogen. The hydrogen atoms are "attached" to one side of the oxygen atom, resulting in a water molecule having a positive charge on the side where the hydrogen atoms are and a negative charge on the other side, where the oxygen atom is. Since opposite electrical charges attract, water molecules tend to attract each other, making water kind of "sticky"—the hydrogen atoms (positive charge) attracts the oxygen side (negative charge) of a different water molecule.

✓ **Important Point**: all these water molecules attracting each other means they tend to clump together. This is why water drops are, in fact, drops! If it wasn't for some of Earth's forces, such as gravity, a drop of water would be ball shaped—a perfect sphere. Even if it doesn't form a perfect sphere on Earth, we should be happy water is sticky.

Along with H_2O molecules, hydrogen (H^+), hydroxyl (OH^-), sodium, potassium, and magnesium, there are other ions and elements present. Additionally, water contains dissolved compounds including various carbonates, sulfates, silicates, and chlorides. Rain water, often assumed to be the equivalent of distilled water, is not immune to contamination that is collected as it descends through the atmosphere. The movement of water across the face of land contributes to its contamination, taking up dissolved gases, such as carbon dioxide and oxygen, and a multitude of organic substances and minerals leached from the soil. Don't let that crystal clear lake or pond fool you. These are not filled with water alone but are composed of a complex medium of chemical ingredients far exceeding the brief list presented here; it is a special medium in which highly specialized life can occur.

How important is water to life? To answer this question all we need do is to take a look at the common biological cell: it easily demonstrates the importance of water to life.

Living cells comprise a number of chemicals and organelles within a liquid substance, the cytoplasm, and the cell's survival may be threatened by changes in the proportion of water in the cytoplasm. This change in proportion of water in the cytoplasm can occur through desiccation (evaporation), oversupply, or the loss of either nutrients or water to the external environment. A cell that is unable to control and maintain homeostasis (i.e., the correct equilibrium or proportion of water) in its cytoplasm may be doomed. It may not survive.

✓ **Important Point:** As mentioned, water is called the "universal solvent" because it dissolves more substances than any other liquid. This means that wherever water goes, either through the ground or through our bodies, it takes along valuable chemicals, minerals, and nutrients.

INFLAMMABLE AIR + VITAL AIR = WATER

In 1783 in England, Henry Cavendish (a brilliant chemist and physicist) was "playing with" electric current. Specifically, Cavendish was passing electric current through a variety of substances to see what appended. Eventually, he got around to water. He filled a tube with water and sent his current through it. The water vanished.

To say that Cavendish was flabbergasted by the results of this experiment would be a mild understatement. "The tube has to have a leak in it," he reasoned.

He repeated the experiment again—same result.

Then again—same result.

The fact is he made the water disappear again and again. Actually, what Cavendish had done was convert the liquid water to its gaseous state—into an invisible gas.

When Cavendish analyzed the contents of the tube, he found it contained a mixture of two gases, one of which was *inflammable air* and the other was a heavier gas. This heavier gas had only been discovered a few years earlier by his colleague Joseph Priestly (English chemist and clergyman) who, finding that it kept a mouse alive and supported combustion, called it *vital air.*

JUST TWO H'S AND ONE O

Cavendish had been able to separate the two main constituents that make up water. All that remained was for him to put the ingredients back together again. He accomplished this by mixing a measured volume of inflammable air with different volumes of its vital counterpart, and setting fire to both. He found that most mixtures burned well enough, but when the proportions were precisely two to one, there was an explosion and the walls of his test tubes were covered with liquid droplets. He quickly identified these as water.

Cavendish made an announcement: Water was not water. Moreover, water is not just an odorless, colorless, and tasteless substance that lay beyond reach of chemical analysis. Water is not an element in its own right, but a compound of two independent elements, one a supporter of combustion and the other combustible. When united, these two elements become the preeminent quencher of thirst and flames.

It is interesting to note that a few year later, the great French genius, Antoine Lavoisier, tied the compound neatly together by renaming the ingredients *hydrogen*—"the water producer"—and *oxygen*. In a fitting tribute to his guillotined corpse (he was a victim of the French Revolution), his tombstone came to carry a simple and telling epitaph, a fitting tribute to the father of a new age in chemistry—*just two H's and one O.*

SOMEWHERE BETWEEN 0° AND 105°

We take water for granted now. Every high school student knows that water is a chemical compound of two simple and abundant elements. And yet scientists continue to argue the merits of rival theories on the structure of water. The fact is we still know little about water. For example, we don't know how water works.

Part of the problems lies with the fact that no one has ever seen a water molecule. It is true that we have theoretical diagrams and equations. We also have a disarmingly simple formula—H_2O. The reality, however, is that water is very complex. X-rays, for example, have shown that the atoms in water are intricately laced.

It has been said over and over again that water is special, strange, and different. Water is also almost indestructible. Sure, we know that electrolysis can separate water atoms, but we also know that once they get together again they must be heated up to more than 2,900° C to separate them again.

Water is also idiosyncratic. This can be seen in the way in which the two atoms of hydrogen in a water molecule (see Figure 25.1) take up a very precise and strange (different) alignment to each other. Not all angles of 45, 60, or 90 degrees—oh no, not water. Remember, water is different. The two hydrogen atoms always come to rest at an angle of approximately 105 degrees from each other, making all diagrams of their attachment to the larger oxygen atom look like Mickey Mouse ears on a very round head (see Figure 24.1; remember that everyone's favorite mouse is mostly water, too).

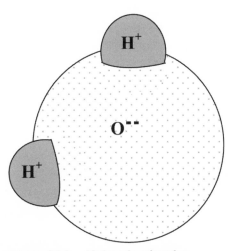

Figure 24.1. Molecule of water.

This 105-degree relationship makes water lopsided, peculiar, and eccentric—it breaks all the rules. You're not surprised are you?

One thing is certain; however, this 105-degree angle is crucial to all life as we know it. Thus, the answer to defining why water is special, strange, different—and vital—lies somewhere between 0 and 105 degrees.

Water's Physical Properties

Water has several unique physical properties:

- Water is unique in that it is the only natural substance that is found in all three states—liquid, solid (ice), and gas (steam)—at the temperatures normally found on Earth. Earth's water is constantly interacting, changing, and in movement.
- Water freezes at 32° F and boils at 212° F (at sea level, but 186.4° F at 14,000 feet). In fact, water's freezing and boiling points are the baseline with which temperature is measured: 0° on the Celsius scale is water's freezing point, and 100° is water's boiling point. Water is unusual in that the solid form, ice, is less dense than the liquid form, which is why ice floats.
- Water has a high specific heat index. This means that water can absorb a lot of heat before it begins to get hot. This is why water is valuable to industries and in your car's radiator as a coolant. The high specific heat index of water also helps regulate the rate at which air changes temperature, which is why the temperature change between seasons is gradual rather than sudden, especially near the oceans.
- Water has a very high surface tension. In other words and as previously mentioned, water is sticky and elastic, and tends to clump together in drops rather than spread out in a thin film. Surface tension is responsible for capillary action (discussed in detail later), which allows water (and its dissolved substances) to move through the roots of plants and through the thin blood vessels in or bodies.

Here's a quick rundown of some of water's properties:

- Weight: 62.416 pounds per cubic foot at 32° F
- Weight: 61.998 pounds per cubic foot at 100° F
- Weight: 8.33 pounds/gallon, 0.036 pounds/cubic inch
- Density: 1 gram (gm) per cubic centimeter (cc) at 39.2° F, 0.95865 gm/cc at 212° F
- 1 gallon = 4 quarts = 8 pints = 128 ounces = 231 cubic inches
- 1 liter = 0.2642 gallons = 1.0568 quart = 61.02 cubic inches
- 1 million gallons = 3.069 acre-feet = 133,685.64 cubic feet

Capillary Action

If we were to mention the term *capillary action* to the person on the street, he or she might instantly nod and respond that the human body is full of them—that capillaries are the tiny blood vessels that connect the smallest arteries and the smallest of the veins. This would be true, of course. But in the context of water science, *capillary* is something different than capillary action in the human body.

Even if you've never heard of capillary action, it is still important in your life. Capillary action is important for moving water (and all of the things that are dissolved in it) around. It is defined as the movement of water within the spaces of a porous material due to the forces of adhesion, cohesion, and surface tension.

Surface tension is a measure of the strength of the water's surface film. The attraction between the water molecules creates a strong film, which, among other common liquids is only surpassed by that of mercury. This surface tension permits water to hold up substances heavier and denser than itself. A steel needle carefully placed on the surface of that glass of water will float. Some aquatic insects such as the water strider rely on surface tension to walk on water.

Capillary action occurs because water is sticky thanks to the forces of cohesion (water molecules like to stay closely together) and adhesion (water molecules are attracted and stick to other substances). So, water tends to stick together, as in a drop, and it sticks to glass, cloth, organic tissues, and soil. Dip a paper towel into a glass of water and the water will "climb" onto the paper towel. In fact, it will keep going up the towel until the pull of gravity is too much for it to overcome.

The Water Cycle

In chapters 7 and 13, we briefly describe the hydrological, or water, cycle. Its importance to our very survival on Earth cannot be overstated. Simply, this natural cycle is one of the most important ongoing processes in nature. The natural *water cycle* or *hydrological cycle* is the means by which water in all three forms—solid, liquid, and vapor—circulates through the biosphere. The water cycle is all about describing how water moves above, on, and through the Earth. Much more water, however, is "in storage" for long periods than is actually moving through the cycle. The storehouses for the vast majority of all water on Earth are the oceans. It is estimated that of the 332,500,000 cubic miles of the world's water supply, about 321,000,000 cubic miles is stored in oceans. That is about 96.5%. It is also estimated that the oceans supply about 90% of the evaporated water that goes into the water cycle.

Water—lost from the Earth's surface to the atmosphere either by evaporation from the surface of lakes, rivers, and oceans or through the transpiration of plants—forms clouds that condense to deposit moisture on the land and sea. Evaporation from the oceans is the primary mechanism supporting the surface-to-atmosphere portion of the water cycle. Note, however, that a drop of water might travel thousands of miles between the time it evaporates and the time it falls to Earth again as rain, sleet, or snow. The water that collects on land flows to the ocean in streams and rivers or seeps into the Earth, joining groundwater. Even groundwater eventually flows toward the ocean for recycling.

The cycle constantly repeats itself, a cycle without end.

✓ **Note:** The length of time that it takes water to return to the atmosphere after falling from the clouds varies tremendously. After a short summer shower, most of the rainfall on land can evaporate into the atmosphere in only a matter of minutes. A drop of rain falling on the ocean may take as long as 37,000 years before it returns to the atmosphere, and some water has been in the ground or caught in glaciers for millions of years.

✓ **Important Point:** Only about 2% of the water absorbed into plant roots is used in photosynthesis. Nearly all of it travels through the plant to the leaves, where transpiration to the atmosphere begins the cycle again.

SPECIFIC WATER MOVEMENTS

After having reviewed the water cycle in very simple terms to provide foundational information, it is important to point out that the actual movement of water on Earth is much more complex. Three different methods of transport are involved in this water movement: *evaporation, precipitation,* and *run-off.*

Evaporation of water is a major factor in hydrologic systems. Evaporation is a function of temperature, wind velocity, and relative humidity. Evaporation (or vaporization) is, as the name implies, the formation of vapor. Dissolved constituents such as salts remain behind when water evaporates. Evaporation of the surface water of oceans provides most of the water vapor. It should be pointed out, however, that water can also vaporize through plants, especially from leaf surfaces. This process is called *evapotranspiration.* Plant transpiration is pretty much an invisible process—since the water is evaporating from the leaf surfaces, you don't just go out and see the leaves "breathe." During a growing season, a leaf will transpire many times more water than its own weight. A large oak tree can transpire 40,000 gallons (151,000 liters) per year (U.S. Geological Survey [USGS], 2006).

USGS (2006) points out that the amount of water that plants transpire varies greatly geographically and over time. There are a number of factors that determine transpiration rates:

- **Temperature:** Transpiration rates go up as the temperature goes up, especially during the growing season, when the air is warmer due to stronger sunlight and warmer air masses.

- **Relative humidity:** As the relative humidity of the air surrounding the plant rises, the transpiration rate falls. It is easier for water to evaporate into dryer air than into more saturated air.
- **Wind and air movement:** Increased movement of the air around a plant results in a higher transpiration rate.
- **Soil-moisture availability:** When moisture is lacking, plants can begin to senesce (i.e., age prematurely, which can result in leaf loss) and transpire less water.
- **Type of plant:** Plants transpire water at different rates. Some plants that grow in arid regions, such as cacti and succulents, conserve precious water by transpiring less water than other plants.

✓ **Interesting Point:** It may surprise you that ice can also vaporize without melting first. However, this *sublimation* process is slower than vaporization of liquid water.

Evaporation rates are measured with evaporation pans. These evaporation pans provide data that indicate the atmospheric evaporative demand of an area and can be used to estimate (1) the rates of evaporation from ponds, lakes, and reservoirs; and (2) evapotranspiration rates. It is important to note that several factors affect the rate of pan evaporation. These factors include the type of pan, type of pan environment, method of operating the pan, exchange of heat between pan and ground, solar radiation, air temperature, wind, and temperature of the water surface (Jones, 1992).

Precipitation includes all forms in which atmospheric moisture descends to Earth—rain, snow, sleet, and hail. Before precipitation can occur, the water that enters the atmosphere by vaporization must first condense into liquid (clouds and rain) or solid (snow, sleet, and hail) before it can fall. This vaporization process absorbs energy. This energy is released in the form of heat when the water vapor condenses. You can best understand this phenomena when you compare it with what occurs when water that evaporates from your skin absorbs heat, making you feel cold.

✓ **Note:** The annual evaporation from ocean and land areas is the same as the annual precipitation.

Run-off is the flow back to the oceans of the precipitation that falls on land. This journey to the oceans is not always unobstructed—flow back may be intercepted by vegetation (from which it later evaporates), a portion is held in depressions, and a portion infiltrates into the ground. A part of the infiltrated water is taken up by plant life and returned to the atmosphere through evapotranspiration, while the remainder either moves through the ground or is held by capillary action. Eventually, water drips, seeps, and flows its way back into the oceans.

Assuming that the water in the oceans and ice caps and glaciers is fairly constant when averaged over a period of years, the water balance of the Earth's surface can be expressed by the relationship: water lost = water gained (Turk & Turk, 1988).

Q and Q Factors

While potable water practitioners must have a clear and complete understanding of the natural water cycle, they must, as previously mentioned, also factor in two major considerations: quality and quantity—the *Q and Q factors*. These are (1) providing a *quality* potable water supply that is clean, wholesome, and safe to drink; and (2) finding a water supply available in adequate *quantities* to meet the anticipated demand.

✓ **Important Point:** A couple of central facts important to our discussion of freshwater supplies are: (1) water is very much a local or regional resource, and (2) problems of its shortage or pollution are equally local problems. Human activities affect the quantity of water available at a locale at any time by changing either the total volume that exists there, or aspects of quality that restrict or devalue it for a particular use. Thus, the total human effects on water supplies is the sum of the separate human effects on the various drainage basins and groundwater aquifers. In the global system, the central, critical fact about water is the natural variation in its availability (Meyer, 1996). Simply put, **not all lands are watered equally**.

Watershed Protection

Watershed protection is one of the barriers in the multiple-barrier approach to protecting source water. In fact, watershed protection is the primary barrier, the first line of defense against contamination of drinking water at its source.

MULTIPLE-BARRIER CONCEPT

On August 6, 1996, during the Safe Drinking Water Act Reauthorization signing ceremony, President Bill Clinton stated, "A fundamental promise we must make to our people is that the food they eat and the water they drink are safe."

No rational person could doubt the importance of the promise made in this statement.

The Safe Drinking Water Act (SDWA), passed in 1974, amended in 1986, and (as stated previously) reauthorized in 1996, gives the United States Environmental Protection Agency (EPA) the authority to set drinking water standards. This document is important for many reasons, primarily because it describes how the EPA establishes these standards.

Drinking water standards are regulations that the EPA sets to control the level of contaminants in the nation's drinking water. These standards are part of the Safe Drinking Water Act's *multiple-barrier approach* to drinking water protection. The multiple-barrier approach includes the following elements:

1. **Assessing and protecting drinking water sources** means doing everything possible to prevent microbes and other contaminants from entering water supplies. Minimizing human and animal activity around our watersheds is one part of this barrier.
2. **Optimizing treatment processes** provides a second barrier. This usually means filtering and disinfecting the water. It also means making sure that the people who are responsible for our water are properly trained and certified, and are knowledgeable of the public health issues involved.
3. **Ensuring the integrity of distribution systems** consists of maintaining the quality of water as it moves through the system on its way to the customer's tap.
4. **Effecting correct cross-connection control procedures** is a critical fourth element in the barrier approach because the greatest potential hazard in water distribution systems is associated with cross-connections to nonpotable waters. There are many connections between potable and nonpotable systems—for example, every drain in a hospital constitutes such a connection—but cross-connections are those through which backflow can occur (Angele, 1974).
5. **Continuous monitoring and testing of the water before it reaches the tap** is a critical element in the barrier approach. It should include specific procedures to follow should potable water ever fail to meet quality standards.

With the involvement of the EPA, local governments, drinking water utilities, and citizens, these multiple barriers ensure that the tap water in the United States and territories is safe to drink. Simply, in the multiple-barrier concept, we employ a holistic approach to water management that begins at the source and continues with treatment through disinfection and distribution.

The bottom line on the multiple-barrier approach to protecting the watershed is best summed up in the following:

> Ideally, under the general concept of "quality in, means quality out," a protected watershed ensures that surface runoff and inflow to the source waters occur within a pristine environment. (Spellman, 2003)

Potable Water Source

Because of huge volume and flow conditions, the quality of natural water cannot be modified significantly within the body of water. Accordingly, humans must augment nature's natural purification processes with physical, chemical, and biological treatment procedures. Essentially, this quality-control approach is directed to the water withdrawn from a source for a specific use, which is treated for safety.

POTABLE WATER

Potable water is water fit for human consumption and domestic use. It is sanitary and normally free of minerals, organic substances, and toxic agents in excess of reasonable amounts for domestic usage in the area served, and is normally adequate in quantity for the minimum health requirements of the persons served.

In regard to a potential potable water supply, the key words, as previously mentioned, are *quality* and *quantity*. Obviously, if we have a water supply that is unfit for human consumption, we have a quality problem. If we do not have an adequate supply of quality water, we have a quantity problem.

In this section, the focus is on surface water and groundwater hydrology and the mechanical components associated with collection and conveyance of water from its source to the public water supply system for treatment. Well supplies are also discussed.

KEY DEFINITIONS

Surface water is the water on the Earth's surface as distinguished from water underground (groundwater).

Groundwater is the subsurface water occupying a saturated geological formation from which wells and springs are fed.

Hydrology is the applied science pertaining to properties, distribution, and behavior of water.

Permeable refers to a material or substance that water can pass through.

Overland flow is the movement of water on and just under the Earth's surface.

Surface runoff is the amount of rainfall that passes over the surface of the Earth

Spring is a surface feature in which, without the help of man, water issues from rock or soil onto the land or into a body of water, the place of issuance being relatively restricted in size.

Precipitation is the process by which atmospheric moisture is discharged onto the Earth's crust. Precipitation takes the form of rain, snow, hail, and sleet.

Water rights are the rights, acquired under the law, to use the water accruing in surface or groundwater, for a specified purpose in a given manner and usually within the limits of a given period.

Drainage basin is an area from which surface runoff or groundwater recharge is carried into a single drainage system. It is also called a *catchment area, watershed,* and *drainage area.*

Watershed is a drainage basin from which surface water is obtained.

Recharge area is an area from which precipitation flows into underground water sources.

Raw water is the untreated water to be used after treatment for drinking water.

Caisson is a large pipe placed in a vertical position.

Impermeable refers to a material or substance water will not pass through.

Contamination refers to the introduction into water of toxic materials, bacteria, or other deleterious agents that make the water unfit for its intended use.

Aquifer is a porous, water-bearing geologic formation.

Water table refers to the average depth or elevation of the groundwater over a selected area. The upper surface of the zone of saturation, except where that surface is formed by an impermeable body.

Unconfined aquifer is an aquifer that sits on an impervious layer, but is open on the top to local infiltration. The recharge for an unconfined aquifer is local. It is also called a *water table aquifer.*

Confined aquifer is an aquifer that is surrounded by formations of less permeable or impermeable material.

Porosity is the ratio of pore space to total volume. It refers to that portion of a cubic foot of soil that is air space and could therefore contain moisture.

Static level is the height to which the water will rise in the well when the pump is not operating.

Pumping level is the level at which the water stands when the pump is operating.

Drawdown is the distance or difference between the static level and the pumping level. When the drawdown for any particular capacity well and rate pump bowls is determined, the pumping level is known for that capacity. The pump bowls are located below the pumping level so that they will always be underwater. When the drawdown is fixed or remains steady, the well is then furnishing the same amount of water as is being pumped.

Cone of depression refers to the fact that as the water in a well is drawn down, the water near the well drains or flows into it. The water will drain farther back from the top of the water table into the well as drawdown increases.

Radius of influence is the distance from the well to the edge of the cone of depression, the radius of a circle around the well from which water flows into the well.

Annular space is the space between the casing and the wall of the hole.

Specific yield is the geologist's method for determining the capacity of a given well and the production of a given water-bearing formation; it is expressed as gallons per minute per foot of drawdown.

Surface Water

Where do we get our potable water from? From what water source is our drinking water provided? To answer these questions, we would most likely turn to one of two possibilities: our public water is provided by a groundwater or surface water source because these two sources are, indeed, the primary sources of most water supplies.

From the earlier discussion of the hydrologic or water cycle, we know that regardless of which of the two sources we obtain our drinking water from, the source is constantly being replenished (we hope) with a supply of fresh water. This water cycle phenomenon was best summed up by Heraclitus of Ephesus, who said, "You could not step twice into the same river; for other waters are ever flowing on to you."

In the following section, we discuss one of the primary duties of the drinking water practitioner (and humankind in general): to find and secure a source of potable water for human use.

LOCATION! LOCATION! LOCATION!

In the real estate business, location is everything. The same can be said when it comes to sources of water. In fact, the presence of water defines "location" for communities. Although communities differ widely in character and size, all have the common concerns of finding water for industrial, commercial, and residential use. Freshwater sources that can provide stable and plentiful supplies for a community don't always occur where we wish. Simply put, on land, the availability of a regular supply of potable water is the most important factor affecting the presence—or absence—of many life forms. A map of the world immediately shows us that surface waters are not uniformly distributed over the Earth's surface. U.S. land holds rivers, lakes, and streams on only about 4% of its surface. The largest populations of any life forms, including humans, are found in regions of the United States (and the rest of the world) where potable water is readily available, because lands barren of water simply won't support large populations. One thing is certain: if a local supply of potable water is not readily available, the locality affected will seek a source. This is readily apparent (absolutely crystal clear), for example, when one studies the history of water "procurement" for the communities located within the Los Angeles Basin.

✓ **Important Point:** The volume of freshwater sources depends on geographic, landscape, and temporal variations, and on the effect of human activities.

HOW READILY AVAILABLE IS POTABLE WATER?

Approximately 326 million cubic miles of water compose Earth's entire water supply. Of this massive amount of water, only about 3% is fresh, although salt water does provide us indirectly with fresh water through evaporation from the oceans. Even most of the minute percentage of fresh water Earth holds is locked up in polar ice caps and in glaciers. The rest is held in lakes, in flows through soil, and in river and stream systems. Only 0.027% of Earth's fresh water is available for human consumption (see Table 24.1 for the distribution percentages of Earth's water supply).

Table 24.1. World Water Distribution

Location	Percent of Total
Land areas	
Freshwater lakes	0.009
Saline lakes and inland seas	0.008
Rivers (average instantaneous volume)	0.0001
Soil moisture	0.005
Groundwater (above depth of 4,000 m)	0.61
Ice caps and glaciers	2.14
Total: Land areas	**2.8**
Atmosphere (water vapor)	0.001
Oceans	97.3
Total all locations (rounded)	**100**

Source: USGS, 2006.

We see from Table 24.1 that the major sources of drinking water are from surface water, groundwater, and from groundwater under the direct influence of surface water (i.e., springs or shallow wells).

Surface waters are not uniformly distributed over the Earth's surface. In the United States, for example, only about 4% of the landmass is covered by rivers, lakes, and streams. The volumes of these freshwater sources depend on geographic, landscape, and temporal variations, and on the effects of human activities.

Again, surface water is that water that is open to the atmosphere and results from overland flow (i.e., runoff that has not yet reached a definite stream channel). Put a different way, surface water is the result of surface runoff.

For the most part, however, *surface* (as used in the context of this text) refers to water flowing in streams and rivers, as well as water stored in natural or artificial lakes, artificial impoundments such as lakes made by damming a stream or river, springs that are affected by a change in level or quantity, shallow wells that are affected by precipitation, wells drilled next to or in a stream or river, rain catchments, and muskeg and tundra ponds.

Specific sources of surface water include:

1. Rivers
2. Streams
3. Lakes
4. Impoundments (artificial lakes made by damming a river or stream)
5. Very shallow wells that receive input via precipitation
6. Springs affected by precipitation (flow or quantity directly dependent on precipitation)
7. Rain catchments (drainage basins)
8. Tundra ponds or muskegs (peat bogs)

Surface water has advantages as a source of potable water. Surface water sources are usually easy to locate—unlike groundwater, finding surface water does not take a geologist or hydrologist. Normally, surface water is not tainted with minerals precipitated from the Earth's strata.

Ease of discovery aside, surface water also presents some disadvantages: surface water sources are easily contaminated (polluted) with microorganisms that can cause waterborne diseases (anyone who has suffered from "hiker's disease" or "hiker's diarrhea" can attest to this), and from chemicals that enter from surrounding runoff and upstream discharges. Water rights can also present problems.

As mentioned, most surface water is the result of surface runoff. The amount and flow rate of this surface water is highly variable, which comes into play for two main reasons: (1) human interferences (influences) and (2) natural conditions. In some cases, surface water runs quickly off land surfaces. From a water resources standpoint, this is generally undesirable, because quick runoff does not provide enough time for the water to infiltrate the ground and recharge groundwater aquifers. Surface water that quickly runs off land also causes erosion and flooding problems. Probably the only good thing that can be said about surface water that runs off quickly is that it usually does not have enough contact time to increase in mineral content. Slow surface water off land has all the opposite effects.

Drainage basins collect surface water and direct it on its gravitationally influenced path to the ocean. The drainage basin is normally characterized as an area measured in square miles, acres, or sections. Obviously, if a community is drawing water from a surface water source, the size of its drainage basin is an important consideration.

Surface water runoff, like the flow of electricity, flows or follows the path of least resistance. Surface water within the drainage basin normally flows toward one primary watercourse (river, stream, brook, creek, etc.), unless some artificial distribution system (canal or pipeline) diverts the flow.

✓ **Important Point:** Many people probably have an overly simplified idea that precipitation falls on the land, flows overland (runoff), and runs into rivers, which then empty into the oceans. That is "overly simplified" because rivers also gain and lose water to the ground. Still, it is true that much of the water in rivers comes directly from runoff from the land surface, which is defined as surface runoff.

Surface water runoff from land surfaces depend on several factors, which include:

- *Rainfall duration:* Even a light, gentle rain, if it lasts long enough, can, with time, saturate soil and allow runoff to take place.
- *Rainfall intensity:* With increases in intensity, the surface of the soil quickly becomes saturated. This saturated soil can hold no more water; as more rain falls and water builds up on the surface, it creates surface runoff.
- *Soil moisture:* The amount of existing moisture in the soil has a definite effect on surface runoff. Soil already wet or saturated from a previous rain causes surface runoff to occur sooner than if the soil were dry. Surface runoff from frozen soil can be up to 100% of snow melt or rain runoff because frozen ground is basically impervious.
- *Soil composition:* The composition of the surface soil directly affects the amount of runoff. For example, hard rock surfaces, obviously, result in 100% runoff. Clay soils have very small void spaces that swell when wet; the void spaces close and do not allow infiltration. Coarse sand possesses large void spaces that allow easy flow through of water, which produces the opposite effect, even in a torrential downpour.
- *Vegetation cover:* Groundcover limits runoff. Roots of vegetation and pine needles, pine cones, leaves, and branches create a porous layer (a sheet of decaying natural organic substances) above the soil. This porous organic sheet readily allows water into the soil. Vegetation and organic waste also act as cover to protect the soil from hard, driving rains, which can compact bare

soils, close off void spaces, and increase runoff. Vegetation and groundcover work to maintain the soil's infiltration and water-holding capacity, and also work to reduce soil moisture evaporation.

- *Ground slope:* When rain falls on steeply sloping ground, up to 80% or more may become surface runoff. Gravity moves the water down the surface more quickly than it can infiltrate the surface. Water flow off flat land is usually slow enough to provide opportunity for a higher percentage of the rainwater to infiltrate the ground.
- *Human influences:* Various human activities have a definite influence on surface water runoff. Most human activities tend to increase the rate of water flow. For example, canals and ditches are usually constructed to provide steady flow, and agricultural activities generally remove groundcover that would work to retard the runoff rate. On the opposite extreme, artificial dams are generally built to retard the flow of runoff.

Paved streets, tarmac, paved parking lots, and buildings are impervious to water infiltration, greatly increasing the amount of storm water runoff from precipitation events. These artificial surfaces, which work to hasten the flow of surface water, often cause flooding, sometimes with devastating consequences. In badly planned areas, even relatively light precipitation can cause local flooding. Impervious surfaces not only present flooding problems, they also do not allow water to percolate into the soil to recharge groundwater supplies—often another devastating blow to a location's water supply.

ADVANTAGES AND DISADVANTAGES OF SURFACE WATER

The biggest advantage of using a surface water supply as a water source is that these sources are readily located; finding surface water sources does not demand sophisticated training or equipment. Many surface water sources have been used for decades and even centuries (in the United States, for example), and considerable data is available on the quantity and quality of the existing water supply. Surface water is also generally softer (not mineral-laden), which makes its treatment much simpler.

The most significant disadvantage of using surface water as a water source is pollution. Surface waters are easily contaminated (polluted) with microorganisms that cause waterborne diseases and chemicals that enter the river or stream from surface runoff and upstream discharges. Another problem with many surface water sources is turbidity, which fluctuates with the amount of precipitation. Increases in turbidity increase treatment cost and operator time. Surface water temperatures can be a problem because they fluctuate with ambient temperature, making consistent water quality production at a waterworks plant difficult. Drawing water from a surface water supply might also present problems; intake structures may clog or become damaged from winter ice, or the source may be so shallow that it completely freezes in the winter. Water rights cause problems too—removing surface water from a stream, lake, or spring requires a legal right. The lingering, seemingly unanswerable question is who owns the water?

Using surface water as a source means that the purveyor is obligated to meet the requirements of the surface water treatment rule (SWTR) and interim enhanced surface water treatment rule. Note that this rule only applies to large public water systems that serve more than 10,000 people. It tightened controls on disinfection byproducts and turbidity and regulates *Cryptosporidium.*

Groundwater Supply

Unbeknownst to most of us, our Earth possesses an unseen ocean. Unlike the surface oceans that cover most of the globe, this ocean is freshwater. It consists of the groundwater that lies contained in aquifers beneath Earth's crust. This gigantic water source forms a reservoir that feeds all the natural fountains and springs of Earth. But how does water travel into the aquifers that lie under Earth's surface?

Groundwater sources are replenished from a percentage of the average approximately three feet of water that falls to Earth each year on every square foot of land. Water falling to Earth as precipitation follows three courses. Some runs off directly to rivers and streams, eventually working back to the sea. Evaporation and transpiration through vegetation takes up about 2 feet. The remaining 6 inches seeps into the ground, entering and filling every interstice, each hollow and cavity. Gravity pulls water toward the center of the Earth. That means that water on the surface will try to seep into the ground below it. Although groundwater composes only one-sixth of the total, (1,680,000 miles of water), if we could spread out this water over the land, it would blanket it to a depth of 1,000 feet.

GROUNDWATER

As mentioned, part of the precipitation that falls on land infiltrates the land surface, percolates downward through the soil under the force of gravity, and becomes groundwater. Groundwater, like surface water, is extremely important to the hydrologic cycle and to our water supplies. Almost half of the people in the United States drink public water from groundwater supplies. Overall, more water exists as groundwater than surface water in the United States, including the water in the Great Lakes. But sometimes pumping it to the surface is not economical, and in recent years, pollution of groundwater supplies from improper disposal has become a significant problem.

We find groundwater in saturated layers called *aquifers* under the Earth's surface. Three types of aquifers exist: unconfined, confined, and springs.

Aquifers are made up of a combination of solid material such as rock and gravel and open spaces called *pores*. Regardless of the type of aquifer, the groundwater in the aquifer is in a constant state of motion. This motion is caused by gravity or by pumping.

The actual amount of water in an aquifer depends on the amount of space available between the various grains of material that make up the aquifer. The amount of space available is called porosity. The ease of movement through an aquifer depends on how well the pores are connected. For example, clay can hold a lot of water and has high porosity, but the pores are not connected, so water moves through the clay with difficulty. The ability of an aquifer to allow water to infiltrate is called *permeability.*

The aquifer that lies just under the Earth's surface is called the *zone of saturation,* an unconfined aquifer (see Figure 24.2). The top of the zone of saturation is the water table. An unconfined aquifer is only contained on the bottom and depends on local precipitation for recharge. This type of aquifer is often called a *water table aquifer.*

Unconfined aquifers are a primary source of shallow well water (see Figure 24.2). These wells are shallow (and not desirable as a public drinking water source). They are subject to local contamination from hazardous and toxic materials—fuel and oil, septic tanks, and agricultural runoff provided increased levels of nitrates and microorganisms. These wells may be classified as groundwater under the direct influence of surface water (GUDISW), and therefore require treatment for control of microorganisms.

A confined aquifer is sandwiched between two impermeable layers that block the flow of water. The water in a confined aquifer is under hydrostatic pressure. It does not have a free water table (see Figure 24.3).

Confined aquifers are called *artesian aquifers.* Wells drilled into artesian aquifers are called *artesian wells* and commonly yield large quantities of high-quality water. An artesian well is any well where the water in the well casing would rise above the saturated strata. Wells in confined aquifers are normally referred to as *deep wells* and are not generally affected by local hydrological events.

A confined aquifer is recharged by rain or snow in the mountains where the aquifer lies close to the surface of the Earth. Because the recharge area is some distance from areas of possible contamination, the possibility of contamination is usually very low. However, once contaminated, confined aquifers may take centuries to recover.

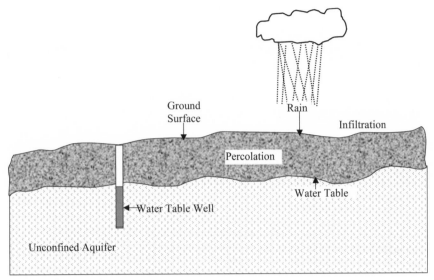

Figure 24.2. Unconfined aquifer.

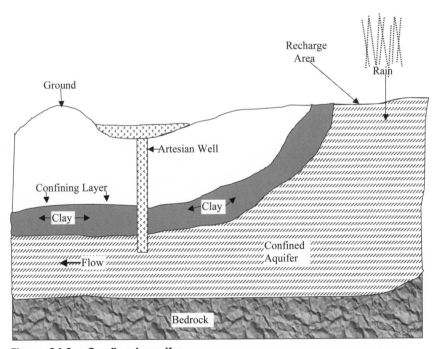

Figure 24.3. Confined aquifer.

Groundwater naturally exits in the Earth's crust in areas called *springs.* The water in a spring can originate from a water table aquifer or from a confined aquifer. Only water from a confined spring is considered desirable for a public water system.

GROUNDWATER QUALITY

Generally, groundwater possesses high chemical, bacteriological, and physical quality. When pumped from an aquifer composed of a mixture of sand and gravel, if not directly influenced by surface water, groundwater is often used without filtration. It can also be used without disinfection if it has a low coliform count. However, as mentioned, groundwater can become contaminated. When septic systems fail, saltwater intrudes, improper disposal of wastes occurs, improperly stockpiled chemicals leach, underground storage tanks leak, hazardous materials spill, fertilizers and pesticides are misplaced, and when mines are improperly abandoned, groundwater can become contaminated.

To understand how an underground aquifer becomes contaminated, you must understand what occurs when pumping is taking place within the well. When groundwater is removed from its underground source (i.e., from the water-bearing stratum) via a well, water flows toward the center of the well. In a water table aquifer, this movement causes the water table to sag toward the well. This sag is called the *cone of depression.* The shape and size of the cone depends on the relationship between the pumping rate and the rate at which water can move toward the well. If the rate is high, the cone is shallow and its growth stabilizes. The area that is included in the cone of depression is called the *cone of influence,* and any contamination in this zone will be drawn into the well.

Groundwater under the Direct Influence of Surface Water

GUDISW is not classified as a groundwater supply. A supply designated as GUDISW must be treated under the state's surface water rules rather than the groundwater rules.

The SWTR of the Safe Drinking Water Act requires each site to determine which groundwater supplies are influenced by surface water (i.e., when surface water can infiltrate a groundwater supply and could contaminate it with *Giardia,* viruses, turbidity, and organic material from the surface water source). To determine whether a groundwater supply is under the direct influence of surface water, the EPA has developed procedures that focus on significant and relatively rapid shifts in water quality characteristics, including turbidity, temperature, and pH. When these shifts can be closely correlated with rainfall or other surface water conditions, or when certain indicator organisms associated with surface water are found, the source is said to be under the direct influence of surface water.

Almost all groundwater is in constant motion through the pores and crevices of the aquifer in which it occurs. The water table is rarely level; it generally follows the shape of the ground surface. Groundwater flows in the downhill direction of the sloping water table. The water table sometimes intersects low points of the ground, where it seeps out into springs, lakes, or streams. Usual groundwater sources include wells and springs that are not influenced by surface water or local hydrologic events.

As a potable water source, groundwater has several advantages over surface water. Unlike surface water, groundwater is not easily contaminated. Groundwater sources are usually lower in bacteriological contamination than surface waters. Groundwater quality and quantity usually remains stable throughout the year. In the United States, groundwater is available in most locations.

As a potable water source, groundwater does present some disadvantages compared with surface water sources. Operating costs are usually higher, because groundwater supplies must be pumped to the surface. Any contamination is often hidden from view. Removing any contaminants is very difficult. Groundwater often possesses high mineral levels, and thus an increased level of hardness, because it is in contact longer with minerals. Near coastal areas, groundwater sources may be subject to salt water intrusion.

✓ **Important Point:** Groundwater quality is influenced by the quality of its source. Changes in source waters or degraded quality of source supplies may seriously impair the quality of the groundwater supply.

Prior to moving onto water use, it is important to point out that our freshwater supplies are constantly renewed through the hydrologic cycle, but the balance between the normal ratio of freshwater to salt water is not subject to our ability to change. As our population grows and we move into lands without ready freshwater supplies, we place ecological strain on those areas, and on their ability to support life.

Communities that build in areas without adequate local water supply are at risk in the event of emergency. Proper attention to our surface and groundwater sources, including remediation, pollution control, and water reclamation and reuse can help to ease the strain, but technology cannot fully replace adequate local freshwater supplies, whether from surface or groundwater sources.

Water Use

In the United States, rainfall averages approximately $4,250 \times 10^9$ gallons a day. About two-thirds of this returns to the atmosphere through evaporation directly from the surface of rivers, streams, and lakes, and from transpiration from plant foliage. This leaves approximately $1,250 \times 10^9$ gallons a day to flow across or through the Earth to the sea.

The USGS (2004) points out that estimates in the United States indicate that about 408 billion gallons per day (one thousand million gallons per day, abbreviated Bgal/d) were withdrawn from all uses during 2000. This total has varied less than 3% since 1985 as withdrawals have stabilized for the two largest uses—thermoelectric power and irrigation. Fresh groundwater withdrawals (83.3 Bgal/d) during 2000 were 14% more than during 1985. Fresh surface-water withdrawals for 2000 were 262 Bgal/d, varying less than 2% since 1985.

About 195 Bgal/d, or 8% of all freshwater and saline-water withdrawals for 2000, were used for thermoelectric power. Most of this water was derived from surface water and used for once-through cooling at power plants. About 52% of fresh surface-water withdrawals and about 96% of saline-water withdrawals were for thermoelectric power use. Withdrawals for thermoelectric power have been relatively stable since 1985.

Irrigation remained the largest use of freshwater in the United States and totaled 137 Bgal/d for 2000. Since 1950, irrigation has accounted for about 65% of total water withdrawals, excluding those for thermoelectric power. Historically, more surface water than groundwater has been used for irrigation. However, the percentage of total irrigation withdrawals from groundwater has continued to increase, from 23% in 1950 to 42% in 2000. Total irrigation withdrawals were 2% more for 2000 than for 1995, because of a 16% increase in groundwater withdrawals and a small decrease in surface-water withdrawals. Irrigated acreage more than doubled between 1950 and 1980, then remained constant before increasing nearly 7% between 1995 and 2000. The number of acres irrigated with sprinkler and microirrigation systems has continued to increase and now composes more than one-half the total irrigated acreage.

Public-supply withdrawals were more than 43 Bgal/d for 2000. Public-supply withdrawals during 1950 were 14 Bgal/d. During 2000, about 85% of the population in the United States obtained drinking water from public suppliers, compared with 62% during 1950. Surface water provided 63 percent of the total during 2000, whereas surface water provided 74% during 1950.

Self-supplied industrial withdrawals totaled nearly 20 Bgal/d in 2000, or 12% less than in 1995. Compared with 1985, industrial self-supported withdrawals declined by 24%. Estimates of industrial water use in the United State were largest during the years from 1965 to 1980, but during 2000, estimates were at the lowest level since reporting began in 1950. Combined withdrawals for self-supplied domestic, livestock, aquaculture, and mining were less than 13 Bgal/d for 2000, and represented about 3% of total withdrawals.

States with the largest surface-water withdrawals were California, which has large withdrawals for irrigation and thermoelectric power, and Texas and Nebraska, which had large withdrawals for irrigation.

In this text, the primary concern with water use is in regard to municipal applications (demand). Municipal water demand is usually classified according to the nature of the user. These classifications are:

1. **Domestic:** Domestic water is supplied to houses, schools, hospitals, hotels, restaurants, and so on for culinary, sanitary, and other purposes. Use varies with the economic level of the consumer, the range being 20 to 100 gallons per capita per day. It should be pointed out that these figures include water used for watering gardens and lawns and washing cars.
2. **Commercial and industrial:** Commercial and industrial water is supplied to stores, offices, and factories. The importance of commercial and industrial demand is based, of course, on whether there are large industries that use water supplied from the municipal system. These large industries demand a quantity of water directly related to the number of persons employed, to the actual floor space or area of each establishment, and to the number of units manufactured or produce. Industry in the United States uses an average of 150 Bgal/d of water each day.
3. **Public use:** Public use water is the water furnished to public buildings and used for public services. This includes water for schools, public buildings, fire protection, and for flushing streets.
4. **Loss and waste:** Water that is lost or wasted (i.e., unaccounted for) is attributable to leaks in the distribution system, inaccurate meter readings, and for unauthorized connections. Loss and waste of water can be expensive. To reduce loss and waste, a regular program that includes maintenance of the system and replacement or recalibration of meters is required (McGhee, 1991).

References and Recommended Reading

Angele, F. J., Sr. (1974). *Cross connections and backflow protection* (2nd ed.). Denver: American Water Association.

Environmental Protection Agency (EPA) (2006). *Watersheds.* Accessed December 2006 from http://www.epa.gov/owow/watershed/whatis.html.

Jones, F. E. (1992). *Evaporation of water.* Chelsea, Michigan: Lewis Publishers.

Lewis, S. A. (1996). *The Sierra Club guide to safe drinking water.* San Francisco: Sierra Club Books.

McGhee, T. J. (1991). *Water supply and sewerage* (6th ed.). New York: McGraw-Hill.

Meyer, W. B. (1996). *Human impact on earth.* New York: Cambridge University Press.

Peavy, H. S., et al. (1985). *Environmental engineering.* New York: McGraw-Hill.

Pielou, E. C. (1998). *Fresh water.* Chicago: University of Chicago Press.

Powell, J. W. (1904). *Twenty-second annual report of the Bureau of American Ethnology to the Secretary of the Smithsonian Institution, 1900–1901.* Washington, D.C.: Government Printing Office.

Spellman, F. R. (2003). *Handbook of water and wastewater treatment plant operations.* Boca Raton, FL: Lewis Publishers.

Turk, J., & Turk, A. (1988). *Environmental science* (4th ed.). Philadelphia: Saunders.

United States Geological Survey (USGS) (2004). *Estimated use of water in the United States in 2000.* Washington, DC: U.S. Geological Survey.

United States Geological Survey (USGS) (2006). *Water science in schools.* Washington, DC: U.S. Geological Survey.

CHAPTER 25

Stream Systems

God is the color of water. Water doesn't have a color.

—James McBride

Stream Genesis and Structure

Early in the spring on a snow- and ice-covered high alpine meadow, the water cycle continues. The cycle's main component, water, has been held in reserve—literally frozen, for the long dark winter months, but with longer, warmer spring days, the sun is higher, more direct, and of longer duration, and the frozen masses of water respond to the increased warmth. The melt begins with a single drop, then two, then more and more. As the snow and ice melts, the drops join a chorus that continues unending; they fall from their ice-bound lip to the bare rock and soil terrain below.

The terrain the snowmelt strikes is not like glacial till, the unconsolidated, heterogeneous mixture of clay, sand, gravel, and boulders, dug-out, ground-out, and exposed by the force of a huge, slow, and inexorably moving glacier. Instead, this soil and rock ground is exposed to the falling drops of snowmelt because of a combination of wind and tiny, enduring force exerted by drops of water as they collide with the thin soil cover season after season, exposing the intimate bones of the Earth.

Gradually, the single drops increase to a small rush—they join to form a splashing, rebounding, helter-skelter cascade, and many separate rivulets that trickle and then run their way down the face of the granite mountain. At an indented ledge halfway down the mountain slope, a pool forms whose beauty, clarity, and sweet iciness provides the visitor with an incomprehensible, incomparable gift—a blessing from Earth.

The mountain pool fills slowly, tranquil under the blue sky, reflecting the pines, snow and sky around and above it, an open invitation to lie down and drink, and to peer into the glass-clear, deep waters, so clear that it seems possible to reach down over fifty feet and touch the very bowels of the mountain. The pool has no transition from shallow margin to depth; it is simply deep and pure. This pool is such a rarity in the modern world that, as it fills with more melt water, we wish to freeze time, to hold this place and this pool in its perfect state forever. But this cannot be. Mother Nature calls, prodding, urging. For a brief instant, the water laps in the breeze against the outermost edge of the ridge, then a trickle flows over the rim. The giant hand of gravity reaches out and tips the overflowing melt onward and it continues the downward journey, following the path of least resistance to its next destination, several thousand feet below.

When the overflow, still high in altitude but with its rock-strewn bed bent downward toward the sea, meets the angled, broken rocks below, it bounces, bursts, and mists its way against steep, V-shaped walls that form a small valley, carved out over time by water and the forces of the Earth.

Within the valley confines, the melt water has grown from drops to rivulets to a small mass of flowing water. It flows through what is at first a narrow opening, gaining strength, speed, and power as the V-shaped valley widens to form a U shape. The journey continues as the water mass picks up speed and tumbles over massive boulders, and then slows again.

At a larger but shallower pool, waters from hillsides, crevices, springs, rills, and mountain creeks at higher elevations have joined the main body. At the influent pool sides all appears peaceful, quiet, and restful, but not far away, at the effluent end of the pool, gravity takes control again. The overflow is flung over the jagged lip, and cascades downward several hundred feet, where the waterfall again brings its load to a violent, mist-filled meeting.

The water separates and joins again and again, forming a deep, furious, wild stream that calms gradually as it continues to flow over lands that are less steep. The waters widen into pools overhung by vegetation, surrounded by tall trees. The pure, crystalline waters have become progressively discolored on their downward journey, stained brown-black with humic acid, and literally filled with suspended sediments; the once-pure stream is now muddy.

The mass divides and flows in different directions, over different landscapes. Small streams divert and flow into open country. Different soils work to retain or speed the waters, and in some places the waters spread out into shallow swamps, bogs, marshes, fens, or mires. Other streams pause long enough to fill deep depressions in the land and form lakes. For a time, the water remains and pauses in its journey to the sea. But this is only a short-term pause, because lakes are only a short-term resting place in the water cycle. The water will eventually move on by evaporation or seepage into groundwater.

Other portions of the water mass stay with the main flow, and the speed of flow changes to form a river, which braids its way through the landscape, heading for the sea. As it changes speed and slows, the river bottom changes from rock and stone to silt and clay. Plants begin to grow, stems thicken, and leaves broaden. The river is now full of life and the nutrients needed to sustain life. But the river courses onward, its destiny met when the flowing rich mass slows at last and finally spills into the sea (Spellman & Whiting, 1998).

Streams

Streams are bodies of running water that carry rock particles (sediment loads) and dissolved ions and flow down slope along a clearly defined path, called a *channel.* Thus, streams may vary in width from a few inches to several miles. Streams are important for several reasons:

- Streams are an important part of the water cycle; they carry most of the water that goes from the land to the sea.
- Streams are one of the main transporters of sediment load from higher to lower elevations.
- Streams carry dissolved ions, the products of chemical weathering, into the oceans and thus make the sea salty.
- Streams (along with weathering and mass wasting) are a major part of the erosional process.
- Most population centers are located next to streams because they provide a major source of water and transportation.

Key Terms

Evapotranspiration (plant water loss) describes the process whereby plants lose water to the atmosphere during the exchange of gases necessary for photosynthesis. Water loss by evapotranspiration constitutes a major flux back to the atmosphere.

Infiltration capacity is the maximum rate soil can absorb rainfall.

Perennial stream is a type of stream in which flow continues during periods of no rainfall.

Gaining streams are typical of humid regions where groundwater recharges the stream.

Losing streams are typical of arid regions where streams can recharge groundwater.

Laminar flow occurs in a stream where parallel layers of water shear over one another vertically.

Turbulent flow occurs in a stream where complex mixing is the result.

Meandering is a stream condition whereby flow follows a winding and turning course.

Thalweg refers to the line of maximum water of channel depth in a stream.

Riffles refers to shallow, high-velocity flow over mixed gravel-cobble (bar-like) substrate.

Sinuosity refers to the bending or curving shape of a stream course.

Characteristics of Stream Channels

A standard rule of thumb states that flowing waters (rivers and streams) determine their own channels, and these channels exhibit relationships attesting to the operation of physical laws—laws that are not, as of yet, fully understood. The development of stream channels and entire drainage networks and the existence of various regular patterns in the shape of channels indicate that streams are in a state of dynamic equilibrium between erosion (sediment loading) and deposition (sediment deposit), and governed by common hydraulic processes. However, because channel geometry is four dimensional with a long profile, cross-section, and depth and slope profile, and because these mutually adjust over a time scale as short as years and as long as centuries or more, cause-and-effect relationships are difficult to establish. Other variables that are presumed to interact as the stream achieves its graded state include width and depth, velocity, size of sediment load, bed roughness, and the degree of braiding (sinuosity).

STREAM PROFILES

Mainly because of gravity, most streams exhibit a downstream decrease in gradient along their length. Beginning at the headwaters, the steep gradient becomes less so as one proceeds downstream, resulting in a concave longitudinal profile. Though diverse geography provides for almost unlimited variation, a lengthy stream that originates in a mountainous area (such as the one described in the chapter opening) typically comes into existence as a series of springs and rivulets; these coalesce into a fast-flowing, turbulent mountain stream, and the addition of tributaries results in a large and smoothly flowing river that winds through the lowlands to the sea.

When studying a stream system of any length, it becomes readily apparent almost from the start of such studies that a body of flowing water varies considerably from place to place along its length. For example, a common variable, the results of which can be readily seen, is that when discharge increases, it causes corresponding changes in the stream's width, depth, and velocity. In addition to physical changes that occur from location to location along a stream's course, there are a legion of biological variables that correlate with stream size and distance downstream. The most apparent and striking changes are in steepness of slope and in the transition from a shallow stream with large boulders and a stony substrate to a deep stream with a sandy substrate.

The particle size of bed material at various locations is also variable along the stream's course. The particle size usually shifts from an abundance of coarser material upstream to mainly finer material in downstream areas.

SINUOSITY

Unless forced by artificial means in the form of heavily regulated and channelized streams, straight channels are uncommon. Stream flow creates distinctive landforms composed of channels that are straight usually in appearance only, meandering, and braided; channel networks; and floodplains. Simply put, flowing water will follow a sinuous course. The most commonly used measure is the sinuosity index (SI). Sinuosity equals 1 in straight channels and more than 1 in sinuous channels.

$$SI = \frac{Channel\ distance}{Down\ valley\ distance}$$

Meandering is the natural tendency for alluvial channels and is usually defined as an arbitrarily extreme level of sinuosity, typically an SI greater than 1.5. Many variables affect the degree of sinuosity, however, and so SI values range from near unity in simple, well-defined channels to four in highly meandering channels (Gordon et al., 1992).

It is interesting to note that even in many natural channel sections of a stream course that appear straight, meandering occurs in the line of maximum water or channel depth (known as the *thalweg*). Keep in mind that streams have to meander because that is how they renew themselves. By meandering, they wash plants and soil from the land into their waters, and these serve as nutrients for the plants in the rivers. If rivers aren't allowed to meander, if they are channelized, the amount of life they can support gradually decreases. That means fewer fish, ultimately—and fewer bald eagle, herons, and other fishing birds (Spellman, 1996).

Meander flow follows a predictable pattern and causes regular regions of erosion and deposition. The streamlines of maximum velocity and the deepest part of the channel lie close to the outer side of each bend and cross over near the point of inflection between the banks. A huge elevation of water at the outside of a bend causes a helical flow of water toward the opposite bank. In addition, a separation of surface flow causes a back eddy. The result is zones of erosion and deposition, and explains why point bars develop in a downstream direction in depositional zones (Morisawa, 1968).

Did You Know?

Meandering channels can be highly convoluted or merely sinuous, but maintain a single thread in curves having definite geometric shape. Straight channels are sinuous but apparently random in occurrence of bends. Braided channels are those with multiple streams separated by bars and islands (Leopold, 1994).

BARS, RIFFLES, AND POOLS

Implicit in the morphology and formation of meanders are bars, riffles, and pools. Bars develop by deposition in slower, less competent flow on either side of the sinuous main stream. Onward moving water, depleted of bed load, regains competence and shears a pool in the meander, reloading the stream for the next bar. Alternating bars migrate to form riffles.

As stream flow continues along its course, a pool–riffle sequence is formed. Basically the riffle is a mound or hillock and the pool is a depression.

FLOODPLAIN

Stream channels influence the shape of the valley floor through which they course. This self-formed, self-adjusted flat area near the stream is the floodplain, which loosely describes the valley floor prone to periodic inundation during over-bank discharges. Valley flooding is a regular and natural behavior of the stream. Many people learn about this natural phenomenon the hard way—that is, whenever their farms, towns, streets, and homes become inundated by a river or stream that is doing nothing more than following its natural periodic cycle, conforming to the master plan designed by the master planner: Mother Nature.

Did You Know?

Floodplain rivers are found where regular floods form lateral plains outside the normal channel, which seasonally become inundated, either as a consequence of greatly increased rainfall or snow melt.

Water Flow in a Stream

Most elementary students learn early in their education process that water on Earth flows downhill (gravity) from land to the sea. However, they may or may not be told that water flows downhill toward the sea by various routes.

For the moment, the "route" (channel, conduit, or pathway) we are concerned with is the surface water route taken by surface runoff. Surface runoff is dependent on various factors. For example, climate, vegetation, topography, geology, soil characteristics, and land use determine how much surface runoff occurs compared with other pathways.

The primary source (input) of water to total surface runoff, of course, is precipitation. This is the case even though a substantial portion of all precipitation input returns directly to the atmosphere by evapotranspiration. Evapotranspiration is a combination process, as the name suggests, whereby water in plant tissue and in the soil evaporates and transpires to water vapor in the atmosphere.

Probably the easiest way to understand precipitation's input to surface water runoff is to take a closer look at this precipitation input.

Again, a substantial portion of precipitation input returns directly to the atmosphere by evapotranspiration. It is also important to point out that when precipitation occurs, some rainwater is intercepted, blocked, or caught by vegetation where it evaporates, never reaching the ground or being absorbed by plants. A large portion of the rainwater that reaches the surface on the ground, in lakes, and in streams also evaporates directly back to the atmosphere. Although plants display a special adaptation to minimize transpiration, plants still lose water to the atmosphere during the exchange of gases necessary for photosynthesis. Notwithstanding the large percentage of precipitation that evaporates, rain- or meltwater that reaches the ground surface follows several pathways in reaching a stream channel or groundwater.

Soil can absorb rainfall to its infiltration capacity (i.e., to its maximum intake rate). During a rain event, this capacity decreases. Any rainfall in excess of infiltration capacity accumulates on the surface. When this surface water exceeds the depression storage capacity of the surface, it moves as an irregular sheet of overland flow. In arid areas, overland flow is likely because of the low permeability of the soil. Overland flow is also likely when the surface is frozen or when human activities have rendered the land surface less permeable. In humid areas, where infiltration capacities are high, overland flow is rare.

In rain events where the infiltration capacity of the soil is not exceeded, rain penetrates the soil and eventually reaches the groundwater—from which it discharges to the stream slowly and over a long period. This phenomenon helps to explain why stream flow through a dry-weather region remains constant: The flow is continuously augmented by groundwater. This type of stream is known as a *perennial stream,* as opposed to an *intermittent* one, because the flow continues during periods of no rainfall.

Streams that course their way in channels through humid regions are fed water via the water table, which slopes toward the stream channel. Discharge from the water table into the stream accounts for flow during periods without precipitation, and also explains why this flow increases, even without tributary input, as one proceeds downstream. Such streams are called *gaining* or *effluent,* as opposed to *losing* or *influent* streams that lose water into the ground. It is interesting to note that the same stream can shift between gaining and losing conditions along its course because of changes in underlying strata and local climate.

Stream Water Discharge

The current velocity (speed) of water in a channel, which is driven by gravitational energy, varies considerably within a stream's cross-section owing to friction with the bottom and sides, with sediment, with obstructions (rocks and logs, etc.) and the atmosphere, and to sinuosity (bending or curving). Highest velocities, obviously, are found where friction is least, generally at or near the surface and near the center of the channel. In deeper streams current velocity is greatest just below the surface because of the friction with the atmosphere; in shallower streams current velocity is greatest at the surface because of friction with the bed. Velocity decreases as a function of depth, approaching zero at the substrate surface. A general and convenient rule of thumb is that the deepest part of the channel occurs where the stream velocity is the highest. Additionally, both width and depth of a stream increase downstream because discharge (the amount of water passing any point in a given time) increases downstream. As discharge increases, the cross-sectional shape will change, with the stream becoming deeper and wider. Velocity is important to discharge because *discharge* (m³/sec) = *cross-sectional area [width × average depth] (m²) × average velocity* (m/sec).

$$Q = A \times V$$

A stream is constantly seeking balance. This can be seen whenever the amount of water in a stream increases: the stream must adjust its velocity and cross-sectional area to reach balance. Discharge increases as more water is added through precipitation, tributary streams, or from groundwater seeping into the stream. As discharge increases, generally width, depth, and velocity of the stream also increase.

Transport of Material (Load)

Water flowing in a channel may exhibit laminar flow (parallel layers of water shear over one another vertically) or turbulent flow (complex mixing). In streams, laminar flow is uncommon, except at boundaries where flow is very low and in groundwater. Thus the flow in streams generally is turbulent. Turbulence exerts a shearing force that causes particles to move along the stream bed by pushing, rolling, and skipping, referred to as *bed load.* This same shear causes turbulent eddies that entrain particles in suspension (called the *suspended load*—particles size under 0.06 mm). Entrainment is the incorporation of particles when stream velocity exceeds the entraining velocity for a particular particle size.

Did You Know?

Entrainment is a natural extension of erosion and is vital to the movement of stationary particles in changing flow conditions. Remember, all sediments ultimately derive from erosion of basin slopes, but the immediate supply usually derives from the stream channel and banks, while the bed load comes from the streambed itself and is replaced by erosion of bank regions.

The entrained particles in suspension (suspended load) also include fine sediment, primarily clays, silts, and fine sands that require only low velocities and minor turbulence to remain in suspension. These are referred to as *wash load* (under 0.002 mm) because this load is "washed" into the stream from banks and upland areas (Gordon et al., 1992; Spellman, 1996).

Thus, the suspended load includes the wash load and coarser materials at lower flows. Together, the suspended load and bed load constitute the solid load. It is important to note that in bedrock streams the bed load will be a lower fraction than in alluvial streams where channels are composed of easily transported material.

A substantial amount of material is also transported as the dissolved load. Solutes (ions) are generally derived from chemical weathering of bedrock and soils, and their contribution is greatest in subsurface flows, and in regions of limestone geology.

The relative amount of material transported as solute rather than solid load depends on basin characteristics, lithology (i.e., the physical character of rock), and hydrologic pathways. In areas of very high runoff, the contribution of solutes approaches or exceeds sediment load, whereas in dry regions, sediments make up as much as 90% of the total load.

Deposition occurs when stream competence (i.e., the largest particle that can be moved as bed load and the critical erosion—competent—velocity is the lowest velocity at which a particle resting on the streambed will move) falls below a given velocity. Simply stated: The size of the particle that can be eroded and transported is a function of current velocity.

Sand particles are the most easily eroded. The greater the mass of larger particles (e.g., coarse gravel), the higher the initial current velocities must be for movement. However, smaller particles (silts and clays) require even greater initial velocities because of their cohesiveness and because they present smaller, stream-lined surfaces to the flow. Once in transport, particles will continue in motion at somewhat slower velocities than initially required to initiate movement, and will settle at still lower velocities.

Particle movement is determined by size, flow conditions, and mode of entrainment. Particles larger than 0.02 mm (medium-coarse sand size) tend to move by rolling or sliding along the channel bed as traction load. When sand particles fall out of the flow, they move by saltation or repeated bouncing. Particles smaller than 0.06 mm (silt) move as suspended load and particles smaller than 0.002 (clay) move indefinitely as wash load. A considerable amount of particle sorting takes place because of the different styles of particle flow in different section of the stream (Richards, 1982; Likens, 1984).

Unless the supply of sediments becomes depleted, the concentration and amount of transported solids increases. However, discharge is usually too low throughout most of the year to scrape, scour, or shape channels, or to move significant quantities of sediment in all but sand-bed streams, which can experience change more rapidly. During extreme events, the greatest scour occurs and the amount of material removed increases dramatically.

Sediment inflow into streams can both be increased and decreased as a result of human activities. For example, poor agricultural practices and deforestation greatly increase erosion.

Artificial structures such as dams and channel diversions can, on the other hand, greatly reduce sediment inflow.

References and Recommended Reading

Giller, P. S., & Jalmqvist, B. (1998). *The biology of streams and rivers.* Oxford: Oxford University Press.

Gordon, N. D., McMahon, T. A., & Finlayson, B. L. (1992). *Stream hydrology: An introduction for ecologists.* Chichester: John Wiley.

Leopold, L. B. (1994). *A view of the river.* Cambridge, Massachusetts: Harvard University Press.

Likens, W. M. (1984). Beyond the shoreline: A watershed ecosystem approach. *Vert Int Ver Theor Angew Liminol 22,* 1–22.

Morisawa, M. (1968). *Streams: Their dynamics and morphology.* New York: McGraw-Hill.

Richards, K. (1982). *Rivers: Form and processes in alluvial channels.* London: Mehuen.

Spellman, F. R. (1996). *Stream ecology and self-purification.* Lancaster, PA: Technomic.

Spellman, F. R., & Whiting, N. (1998). *Environmental science and technology.* Rockville, MD: Government Institutes.

U.S. Geological Survey (USGS) (2006). *Sinkholes.* Accessed 7/06/08 from http://ga.wwater.usgs.gov/edu/earthqwsinkholes.htm

CHAPTER 26

Lakes

As a lotus flower is born in water, grows in water and rises out of water to stand above it unsoiled, so I, born in the world, raised in the world having overcome the world, live unsoiled by the world.

—Buddha (563–483 BCE)

Lentic Habitat

When we look at a globe of the Earth, we quickly perceive that ours is a water planet. Water covers most of Earth's surface (about three quarters of it). With all of Earth's water, the irony is that if all of Earth's 325 trillion gallons were squeezed into a gallon container and we poured off what was not drinkable (polluted, salty, or frozen) we would be left with one drop. A small percentage of this drop represents all the freshwater contained in all of the Earth's lakes.

Lakes and ponds are found in many parts of the world and many of them are of great importance to human welfare. The total volume of freshwater contained in lakes is 91,000 cubic kilometers. More specifically, freshwater contained in lakes represent 0.007 percent of Earth's total water and 0.26 percent of Earth's total freshwater supply (Spellman, 2008).

Before briefly discussing the geology of lakes and ponds, it is important to be familiar with a few key terms.

- **Lake** may be defined as a body of standing surface-water runoff (and maybe some groundwater seepage) occupying a depression in the land.
- **Limnology** is the study of freshwater ecology, which is divided into two classes: lentic and lotic.

- **Lentic class** (calm zone) refers to lakes, ponds, and swamps. These are composed of four zones: littoral, limnetic, profundal, and benthic zones.
  - The littoral zone is the outermost shallow region of the lentic class, which has light penetration to the bottom.
  - The limnetic zone is the open-water zone of the lentic class to a depth of effective light penetration.
  - The euphotic zone refers to all lighted regions (light penetration) formed of the littoral and limnetic zones.
  - The profundal zone is a deep-water region beyond light penetration of the lentic class.
  - The benthic zone is the bottom region of a lake.
- **Lotic class** (washed) consists of rivers and streams and is composed of two zones: rapids and pools.
  - In the rapids zone, the stream velocity prevents sedimentation, with a firm bottom provided for organisms specifically adapted to live attached to the substrate.
  - The pool area is a deeper region with a velocity slow enough to allow sedimentation. The bottom is soft due to silts and settleable solids that cause lowered dissolved oxygen (DO) due to decomposition.
- **Pond** is smaller than a lake. Ponds may be natural or artificial in origin.
- **Stratified lake** can be divided into three horizontal layers: epilimnion (upper, usually oxygenated layer); mesolimnion or thermocline (middle layer of rapidly changing temperature); and hypolimnion (lowest layer, which is subject to deoxygenation).

Did You Know?

It's not that water that forms lakes get trapped, but that water entering a lake comes in faster than it can escape, either via outflow in a river, seepage into the ground, or by evaporation (U.S. Geological Survey, 2008).

Lakes and ponds range in size of just a few square feet (ponds are generally 2 to 8 hectares) to thousands of square miles. Scattered throughout the Earth, many of the first lakes evolved during the Pleistocene Ice Age. Lakes are found at all altitudes. Lake Titicaca (Peru and Chile) is 12,500 feet above sea level. At the other extreme, the Dead Sea in Israel and Jordan is almost 1,300 feet below sea level. Many ponds are seasonal, just lasting a couple of months, such as sessile pools, while lakes last many years (none—even the Great Lakes—will last forever).

Lakes and ponds are divided into four different "zones" usually determined by depth and distance from the shoreline. The four distinct zones—littoral, limnetic, profundal, and benthic—are shown in Figure 26.1. Miller (1998) points

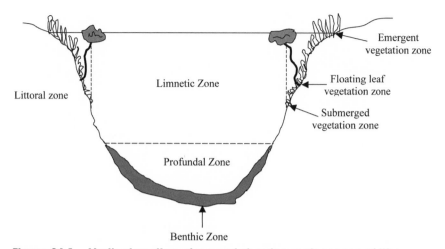

Figure 26.1.   Vertical section of a pond showing major zones of life.
Modified from Enger, Kormelink, Smith, and Smith, 1989, *Environmental Science: An Introduction.* Dubuque, Iowa: Wm. C. Brown Publications, p. 77

out that each zone provides a variety of ecological niches for different species of plant and animal life.

The littoral zone is the topmost zone near the shores of the lake or pond with light penetration to the bottom. It provides an interface zone between the land and the open water of lakes. This zone contains rooted vegetation such as grasses, sedges, rushes, water lilies and waterweeds, and a large variety of organisms. The littoral zone is further divided into concentric zones, with one group replacing the other as the depth of water changes. Figure 26.1 also shows these concentric zones: emergent vegetation, floating leaf vegetation, and submerged vegetation zones, proceeding from shallow to deeper water.

The littoral zone is the warmest zone since it is the area that light hits. It contains flora such as rooted and floating aquatic plants, and a very diverse community, which can include several species of algae (like diatoms), grazing snails, clams, insects, crustaceans, fish, and amphibians. The aquatic plants aid in providing support by establishing excellent habitats for photosynthetic and heterotrophic (requires organic food from the environment) microflora as well as many zooplankton and larger invertebrates (Wetzel, 1983).

From Figure 26.2 it can be seen that the limnetic zone is the open-water zone up to the depth of effective light penetration; that is, the open water away from the shore. The community in this zone is dominated by minute suspended organisms, the plankton, such as phytoplankton (plants) and zooplankton (animals), and some consumers such as insects and fish. Plankton are small organisms that can feed and reproduce on their own and serve as food for small chains.

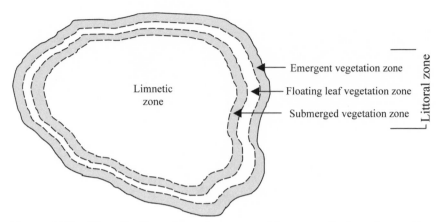

Emergent vegetation zone
Floating leaf vegetation zone
Submerged vegetation zone

Limnetic zone

Littoral zone

Figure 26.2. View looking down on concentric zones that make up the littoral zone.

Did You Know?

Without plankton in the water, there would not be any living organisms in the world, including humans.

In the limnetic zone, the population density of each species is quite low. The rate of photosynthesis is equal to the rate of respiration; thus, the limnetic zone is at compensation level. Small shallow ponds do not have this zone; they have a littoral zone only. When all lighted regions of the littoral and limnetic zones are discussed as one, the term *euphotic* is used for both, designating these zones as having sufficient light for photosynthesis and the growth of green plants to occur.

The small plankton do not live for a long time. When they die, they fall into the deep-water part of the lake or pond, the profundal zone. The profundal zone, because it is the bottom or deep-water region, is not penetrated by light. This zone is primarily inhabited by heterotrophs, adapted to its cooler, darker water and lower oxygen levels.

The final zone, the benthic zone, is the bottom region of the lake. It supports scavengers and decomposers that live on sludge. The decomposers are mostly large numbers of bacteria, fungi, and worms, which live on dead animal and plant debris and wastes that find their way to the bottom.

Classification of Lakes

It is our natural tendency to classify things. Kevern et al. (1999) classify lakes in three ways. One classification is based on productivity of the lake (or its relative richness). This is the trophic basis of classification. A second classification is based on the times during the year that the water of a lake becomes mixed and the extent to which the water is mixed. And a third classification is based on the fish community of lakes.

For the purpose of this text, a somewhat different classification scheme is used. That is, lakes are classified based on eutrophication, special types of lakes, and impoundments.

Eutrophication is a natural aging process that results in organic material being produced in abundance due to a ready supply of nutrients accumulated over time. Through natural succession (i.e., the process by which biological communities replace each other in a relatively predictable sequence), eutrophication cause a lake ecosystem to turn into a bog and eventually to a terrestrial ecosystem. Eutrophication has received a great amount of publicity lately. In recent years, humans have accelerated the eutrophication of many surface waters by the addition of wastes containing nutrients. This accelerated process is called *cultural eutrophication*. Sources of human wastes and pollution are sewage, agricultural runoff, mining, industrial wastes, urban runoff, leaching from cleared land, and landfills.

CLASSIFICATION OF LAKES BASED ON EUTROPHICATION

Lakes can be classified into three types based on their eutrophication stage.

1. **Oligotrophic lakes (few foods):** Oligotrophic lakes are young, deep, crystal-clear water, nutrient-poor lakes with little biomass productivity. Only a small quantity of organic matter grows in an oligotrophic lake; the phytoplankton, the zooplankton, the attached algae, the macrophytes (aquatic weeds), the bacteria, and the fish are all present as small populations. It's like planting corn in sandy soil: there won't be much growth (Kevern et al., 1999). Lake Superior is an example from the Great Lakes.

2. **Mesotrophic lakes:** It is hard to draw distinct lines between oligotrophic and eutrophic lakes, and often the term *mesotrophic* is used to describe a lake that falls somewhere between the two extremes. Mesotrophic lakes develop with the passage of time. Nutrients and sediments are added through runoffs, and the lake becomes more productive biologically. There is a great diversity of species with very low populations at first, but a shift toward higher and higher populations with fewer and fewer species. Sediments and solids contributed by runoffs and organisms make the lake shallower. At an advanced mesotrophic stage, a lake has undesirable odors and colors in certain parts. Turbidity increases and the bottom has organic deposits. Lake Ontario has reached this stage.

3. **Eutrophic lakes (good foods):** Eutrophic lakes have a large or excessive supply of nutrients. As the nutrients continue to be added, large algal blooms occur,

fish types change from sensitive to more pollution-tolerant ones, and biomass productivity becomes very high. Populations of a small number of species become very high. The lake takes on undesirable characteristics such as offensive odors, very high turbidity, and a blackish color. This high level of turbidity can be seen in studies of Lake Washington in Seattle, Washington. Laws (1993) reports that Secchi depth (measure of turbidity of water) measurements made in Lake Washington from 1950 to 1979 show an almost fourfold reduction in water clarity. Along with the reduction in turbidity, the lake becomes very shallow. Lake Erie is at this stage. Over time, a lake eventually becomes filled with sediments as it evolves into a swamp and finally into a land area.

SPECIAL TYPES OF LAKES

Odum (1971) refers to several special lake types.

1. **Dystrophic (like bog lakes):** Dystrophic lakes develop from the accumulation of organic matter from outside of the lake. In this case the watershed is often forested and there is an input of organic acids (e.g., humic acids) from the breakdown of leaves and evergreen needles. There follows a rather complex series of events and processes resulting finally in a lake that is usually low in pH (acid) and often is moderately clear, but color ranges from yellow to brown. Dissolved solids, nitrogen, phosphorus, and calcium are low and humic matter is high. These lakes are sometimes void of fish fauna; other organisms are limited. When fish are present, production is usually poor. They are typified by the bog lakes of northern Michigan.
2. **Deep ancient lakes:** Deep ancient lakes contain animals found nowhere else (endemic fauna). An example is Lake Baikal in Russia.
3. **Desert salt lakes:** Desert salt lakes are specialized environments like the Great Salt Lake in Utah where evaporation rates exceed precipitation rates, resulting in salt accumulation.
4. **Volcanic lakes:** Volcanic lakes occur on volcanic mountain peaks as in Japan and the Philippines.
5. **Chemically stratified lakes:** Examples of chemically stratified lakes include Big Soda Lake in Nevada. These lakes are stratified due to different densities of water caused by dissolved chemicals. They are *meromictic,* which means "partly mixed."
6. **Polar lakes:** Lakes in the polar regions have a surface water temperature mostly below 4° C.
7. **Marl lakes:** According to Kevern et al. (1999), Marl lakes are different in that they generally are very unproductive; yet, they may have summer-time depletion of dissolved oxygen in the bottom waters and very shallow Secchi

disk depths, particularly in the late spring and early summer. These lakes gain significant amounts of water from springs that enter at the bottom of the lake. When rainwater percolates through the surface soils of the drainage basin, the leaves, grass, and other organic materials incorporated in these soils are attacked by bacteria. These bacteria extract the oxygen dissolved in the percolating rainwater and add carbon dioxide. The resulting concentrations of carbon dioxide can get quite high and when they interact with the water, carbonic acid is formed.

As this acid-rich water percolates through the soils, it dissolves limestone. When such groundwater enters a lake through a spring, it contains very low concentrations of dissolved oxygen and is super-saturated with carbon dioxide. The limestone that was dissolved in the water reforms very small particles of solid limestone in the lake as the excess carbon dioxide is given off from the lake to the atmosphere. These small particles of limestone are marl and, when formed in abundance, cause the water to appear turbid, yielding a shallow Secchi disk depth. The low dissolved oxygen in the water entering from the springs produces low dissolved oxygen concentrations at the lake bottom.

IMPOUNDMENTS (SHUT-INS)

Impoundments are artificial lakes made by trapping water from rivers and watersheds. They vary in their characteristics according to the region and nature of drainage. They have high turbidity and a fluctuating water level. The biomass productivity, particularly of benthos, is generally lower than that of natural lakes (Odum, 1971).

References and Recommended Reading

Kevern, N. R., King, D. L., & Ring, R. (1999). Lake classification systems, part I, *The Michigan Riparian*, p. 1.

Laws, E. A. (1993). *Aquatic pollution: An introductory text.* New York: John Wiley & Sons, p. 59.

Miller, G. T. (1998). *Environmental science: An introduction.* Belmont, CA: Wadsworth, p. 77.

Odum, E. P. (1971). *Fundamentals of ecology.* Philadelphia: Saunders College Publishing, pp. 312–313.

Spellman, F. R. (2008). *The science of water* (2nd ed.). Boca Raton, FL: CRC.

U.S. Geological Survey (USGS) (2008). *Earth's water: Lakes and reservoirs.* Accessed 09/14/08 from http://ga.water.usgs.gov/edu/earthlakes.html.

Wetzel, R. G. (1983). *Limnology.* New York: Harcourt Brace Jovanovich College.

CHAPTER 27

Oceans and Their Margins

Blue, green, grey, white, or black; smooth, ruffled, or
mountainous; that ocean is not silent.

—H. P. Lovecraft

Oceans

Oceans are the storehouse of Earth's water. Oceans cover about 71% of Earth's
surface. The average depth of Earth's oceans is about 3,800 m with the greatest
ocean depth recorded at 11,036 m in the Mariana Trench. At the present time,
the oceans contain a volume of about 1.35 billion cubic kilometers (96.5% of
Earth's total water supply), but the volume fluctuates with the growth and melting of glacial ice.

Composition of ocean water has remained constant throughout geologic
time. The major constituents dissolving in ocean water (from rivers and precipitation and the result of weathering and degassing of the mantle by volcanic
activity) is composed of about 3.5 percent, by weight, of dissolved salts including
chloride (55.07%), sodium (30.62%), sulfate (7.72%), magnesium (3.68%),
calcium (1.17%), potassium (1.10%), bicarbonate (0.40%), bromine (0.19%),
and strontium (0.02%).

The most significant factor related to ocean water that everyone is familiar
with is the salinity of the water—how salty it is. *Salinity,* a measure of the amount
of dissolved ions in the oceans, ranges between 33 and 37 parts per thousand.
Often the concentration is the amount (by weight) of salt in water, as expressed
in parts per million (ppm). Water is saline if it has a concentration of more than
1,000 ppm of dissolved salts; ocean water contains about 35,000 ppm of salt (U.S.

Geological Survey [USGS], 2007). Chemical precipitation, absorption onto clay minerals, and plants and animals prevent seawater from containing even higher salinity concentrations. However, salinity does vary in the oceans because surface water evaporates, rain and stream water is added, and ice forms or thaws.

Did You Know?

Salinity is higher in mid-latitude oceans because evaporation exceeds precipitation. Salinity is also higher in restricted areas of the oceans like the Red Sea (up to 40 parts per thousand). Salinity is lower near the equator because precipitation is higher and is lower near the mouths of major rivers because of input of fresh water.

Along with salinity, another important property of seawater includes temperature. The temperature of surface seawater varies with latitude, from near 0° C near the poles to 29° C near the equator. Some isolated areas can have temperatures up to 37° C. Temperature decreases with ocean depth.

THE OCEAN FLOOR

The bottoms of the oceans' basins (ocean floors) are marked by mountain ranges, plateaus, and other relief features similar to (although not as rugged as) those on the land.

As shown in Figure 27.1, the floor of the ocean has been divided into four divisions: the continental shelf, continental slope, continental rise, and deep-sea floor or abyssal plain.

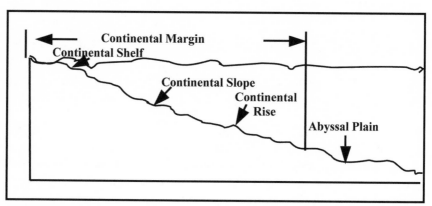

Figure 27.1. Cross section of ocean floor showing major elements of topography.

- **Continental shelf** is the flooded, nearly flat true margins of the continents. Varying in width to about 40 miles and a depth of approximately 650 feet, continental shelves slope gently outward from the shores of the continents (see Figure 27.1). Continental shelves occupy approximately 7.5% of the ocean floor.
- **Continental slope** is a relatively steep slope descending from the continental shelf (see Figure 27.1) rather abruptly to the deeper parts of the ocean. These slopes occupy about 8.5% of the ocean floor.
- **Continental rise** is a broad gentle slope below the continental slope containing sediment that has accumulated along parts of the continental slope.
- **Abyssal plain** is a sediment-covered deep-sea plain about 12,000–18,000 feet below sea level. This plane makes up about 42% of the ocean floor.

The deep ocean floor does not consist exclusively of the abyssal plain. In places there are areas of considerable relief. Among the more important such features are:

- **Seamounts:** isolated mountain-shaped elevations more than 3,000 feet high.
- **Mid-oceanic ridge:** submarine mountains, extending more than 37,000 miles through the oceans, generally 10,000 feet above the abyssal plain.
- **Trench:** a deep, steep-sided trough in an abyssal plain.
- **Guyots:** a seamount that is flat-topped and was once a volcano. They rise from the ocean bottom and usually are covered by 3,000 to 6,000 feet of water.

Ocean Tides, Currents, and Waves

Water is the master sculptor of Earth's surfaces. The ceaseless, restless motion of the sea is an extremely effective geologic agent. Besides shaping inland surfaces, water sculpts the coast. Coasts include sea cliffs, shores, and beaches. Seawater set in motion erodes cliffs, transports eroded debris along shores, and dumps it on beaches. Therefore, most coasts retreat or advance. In addition to the unceasing causes of motion—wind, density of sea water, and rotation of the Earth—the chief agents in this process are tides, currents, and waves.

TIDES

The periodic rise and fall of the sea (once every 12 hours and 26 minutes) produces the tides. Tides are caused by the gravitational attraction of the Moon and to a lesser extent, the sun on the Earth. The Moon has a larger effect on tides and causes the Earth to bulge toward the Moon. It is interesting to note that at the same time the Moon causes a bulge on Earth, a bulge occurs on the opposite side of the Earth due to inertial forces (further explanation is beyond the scope of this text). The effect of the tides is not too noticeable in the open sea, the difference between high and low tide amounting to about 2 feet. The tidal range may be considerably greater near shore, however. It may range from less than 2 feet to as much as 50 feet. The tidal range will vary according to the phase of the Moon and the distance of the Moon from the Earth. The type of shoreline and the physical configuration of the ocean floor also affects the tidal range.

CURRENTS

The oceans have localized movements of masses of seawater called *ocean currents*. These are the result of drift of the upper 50 to 100 m of the ocean due to drag by wind. Thus, surface ocean currents generally follow the same patterns as atmospheric circulation with the exception that atmospheric currents continue over the land surface, while ocean currents are deflected by the land. Along with wind action, current may also be caused by tides, variation in salinity of the water, rotation of the Earth, and concentrations of turbid or muddy water. Temperature changes in water affect water density which, in turn, causes currents—theses currents cause seawater to circulate vertically.

WAVES

Waves, varying greatly in size, are produced by the friction of wind on open water. Wave height and power depend on wind strength and *fetch*—the amount of unobstructed ocean over which the wind has blown. In a wave, water travels in loops. Essentially an up-and-down movement of the water, the diameter of the loops decreases with depth. The diameter of loops at the surface is equal to wave height (h). Breakers are formed when the wave comes into shallow water near the shore. The lower part of the wave is retarded by the ocean bottom, and the top, having greater momentum, is hurled forward causing the wave to break. These breaking waves may do great damage to coastal property as they race across coastal lowlands driven by winds or gale or hurricane velocities.

Coastal Erosion, Transportation, and Deposition

The geologic work of the sea, like previously discussed geologic agents, consist of erosion, transportation, and deposition. The sea accomplishes its work of coastal landform sculpting largely by means of waves and wave-produced currents; their effect on the seacoast may be quite pronounced. The coast and accompanying coastal deposits and landform development represent a balance between wave energy and sediment supply.

WAVE EROSION

Waves attack shorelines and erode by a combination of several processes. The resistance of the rocks composing the shoreline and the intensity of wave action to which it is subjected are the factors that determine how rapidly the shore will be eroded. Wave erosion works chiefly by hydraulic action, corrosion, and attrition. As waves strike a sea cliff, *hydraulic action* crams air into rock crevices, putting tremendous pressure on the surrounding rock; as waves retreat, the explosively expanding air enlarges cracks and breaks off chunks of rock (*scree*). Chunks hurled by waves against the cliff break off more scree via a sandpapering action—a process called *corrasion*. When the sea rubs and grinds rocks together, it forms a scree that is thrown into the cliffs, which reduces broken rocks to pebbles and sand grains. This process is called *attrition* (Lambert, 2007).

Several features are formed by marine erosion. Different combinations of wave action, rock type, and rock beds produce these features. Some of the more typical erosion-formed features of shorelines are discussed in the following list.

- **Sea cliffs or wave-cut cliffs** are formed by wave erosion of underlying rock followed by the caving-in of the overhanging rocks. As waves eat farther back inland, they leave a wave-cut beach or platform. Such cliffs are essentially vertical and are common at certain localities along the New England and Pacific coasts of North America.
- **Wave-cut benches** are the result of wave action not having enough time to lower the coastline to sea level. Because of the resistance to erosion, a relatively flat wave-cut bench develops. If subsequent uplift of the wave-cut bench occurs, it may be preserved above sea level as a wave-cut bench.
- **Headlands** are finger-like projections of resistant rock extending out into the water. Indentations between headlands are termed *coves*.
- **Sea caves, sea arches, and stacks** are formed by continued wave action on a sea cliff. Wave action hollows out cavities or caves in the sea cliffs. Eventually, waves may cut completely through a headland to form a sea arch; if the roof of the arch collapses, the rock left separated from the headland is called a *stack*.

MARINE TRANSPORTATION

Waves and currents are important transporting agents. Rip currents and undertow carry rock particles back to the sea, and long-shore currents will pick up sediments (some of it in solution), moving them out from shore into deeper water. Materials carried in solution or suspension may drift seaward for great distances and eventually be deposited far from shore. During the transportation process, sediments undergo additional erosion, becoming reduced in size.

MARINE DEPOSITION

Marine deposition takes place whenever currents and waves suffer reduced velocity. Some rocks are thrown up on the shore by wave action. Most of the sediments thus deposited consist of rock fragments derived from the mechanical weathering of the continents, and they differ considerably from terrestrial or continental deposits. Due to input of sediments from rivers, deltas may form. As a result of beach drift, such features as spits and hooks, bay barriers, and tombolos may form. Depositional features along coasts are discussed in the following list.

- **Beaches** are transitory coastal deposits of debris that lie above the low-tide limit in the shore zone.
- **Barrier islands** are long, narrow accumulations of sand lying parallel to the shore and separated from the shore by a shallow lagoon.
- **Spits and hooks** are elongated, narrow embankments of sand and pebble extending out into the water but attached by one end to the land.
- **Tombolos** are bars of sand or gravel connecting an island with the mainland or another island.
- **Wave-built terraces** are structures built up from sediments deposited in deep water beyond a wave-cut terrace.
- **Deltas** form where sediment supply is greater than ability of waves to remove sediment.

References and Recommended Reading

Amos, W. H. (1969). *Limnology: An introduction to the fresh water environment.* Chestertown, MD: LaMotte Company.

Davis, M. L., & Cornwell, D. A. (1991). *Introduction to environmental engineering.* New York: McGraw-Hill.

Environmental Protection Agency (EPA) (2011). *Classifying lakes and ponds.* Accessed 06/07/11 from http://www.epa.gov/bioweb1/aquatic/classify.html.

Environmental Protection Agency (EPA). (1989). Surface water treatment regulations. *Federal Register,* title 54, part 124, June 29, p. 27486.

Gross, G. M. (1995). *Oceanography: A view of the earth.* Englewood Cliffs: Prentice-Hall.

Lambert D. (2007). *The field guide to geology.* New York: Checkmark Books.

Laws, E. A. (1993). *Aquatic pollution: An introductory text.* New York: John Wiley & Sons.

Masters, G. M. (1991). *Introduction to environmental engineering and science.* Englewood Cliffs, NJ: Prentice-Hall.

McGhee, T. J. (1991). *Water supply and sewerage.* New York: McGraw-Hill.

Nalco. (1988). *The Nalco water handbook* (2nd ed.). New York: McGraw-Hill.

Odum, E. P. (1971). *Fundamentals of ecology.* Philadelphia: Saunders College.

Pinet, P. R. (1996). *Invitation to oceanography.* St. Paul, MN: West.

Smith, R. L. (1974). *Ecology and field biology.* New York: Harper & Row Publishers.

Spellman, F. R. (1996). *Stream ecology and self-purification: An introduction for wastewater and water specialists.* Lancaster, PA: Technomic.

Tchobanoglous, G., & Schroeder, E. D. (1987). *Water quality.* Reading, MA: Addison-Wesley.

U.S. Geological Survey (USGS) (2007). *The water cycle: Water storage in oceans.* Accessed 7/11/08 from http://ga.water.usgs.gov/edu/watercycleoceans.html.

Watson, L. (1988). *The water planet: Celebration of the wonder of water.* New York: Crown.

Wetzel, R. G. (1983). *Limnology.* New York: Harcourt Brace Jovanovich College.

Part V

PEDOSPHERE

CHAPTER 28

Soil: Earth's Skin and Tissue

An oyster that went to bed x-million years ago,
tucked itself into a sand-bottom, yawned (so to speak)
and woke a mile high in the Grand Canyon of the Colorado.

—John Ciardi (1961)

The mountain, and the deep and gloomy wood,
Their colours and their forms, were then to me
An appetite; a feeling and a love,
That had no need of a remoter charm,
By thought supplied, nor any interest
Unborrowed from the eye.

—William Wordsworth (1798)

We take soil for granted. It's always been there, with the implied corollary that it will always be there. But where does soil come from?

Of course soil was formed and in a never-ending process it is still being formed. However, soil formation is a slow process—one at work over the course of millennia as mountains are worn away to dust through bare rock succession.

Any activity, human or natural, that exposes rock to air begins the process. Through the agents of physical and chemical weathering, through extremes of heat and cold, through storms and earthquake and entropy, bare rock is gradually broken, reduced, and worn away. As its exterior structures are exposed and weakened, plant life appears to speed the process along.

Lichens cover the bare rock first, growing on the rock's surface, etching it with mild acids and collecting a thin film of soil that is trapped against the rock and clings. This changes the conditions of growth so much that the lichens can no longer survive, and are replaced by mosses.

The mosses establish themselves in the soil trapped and enriched by the lichens, and collect even more soil (Figure 28.1a and 28.1b). They hold moisture to the surface of the rock, setting up another change in environmental conditions.

Well-established mosses hold enough soil to allow herbaceous plant seeds to invade the rock. Grasses and small flowering plants move in, sending out fine root systems that hold more soil and moisture and work their way into minute fissures in the rock's surface. More and more organisms join the increasingly complex community.

Weedy shrubs are the next invaders, with heavier root systems that find their way into every crevice (Figure 28.2). Each stage of succession affects the decay of the rock's surface and adds its own organic material to the mix. Over the course of time, mountains are worn away, eaten away to soil, as time, plants, and extremes of weather work on them.

The parent material, the rock, becomes smaller and weaker as the years, decades, centuries, and millennia go by, creating the rich, varied, and valuable mineral resource we call *soil*.

Figure 28.1a. Rock face with lichen and moss growth. Avalanche Trail Glacier National Park, Montana.
Photo by Frank R. Spellman

Figure 28.1b. Moss-covered rock. Columbia River Basin, Oregon.
Photo by Frank R. Spellman

Figure 28.2. Plant life growing in lava bed. Craters of the Moon National Park, Idaho.
Photo by Frank R. Spellman

Soil: What Is It?

Perhaps no term causes more confusion in communication between various groups of average persons, soil scientists, soil engineers, and Earth scientists than the word *soil* itself. In simple terms, *soil* can be defined as the topmost layer of decomposed rock and organic matter, which usually contains air, moisture, and nutrients and can therefore support life. Most people would have little difficulty in understanding and accepting this simple definition. Then why are various groups confused about the exact meaning of the word *soil?* Quite simply, confusion reigns because soil is not simple—it is quite complex. In addition, the term *soil* has different meanings to different groups. Like *pollution*, the exact definition of *soil* is a personal judgment call. Let's take a look at how some of these different groups view soil.

The average person seldom gives soil a thought because it usually doesn't directly affect their lives. They seldom think about soil as soil, but they might think of soil in terms of dirt.

First of all, soil is not dirt. Dirt is misplaced soil—soil where we don't want it, such as on our hands, clothes, automobiles, or floors. We try to clean up dirt and keep it out of our living environments.

Second, soil is too special to be called *dirt*, because soil is mysterious and, whether we realize it or not, essential to our existence. Because we think of it as common, we relegate soil to an ignoble position. As our usual course of action, we degrade it, abuse it, throw it away, contaminate it, and ignore it. We treat it like dirt; only feces hold a more lowly status.

Soil deserves better.

Why?

Because soil is not dirt. Moreover, it is not filth or grime or squalor. Soil is composed of clay, air, water, sand, loam, and organic detritus of former life forms. If water is Earth's blood and air is Earth's breath, then soil is its flesh and bone and marrow.

Soil scientists (or pedologists) are people interested in soils as a medium for plant growth. Their focus is on the upper meter or so beneath the land surface (the *weathering zone,* which contains the organic-rich material that supports plant growth) directly above the unconsolidated *parent material.* Soil scientists have developed a classification system for soils based on the physical, chemical, and biological properties that can be observed and measured in the soil.

Soils engineers are typically soils specialists who look at soil as a medium that can be excavated using tools. Soils engineers are not concerned with the plant-growing potential of a particular soil, but rather are concerned with a particular soil's ability to support a load. They attempt to determine through examination and testing a soil's particle size, particle-size distribution, and the plasticity of the soil.

Earth scientists (or geologists) have a view that typically falls between pedologists and soils engineers—they are interested in soils and the weathering processes as past indicators of climatic conditions and in relation to the geologic formation of useful materials ranging from clay deposits to metallic ores.

To gain new understanding of soil, go out to a plowed farm field, pick a handful of soil, and look at it very closely. What are you holding in your hand? Read the two descriptions that follow to gain a better understanding of what soil actually is and why it is critically important to us all.

1. A handful of soil is alive, a delicate living organism—as lively as an army of migrating caribou and as fascinating as a flock of egrets. Literally teeming with life of incomparable forms, soil deserves to be classified as an independent ecosystem, or, more correctly stated, as many ecosystems.
2. When we reach down and pick up a handful of soil, it should remind us (and maybe startle some of us) to the realization that without its thin, living soil layer, Earth is a planet as lifeless as our own Moon (Spellman & Whiting, 2006).

In our attempt to define soil and differentiate soil from dirt, in trying to make the point that soil is vital to all of us, that it is arguably the most valuable of all mineral resources on Earth, we would be hard-pressed to do so in a more succinct but complete fashion than to quote E. L. Konigsburg (1996), who states that soil "is the working layer of the earth" (p. 64).

All about Soil

Weekend gardeners tend to think of soil as the first few inches below the Earth's surface—the thin layer that needs to be weeded and that provides a firm foundation for plants. But the soil actually extends from the surface down to the Earth's hard, rocky crust. It is a zone of transition and, as in many of nature's transition zones, the soil is the site of important chemical and physical processes. In addition, because plants need soil to grow, it is arguably the most valuable of all the mineral resources on Earth (Beazley, 1992).

Before we begin a journey that takes us through the territory that is soil to examine soil from the micro to macro levels, we need to stop for a moment and discuss why, beyond the obvious, soil is so important to us, to our environment, to our very survival.

FUNCTIONS OF SOIL

We normally relate soil to our backyards, to farms, to forests, or to a regional watershed. We think of soil as the substance on which plants grow. Soils play other roles, though. They have five main functions important to us: (1) soil is a medium for plant growth, (2) soils regulate our water supplies, (3) soils are recyclers of raw materials, (4) soils provide a habitat for organisms, (5) soils are used as an engineering medium, and (6) soils provide materials. Let's take a closer look at each of the functions of soils.

Soil: A Plant Growth Medium

We are all aware of the primary function of soil: soil serves as a plant growth medium, a function that becomes more important with each passing day as Earth's population continues to grow. However, while true that soil is a medium for plant growth, soil is actually alive as well. Soil exists in paradox: we depend on soil for life, and at the same time, soil depends on life. Its very origin, its maintenance, and its true nature are intimately tied to living plants and animals. What does this mean? Let's take a look at how one renowned environmental writer, Rachel Carson, whose elegant prose brought this point to the forefront, explained this paradox.

> The soil community . . . consists of a web of interwoven lives, each in some way related to the others—the living creatures depending on the soil, but the soil in turn is a vital element of the earth only so long as this community within it flourishes (1962, p. 56).

The soil might say to us if it could, "Don't kill off the life within me and I will do the best I can to provide life that will help to sustain your life." What we have here is a tradeoff—one vitally important both to soil and to ourselves. Remember that most of Earth's people are tillers of the soil. The soil is their source of livelihood, and those soil tillers provide food for us all.

As a plant-growth medium, soil provides vital resources and performs important functions for the plant. To grow in soil, plants must have water and nutrients—soil provides these. To grow and to sustain growth, a plant must have a root system—soil provides pore spaces for roots. To grow and maintain growth, a plant's roots must have oxygen for respiration and carbon dioxide exchange and ultimate diffusion out of the soil—soil provides the air and pore spaces (the soil's ventilation system) for this. To continue to grow, a plant must have support—soil provides this support.

For growth to occur, a seed planted in a soil must be exposed to the proper amount of sunlight and the soil must provide nutrients through a root system that has space to grow, a continuous stream of water (it requires about 500 grams of water to produce 1 gram of dry plant material) for root nutrient transport and plant cooling, and a pathway for both oxygen and carbon dioxide transfer. Just as important, soil water provides the plant with its normal fullness or tension (turgor) it needs to stand—the structural support it needs to face the sun for photosynthesis to occur.

As well as the functions listed previously, soil is also an important moderator of temperature fluctuations. If you have ever dug in a garden on a hot summer day, you probably noticed that the soil was warmer (even hot) on the surface but much cooler just a few inches below the surface.

Soil: Regulator of Water Supplies

When we walk on land, few of us probably realize that we are actually walking across a bridge. This bridge in many areas transports us across a veritable ocean of water below us, deep—or not so deep—under the surface of the Earth.

Consider what happens to rain. Where does the rain water go? Some, falling directly over water bodies, becomes part of the water body again, but an enormous amount falls on land. Some of the water, obviously, runs off, always following the path of least resistance. In modern communities, storm-water runoff is a hot topic. Cities have taken significant steps to try to control runoff and send it where it can be properly handled to prevent flooding.

Let's take a closer look at precipitation and the "sinks" it "pours" into, then relate this usually natural operation to soil water. We begin with surface water, then move onto that ocean of water below the soil's surface: groundwater.

Surface water (water on the Earth's surface as opposed to subsurface water—groundwater) is mostly a product of precipitation: rain, snow, sleet, or hail. Surface water is exposed or open to the atmosphere and results from the movement of water on and just under the Earth's surface (overland flow). This overland flow is the same thing as surface runoff, which is the amount of rainfall that passes over the Earth's surface. Specific sources of surface water include rivers, streams, lakes, impoundments, shallow wells, rain catchments, and tundra ponds or muskegs (peat bogs).

Most surface water is the result of surface runoff. The amount and flow rate of surface runoff is highly variable. This variability stems from two main factors: (1) human interference (influences) and (2) natural conditions. In some cases, surface water runs quickly off land. Generally, this is undesirable (from a water resources standpoint) because it does not provide enough time for water to infiltrate into the ground and recharge groundwater aquifers. Other problems associated with quick surface water runoff are erosion and flooding. Probably the only good thing that can be said about surface water that quickly runs off land is that it does not have enough time (normally) to become contaminated with high mineral content. Surface water running slowly off land may be expected to have all the opposite effects.

Surface water travels over the land to what amounts to a predetermined destination. What factors influence how surface water moves? Surface water's journey over the face of the Earth typically begins at its drainage basin, sometimes referred to as its *drainage area, catchment,* or *watershed.* For a groundwater source, this is known as the *recharge area*—the area from which precipitation flows into an underground water source.

A surface water drainage basin is usually an area measured in square miles, acres, or sections, and if a city takes water from a surface water source, how large (and what lies within) the drainage basin is essential information for the assessment of water quality.

We know that water doesn't run uphill. Instead, surface water runoff (like the flow of electricity), follows along the path of least resistance. Generally speaking, water within a drainage basin will naturally (by the geological formation of the area) be shunted toward one primary watercourse (a river, stream, creek, or brook) unless some manmade distribution system diverts the flow.

Various factors directly influence the surface water's flow over land. The principal factors are:

1. Rainfall duration: Length of the rainstorm affects the amount of runoff. Even a light, gentle rain will eventually saturate the soil if it lasts long enough. Once the saturated soil can absorb no more water, rainfall builds up on the surface and begins to flow as runoff.
2. Rainfall intensity: The harder and faster it rains, the more quickly soil becomes saturated. With hard rains, the surface inches of soil quickly become inundated, and with short, hard storms, most of the rainfall may end up as surface runoff, because the moisture is carried away before significant amounts of water are absorbed into the Earth.
3. Soil moisture: If the soil is already laden with water from previous rains, the saturation point will be reached sooner than if the soil were dry. Frozen soil also inhibits water absorption: up to 100% of snow melt or rainfall on frozen soil will end up as runoff because frozen ground is impervious.
4. Soil composition: Runoff amount is directly affected by soil composition. Hard rock surfaces will shed all rainfall, obviously, but so will soils with heavy clay composition. Clay soils possess small void spaces that swell when wet. When the void spaces close, they form a barrier that does not allow additional absorption or infiltration. On the opposite end of the spectrum, course sand allows easy water flow-through, even in a torrential downpour.

5. Vegetation cover: Runoff is limited by ground cover. Roots of vegetation and pine needles, pine cones, leaves, and branches create a porous layer (sheet of decaying natural organic substances) above the soil. This porous "organic" sheet (ground cover) readily allows water into the soil. Vegetation and organic waste also act as a cover to protect the soil from hard, driving rains. Hard rains can compact bare soils, close off void spaces, and increase runoff. Vegetation and ground cover work to maintain the soil's infiltration and water-holding capacity. Note that vegetation and groundcover also reduce evaporation of soil moisture as well.

6. Ground slope: Flat land water flow is usually so slow that large amounts of rainfall can infiltrate the ground. Gravity works against infiltration on steeply sloping ground where up to 80% of rainfall may become surface runoff.

7. Human influences: Various human activities have a definite effect on surface water runoff. Most human activities tend to increase the rate of water flow. For example, canals and ditches are usually constructed to provide steady flow, and agricultural activities generally remove ground cover that would work to retard the runoff rate. On the opposite extreme, artificial dams are generally built to retard the flow of runoff.

Human habitations, with their paved streets, tarmac, paved parking lots, and buildings create surface runoff potential, since so many surfaces are impervious to infiltration. All these surfaces hasten the flow of water, and they also increase the possibility of flooding, often with devastating results. Because of urban increases in runoff, a whole new field (industry) has developed: storm-water management.

Paving over natural surface acreage has another serious side effect. Without enough area available for water to infiltrate the ground and percolate through the soil to eventually reach and replenish—recharge—groundwater sources, those sources may eventually fail, with devastating effects on local water supply.

Now let's shift gears and take a look at groundwater. Water falling to the ground as precipitation normally follows three courses. Some runs off directly to rivers and streams, some infiltrates to ground reservoirs, and the rest evaporates or transpires through vegetation. The water in the ground (groundwater) is "invisible," and may be thought of as a temporary natural reservoir (American Society for Testing and Materials, 1969; Spellman, 2008). Almost all groundwater is in constant motion toward rivers or other surface water bodies.

Groundwater is defined as water below the Earth's crust but above a depth of 2,500 feet. Thus, if water is located between the Earth's crust and the 2,500-foot level, it is considered usable (potable) fresh water. In the United States, it is estimated "that at least 50% of total available fresh water storage is in underground aquifers" (Kemmer, 1979).

In this text, we are concerned with that amount of water retained in the soil to ensure plant life and growth. Recall that earlier we stated that producing 1 gram of dry plant material requires about 500 grams of water. Note that about 5 grams of this water becomes an integral part of the plant. Unless rainfall is frequent, you don't have to be a rocket scientist to figure out that the ability of soil to hold water against the force of gravity is very important. Thus, one of the vital functions of soil is to regulate the water supply to plants.

Soil: Recycler of Raw Materials

Can you imagine what it would be like to step out into the open air and be hit by a stench that would not only offend your olfactory sense but could almost reach out and grab you? You look out upon the cluttered fields in front of your domicile and see nothing but stack upon stack upon stack of the sources of horrible, putrefied, foul, decaying, gagging, choking, retching stench. We are talking about plant and animal remains and waste, mountains of it, reaching toward the sky, surrounded by colonies of landing and spiting flies of all varieties. "Impossible," you say. Thankfully, in most cases you are right. However, if it were not for the power of the soil to recycle waste products, this scene or something like it might be imaginable. Putrid garbage would build a mountain toward the Moon. Of course this contingency is impossible because under these conditions, there would be no life to die and to stack up in the first place.

Soil is a recycler, probably the premier recycler on Earth. The simple fact is that if it were not for soil's incredible recycling ability, plants and animals would have run out of nourishment long ago. Soil recycles in other ways. For example, consider the geochemical cycles (i.e., the chemical interactions between soil, water, air, and life on Earth) in which soil plays a major role.

Soil possesses the incomparable ability and capacity to assimilate great quantities of organic wastes and turn them into beneficial organic matter (humus), then to convert the nutrients in the wastes to forms that can be utilized by plants and animals. In turn, the soil returns carbon to the atmosphere as carbon dioxide, where it again will eventually become part of living organisms through photosynthesis. Soil performs several different recycling functions, most of them good, some of them not so good.

Consider one recycling function of soil that might not be so good. Soils have the capacity to accumulate large amounts of carbon as soil organic matter, which can have a major effect on global change such as the greenhouse effect. Moreover, it is important that wastes be applied in appropriate amounts and not contain toxic and environmentally harmful elements or compounds that could poison soils, wastes, and plants.

Soil: Habitat for Soil Organisms

> Life not only formed the soil, but other living things of
> incredible abundance and diversity now exist within it; if
> this were not so the soil would be a dead and sterile thing
>
> —Rachel Carson, *Silent Spring,* 1962

One thing is certain: most soils are not dead and sterile things. The fact is, a handful of soil is an ecosystem. It may contain billions of organisms belonging to thousands of species. Table 28.1 lists a very few of these organisms. Obviously, communities of living organisms inhabit the soil. What is not so obvious is that they are as complex and intrinsically valuable as are those organisms that roam the land surface and waters of Earth.

Table 28.1. Soil Organisms (Representative Sample)

Microorganisms (protists)
    Bacteria
    Fungi
    Actinomycetes
    Algae
    Protozoa

Nonarthropod animals
    Nematodes
    Earthworms and potworms

Arthropod animals
    Springtails
    Mites
    Millipedes and centipedes
    Harvestman
    Ants
    Diplopoda
    Diptera
    Crustacea

Vertebrates
    Mice, moles, voles
    Rabbits, gophers, squirrels

Soil: An Engineering Medium

We usually think of soil as being firm and solid (solid ground, *terra firma*). As solid ground, soil is usually a good substrate on which to build highways and other structures. However, not all soils are firm and solid—some are not as stable as others. While construction of buildings and highways may be suitable in one location on one type of soil, it may be unsuitable in another location with different soil. To construct structurally sound, stable highways and buildings, construction on soils and with soil materials requires knowledge of the diversity of soil properties.

Note that working with manufactured building materials that have been engineered to withstand certain stresses and forces is much different than working with natural soil materials, even though engineers have the same concerns about soils as they do with artificial building materials (concrete and steel). It is much more difficult to make these predictions or determinations for soil's ability to resist compression, to remain in place, its bearing strength, shear strength, and stability, than it is to make the same determinations for manufactured building materials.

Soil: Source of Materials

In addition to providing valuable minerals for various purposes, soil is commonly used to provide road building and dam construction materials.

CONCURRENT SOIL FUNCTIONS

According to the U.S. Department of Agriculture (USDA, 2009), soils perform specific critical functions no matter where they are located, and they perform more than one function at the same time, as described in the following list.

- Soils act like *sponges,* soaking up rainwater and limiting runoff. Soils also affect groundwater recharge and flood-control potentials in urban areas.
- Soils act like *faucets,* storing and releasing water and air for plants and animals to use.
- Soils act like *supermarkets,* providing valuable nutrients as well as air and water to plants and animals. Soils also store carbon and prevent its loss into the atmosphere.
- Soils act like *strainers* or *filters,* filtering and purifying water and air that flow through them.
- Soils *buffer, degrade, immobilize, detoxify,* and *trap pollutants,* such as oil, pesticides, herbicides, and heavy metals, and keep them from entering groundwater supplies. Soils also store nutrients for future use by plants and animals above ground and by microbes within soils.

Soil Properties

Soil is the layer of bonded particles of sand, silt, and clay that covers the land surface of the Earth. Most soils develop in multiple layers. The topmost layer (topsoil) is the layer of soil moved in cultivation, and in which plants grow. This topmost layer is actually an ecosystem composed of both biotic and abiotic components—inorganic chemicals, air, water, decaying organic material that provides vital nutrients for plant photosynthesis, and living organisms. Below the topmost layer is the subsoil, the part of the soil below the plow level, usually no more than 1 meter in thickness. Subsoil is much less productive, partly because it contains much less organic matter. Below that is the parent material, the unconsolidated (and more or less chemically weathered) bedrock or other geologic material from which the soil is ultimately formed. The general rule of thumb is that it takes about 30 years to form 1 inch of topsoil from subsoil; it takes much longer than that for subsoil to be formed from parent material—the length of time depends on the nature of the underlying matter (Franck & Brownstone, 1992).

PHYSICAL PROPERTIES

From the soil pollution technologist's point of view regarding land conservation and methodologies for contaminated soil remediation through reuse and recycling, five major physical properties of soil are of interest. They are soil texture, slope, structure, organic matter, and soil color. Soil texture (the relative proportions of the various soil separates in a soil) is a given and cannot be easily or practically changed significantly. It is determined by the size of the rock particles (sand, silt, and clay particles) or the soil separates within the soil. The largest soil particles are gravel, which consist of fragments larger than 2 mm in diameter.

Particles between 0.05 and 2 mm are classified as sand. Silt particles range from 0.002 to 0.05 mm in diameter, and the smallest particles (clay particles) are less than 0.002 mm in diameter. Though clays are composed of the smallest particles, those particles have stronger bonds than silt or sand, though once broken apart, they erode more readily. Particle size has a direct effect on erodibility. Rarely does a soil consist of only one single size of particle—most are a mixture of various sizes.

The slope (or steepness of the soil layer) is another given, which is important because the erosive power of runoff increases with the steepness of the slope. Slope also allows runoff to exert increased force on soil particles, which breaks them apart more readily and carries them farther away.

Soil structure (tilth) should not be confused with soil texture—they are different concepts. In fact, in the field, the properties determined by soil texture may be considerably modified by soil structure. Soil structure refers to the combination or arrangement of primary soil particles into secondary particles (units or peds). Simply stated, soil structure refers to the way various soil particles clump together. The size, shape, and arrangement of clusters of soil particles called *aggregates* naturally form larger clumps called *peds*. Sand particles do not clump because sandy soils lack structure. Clay soils tend to stick together in large clumps. Good soil develops small, friable (easily crumbled) clumps. Soil develops a unique, fairly stable structure in undisturbed landscapes, but agricultural practices break down the aggregates and peds, lessening erosion resistance.

The presence of decomposed or decomposing remains of plants and animals (organic matter) in soil helps not only fertility but also soil structure, especially the soil's ability to store water. Live organisms such as protozoa, nematodes, earthworms, insects, fungi, and bacteria are typical inhabitants of soil. These organisms work to either control the population of organisms in the soil or to aid in the recycling of dead organic matter. All soil organisms in one way or another work to release nutrients from the organic matter, changing complex organic materials into products that can be used by plants.

Just about anyone who has looked at soil has probably noticed that soil color is often different from one location to another. Soil colors range from very bright to dull grays, to a wide range of reds, browns, blacks, whites, yellows, and even greens. Soil color depends primarily on the quantity of humus and the chemical form of iron oxides present.

Soil scientists use a set of standardized color charts (the *Munsell Color Book*) to describe soil colors. They consider three properties of color—hue, value, and chroma—in combination to come up with a large number of color chips to which soil scientists can compare the color of the soil being investigated.

SOIL SEPARATES

As pointed out in the previous section, soil particles have been divided into groups based on their size termed *soil separates*—sand, silt, and clay—by the International Soil Science Society System, the United States Public Roads Administration, and the USDA. In this text, we use the classification established by the USDA. The size ranges in these separates reflect major changes in how the particles behave and in the physical properties they impart to soils.

In Table 28.2, the names of the separates are given, together with their diameters and the number of particles in 1 gram of soil, according to the USDA.

Sand ranges in diameter from 2 mm to 0.05 mm, and is divided into five classes (see Table 29.2). Sand grains are more or less spherical (rounded) in shape, with variable angularity, depending on the extent to which they have been worn down by abrasive processes such as rolling around by flowing water during soil formation.

Sand forms the framework of soil and gives it stability when in a mixture of finer particles. Sand particles are relatively large, which allows voids that form

Table 28.2. Characteristics of Soil Separates (USDA)

Separate	Diameter (mm)	Number of Particles/Gram
Very coarse sand	2.00–1.00	90
Coarse sand	1.00–0.50	720
Medium sand	0.50–0.25	5,700
Fine sand	0.25–0.10	46,000
Very fine sand	0.10–0.05	722,000
Silt	0.05–0.002	5,776,000
Clay	Below 0.002	90,260,853,000

between each grain to also be relatively large. This promotes free drainage of water and the entry of air into the soil. Sand is usually composed of a high percentage of quartz because it is most resistant to weathering and its breakdown is extremely slow. Many other minerals are found in sand, depending on the rocks from which the sand was derived. In the short term (on an annual basis), sand contributes little to plant nutrition in the soil. However, in the long term (thousands of years of soil formation), soils with a lot of weatherable minerals in their sand fraction develop a higher state of fertility.

Silt (essentially microsand), though spherically and mineralogically similar to sand is smaller—too small to be seen with the naked eye. It weathers faster and releases soluble nutrients for plant growth faster than sand. Too fine to be gritty, silt imparts a smooth feel (like flour) without stickiness. The pores between silt particles are much smaller than those in sand (sand and silt are just progressively finer and finer pieces of the original crystals in the parent rocks). In flowing water, silt is suspended until it drops out when flow is reduced. On the land surface, silt, if disturbed by strong winds, can be carried great distances and is deposited as loess.

The clay soil separate is, for the most part, much different from sand and silt. Clay is composed of secondary minerals that were formed by the drastic alteration of the original forms, or by the recrystallization of the products of their weathering. Because clay crystals are plate-like (sheeted) in shape, they have a tremendous surface area–to–volume ratio, giving clay a tremendous capacity to absorb water and other substances on its surfaces. Clay actually acts as a storage reservoir for both water and nutrients. There are many kinds of clay, each with different internal arrangements of chemical elements, which give them individual characteristics.

Soil Formation

Everywhere on Earth's land surface is either rock formation or exposed soil. When rocks formed deep in the Earth are thrust upward and exposed to the

Earth's atmosphere, the rocks adjust to the new environment, and soil formation begins. Soil is formed as a result of physical, chemical, and biological interactions in specific locations. Just as vegetation varies among biomes, so do the soil types that support that vegetation. The vegetation of the tundra and rain forest differ vastly from each other and from vegetation of the prairie and coniferous forest; soils differ in similar ways.

In the soil-forming process, two related—but fundamentally different—processes are occurring simultaneously. The first is the formation of soil parent materials by weathering of rocks, rock fragments, and sediments. This set of processes is carried out in the zone of weathering. The end point is to produce parent material for the soil to develop in, and is referred to as *C-horizon material* (see Figure 28.3). It applies in the same way for glacial deposits as for rocks. The second set of processes is the formation of the soil profile by soil forming processes, which gradually changes the C-horizon material into A, E, and B horizons. Figure 28.3 illustrates two soil profiles, one on hard granite and one on a glacial deposit.

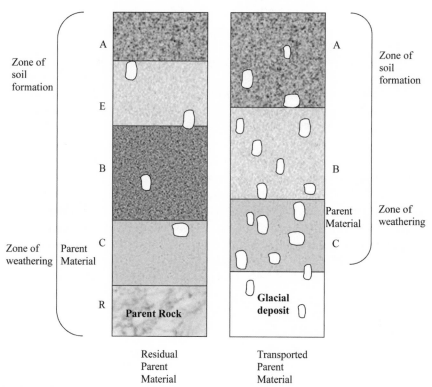

Figure 28.3.   Soil profiles on residual and transported parent materials.

WEATHERING

As mentioned in Chapter 18, soil development takes time and is the result of two major processes: weathering and morphogenesis (morphogenesis was described earlier as bare-rock succession). *Weathering* (the breaking down of bedrock and other sediments that have been deposited on the bedrock by wind, water, volcanic eruptions, or melting glaciers) happens either physically or chemically, or via a combination of both. Weathering is the first step in the erosion process; again, it causes the breakdown of rocks, either to form new minerals that are stable on the surface of Earth, or to break the rocks down to smaller particles. Simply, *weathering* (which projects itself on all surface material above the water table) is the general term used for all the ways in which a rock may be broken down.

Soil Characterization

> Classification schemes of natural objects seek to organize knowledge so that the properties and relationships of the objects may be most easily remembered and understood for some specific purpose. The ultimate purpose of soil classification is maximum satisfaction of human wants that depend on use of the soil. This requires grouping soils with similar properties so that lands can be efficiently managed for crop production. Furthermore, soils that are suitable or unsuitable for pipelines, roads, recreation, forestry, agriculture, wildlife, building sites, and so forth can be identified.
>
> —H. D. Foth (1992)

When people become ill, they may go to a doctor to seek a diagnosis of the cause of the illness and, hopefully, a prognosis of how long before they feel well again.

What do diagnosis and prognosis have to do with soil? Quite a lot. The diagnostic techniques used by a physician to identify the causative factors leading to a particular illness are analogous to the soil practitioner who uses diagnostic techniques to identify a particular soil. Sound far-fetched? It shouldn't because it isn't. Soil scientists must be able to determine the type of soil they study or work with.

Determining the type of soil makes sense, but what does prognosis have to do with all this? Soil practitioners not only need to be able to identify or classify a soil type, but this information allows them to correctly predict how a particular pollutant will react or respond when spilled in that type of soil. The fate of the pollutant is important in determining the possible damage incurred to the environment—

soil, groundwater, and air—because, ultimately, a spill could easily affect all three. Thus, the soil practitioner must not only use diagnostic tools in determining soil type, but must also be familiar with the soil type to judge how a particular pollutant or contaminant will respond when spilled in the soil type.

Let's take a closer look at the genesis of soil classification. From the time humans first advanced from hunter-gatherer status to cultivators of crops, they noticed differences in productive soils and unproductive soils. The ancient Chinese, Egyptians, Romans, and Greeks all recognized and acknowledged the differences in soils as media for plant growth. These early soil classification practices were based primarily on texture, color, and wetness.

Soil classification as a scientific practice did not gain a foothold until the later 18th and early 19th centuries when the science of geology was born. Terms with an obvious geological connotation such as *limestone soils* and *lake-laid soils*, as well as *clayey* and *sandy soils,* came into being. The Russian scientist Dokuchaev was the first to suggest a generic classification of soils—that soils were natural bodies. Dokuchaev's classification work was then further developed by Europeans and Americans. The system is based on the theory that each soil has a definite form and structure (morphology) related to a particular combination of soil-forming factors. This system was used until 1960, when the USDA published *Soil Classification: A Comprehensive System.* This classification system places major emphasis on soil morphology and gives less emphasis to genesis or the soil forming factors as compared with previous systems. In 1975, *Soil Classification: A Comprehensive System* was replaced by *Soil Taxonomy* (USDA, 1999), which classifies objects according to their natural relationships. Soils are classified based on measurable properties of soil profiles.

Note that no clear delineation or line of demarcation can be drawn between the properties of one soil and those of another. Instead, a gradation (sometimes quite subtle, such as from one shade of white to another) occurs in soil properties as one moves from one soil to another. Brady and Weil (2007) point out that "the gradation in soil properties can be compared to the gradation in the wavelengths of light as you move from one color to another. The changing is gradual, and yet we identify a boundary that differentiates what we call 'green' from what we call 'blue'" (p. 58).

To properly characterize the primary characteristics of a soil, a soil must be identified down to the smallest three-dimensional characteristic sample possible. However, to accurately perform a particular soil sample characterization, a sampling unit must be large enough so that the nature of its horizons can be studied, and the range of its properties identified. The *pedon* (rhymes with "head-on") is this unit. The pedon is roughly polygonal in shape, and designates the smallest characteristic unit that can still be called a *soil.*

Because pedons occupy a very small space (from approximately 1 to 10 m²), they cannot be used as the basic unit for a workable field soil-classification system. To solve this problem, a group of pedons, termed a *polypedon,* is of sufficient size to serve as a basic classification unit (commonly called a *soil individual*). In the United States, these groupings have been called a *soil series.*

There is a difference between *a soil* and *the soil.* This difference is important in the soil classification scheme. A soil is characterized by a sampling unit (pedon), which as a group (polypedons) form a soil individual. The soil, on the other hand, is a collection of all these natural ingredients, and is distinguishable from other bodies such as water, air, solid rock, and other parts of the Earth's crust. By incorporating the difference between a soil and the soil, a classification system has been developed that is effective and widely used.

DIAGNOSTIC HORIZONS AND TEMPERATURE AND MOISTURE REGIMES

Soil taxonomy uses a strict definition of soil horizons called diagnostic horizons, which are used to define most of the orders. Two kinds of diagnostic horizons are recognized: surface and subsurface. The surface diagnostic horizons are called *epipedons* (Greek *epi,* "over"; *pedon,* "soil"). The epipedon includes the dark (organic rich) upper part of the soil, the upper eluvial horizons, or sometimes both. Those soils beneath the epipedons are called *subsurface diagnostic horizons.* Each of these layers is used to characterize different soils in soil taxonomy.

In addition to using diagnostic horizons to strictly define soil horizons, soil moisture regime classes can also be used. A *soil moisture regime* refers to the presence of plant-available water or groundwater at a sufficiently high level. The control section of the soil (ranging from 10 to 30 cm for clay and from 30 to 90 cm for sandy soils) designates that section of the soil where water is present or absent during given periods in a year. The control section is divided into sections: upper and lower portions. The upper portion is defined as the depth to which 2.5 cm of water will penetrate within 24 hours. The lower portion is the depth that 7.5 cm of water will penetrate.

Six soil moisture regimes are identified:

- Aridic—characteristic of soils in arid regions
- Xeric—characteristic of having long periods of drought in the summer
- Ustic—characteristic of soils with moisture generally high enough to meet plant needs during growing season
- Udic—common soil in humid climatic regions

- Perudic—an extremely wet moisture regime annually
- Aquic—soil saturated with water and free of gaseous oxygen

Table 28.3 lists the moisture regime classes and the percentage distribution of areas with different soil moisture regimes.

Table 28.3. Soil Moisture Regimes (Percent of Global Area Occupied by Each)

Moisture Regime	Percent of Soils
Aridic	35.9
Xeric	3.5
Ustic	18.0
Udic	33.1
Perudic	1.0
Aquic	8.3

Source: Adaptation from H. Eswaran (1993).

In soil taxonomy, several soil temperature regimes are also used to define classes of soils. Based on mean annual soil temperature, mean summer temperature, and the difference between mean summer and winter temperatures, soil temperature regimes are shown in Table 28.4.

Table 28.4. Soil Temperature Regimes (Percent of Global Areas Occupied by Each)

Soil Temperature Regimes (Mean Annual Temperature [°C])	Percent
Pergelic (0° C)	10.9
Cryic (0–8° C)	13.5
Frigid (0–8° C)	1.2
Mesic (8–15° C)	12.5
Thermic (15–22° C)	11.4
Hyperthermic (>22° C)	18.5
Isofrigid (0–8° C)	0.1
Isomesic (8–15° C)	0.3
Isothermic (15–22° C)	2.4
Isohyperthermic (>22° C)	26.0
Water (NA)	1.2
Ice (NA)	1.4

Source: Adaptation from H. Eswaran (1993).

The diagnostic horizons and moisture and temperature regimes just discussed are the main criteria used to define the various categories in soil taxonomy.

SOIL TAXONOMY

The U.S. Soil Conservation Service's soil classification system, Soil Taxonomy (which is based on measurable properties of soil profiles), places soils in categories (see Table 28.5). Let's take a closer look at each one of these categories.

Table 28.5. Subdivision of Soil Taxonomy Classification System (in Hierarchical Order)

Category	Number of Taxa
Order	11
Suborder	55
Great group	Approximately 230
Subgroup	Approximately 1,200
Family	Approximately 7,500
Series	Approximately 18,500 in U.S.

- **Order:** Soils are not too dissimilar in their genesis. There are 11 soil orders in soil taxonomy. The names and major characteristics of each soil order are shown in Table 28.6.
- **Suborder:** Contains 55 subdivisions of order that emphasize properties that suggest some common features of soil genesis.
- **Great group:** Diagnostic horizons are the major bases for differentiating approximately 230 great groups.
- **Subgroup:** Contains approximately 1,200 subdivisions of the great groups.
- **Family:** Contains approximately 7,500 soils with subgroups having similar physical and chemical properties.
- **Series:** Series is a subdivision of the family, and the most specific unit of the classification system. More than 18,000 soil series are recognized in the United States.

SOIL ORDERS

As stated earlier, 11 soil orders are recognized; they constitute the first category of the classification. The classification of the orders is illustrated in Table 28.6.

SOIL SUBORDERS

Soil orders are further divided into 55 suborders, based primarily on the chemical and physical properties that reflect either the presence or absence of water logging or genetic differences caused by climate and vegetation to give the class

Table 28.6.   Soil Orders (with Simplified Definitions)

Alfisol	Mild forest soil with gray to brown surface horizon, medium to high base supply (refers to amount of interchangeable cations that remain in soil), and a subsurface horizon of clay accumulation.
Andisol	Formed on volcanic ash and cinders, and lightly weathered.
Aridsol	Dry soil with pedogenic (soil forming) horizon, low in organic matter.
Entisol	Recent soil without pedogenic horizons.
Histosol	Organic (peat or bog) soil.
Inceptisol	Soil at the beginning of the weathering process with weakly differentiated horizons.
Mollisol	Soft soil with a nearly black, organic-rich surface horizon and high base supply.
Oxisol	Oxide-rich soil principally a mixture of kaolin, hydrated oxides, and quartz.
Spodosol	Soil that has an accumulation of amorphous materials in the subsurface horizons.
Ultisol	Soil with a horizon of silicate clay accumulation and low base supply.
Vertisol	Soil with high activity clays (cracking clay soil).

Source: USDA Soil Survey Staff, 1960, 1994.

the greatest genetic homogeneity. Thus, the aqualfs (formed under wet conditions) are "wet" (*aqu* for aqua); alfisols become saturated with water sometime during the year. The suborder names all have two syllables, with the first syllable indicating the order, such as *alf* for alfisol and *moll* for mollisol.

SOIL GREAT GROUPS AND SUBGROUPS

Suborders are divided into great groups. They are defined largely by the presence or absence of diagnostic horizons, and the arrangements of those horizons. Great group names are coined by prefixing one or more additional formative elements to the appropriate suborder name. More than 230 great groups are identified.

Subgroups are subdivisions of great groups. Subgroup names indicate to what extent the central concept of the great group is expressed. A *Typic fragiaqualf* is a soil that is typical for the fragiaqualf great group.

SOIL FAMILIES AND SERIES

The family category of classification is based on features that are important to plant growth such as texture, particle size, mineralogical class, and depth. Terms such as *clayey, sandy, loamy,* and others are used to identify textural classes. Terms used to describe mineralogical classes include *mixed, oxidic, carbonatic,* and others. For temperature classes, terms such as *hypothermic, frigid, cryic,* and others are used.

The soil series (subdivided from soil family) gets down to the individual soil, and the name is that of a natural feature or place near where the soil was first recognized. Familiar series names include Amarillo (Texas), Carlsbad (New Mexico), and Fresno (California). In the United States there are more than 18,000 soil series.

Soil Mechanics and Physics

Why does the Leaning Tower of Pisa lean? The tower leans because it was built on a nonuniform consolidation of a clay layer beneath the structure. This process is ongoing (by about 1/25 of an inch per year) and may eventually lead to failure of the tower.

The factors that caused the Leaning Tower of Pisa to lean (and affect using soil as foundational and building materials) are what this chapter is all about. The mechanics and physics of soil are important factors in making the determination as to whether a particular building site is viable for building. Simply put, these two factors can help to answer the question: will the soils present support buildings? This concerns us because wherever humans build, the opportunity for anthropogenic pollutants follows, and to clean up those pollutants we have to excavate below the surface of the soil again.

SOIL MECHANICS

The *mechanics* of soil are physical factors important to engineers because their focus is on soil's suitability as a construction material. Simply put, the engineer must determine the response of a particular volume of soil to internal and external mechanical forces. Obviously, this is important in determining the soil's suitability to withstand the load applied by structures of various types.

By studying soil survey maps and reports, and by checking with soil scientists and other engineers familiar with the region and the soil types of that region, an engineer can determine the suitability of a particular soil for whatever purpose. Conducting field sampling to ensure that the soil product possesses the soil characteristics for its intended purpose is also essential.

The soil characteristics important for engineering purposes include soil texture, kinds of clay present, depth to bedrock, soil density, erodibility, corrosivity, surface geology, plasticity, content of organic matter, salinity, and depth to seasonal water table. Engineers also need to know the soil's space and volume (weight-volume relationships), stress-strain, slope stability, and compaction. Because these concepts are also of paramount importance to determining the fate of materials that are carried through soil, we present these concepts in this section.

Weight-Volume or Space and Volume Relationships

As we mentioned earlier, all natural soil consists of at least three primary components (or phases), solid mineral particles, water, and air (within void spaces between the solid particles). Examining the physical relationships (for soils in particular) between these phases is essential (see Figure 28.4). For convenience and clarity, in Figure 28.4 the mass of soil is represented as a block diagram. Each phase shown in the diagram is a separate block, and each major component has been reduced to a concentrated commodity within a unit volume.

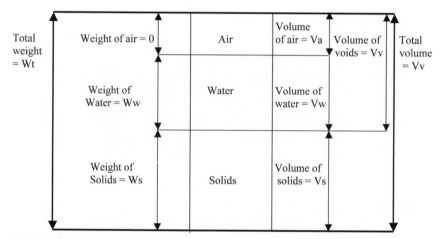

Figure 28.4. Weight/volume relationship of soil.

Note that the proportions of the components shown in Figure 28.4 will vary, sometimes widely, between and within various soil types. Remember that all water that is not chemically attached acts as a void filler. The relationship between free water and void spaces depends on available water (moisture).

The volume of the soil mass is the sum of the volumes of the three components, or

$$V_T = V_a + V_w + V_s \qquad (28.1)$$

The volume of the voids is the sum of V_a and V_w. However, the weight of the voids is only W_w, the weight of the water. Because weighing air in soil voids must be done within the Earth's atmosphere, the weight of air in the soil is factored in at zero. The total weight is expressed as the sum of the weights of the soil solids and water:

$$W_T = W_s + W_w \qquad (28.2)$$

The relationship between weight and volume can be expressed as

$$W_m = V_m G_m Y_w \qquad (28.3)$$

where

W_m = weight of the material (solid, liquid, or gas)
V_v = volume of the material
G_m = specific gravity of the material (dimensionless)
Y_w = unit weight of water

We can solve a few useful problems with the relationships described previously. More importantly, this information about a particular location's soil allows engineers to mechanically adjust the proportions of the three major components by reorienting the mineral grains by compaction or tilling. In remediation, a decision to blend soil types to alter the proportions (such as increasing or decreasing the percentage of void space) may be part of a site cleanup process.

Relationships between volumes of soil and voids are described by the *void ratio, e,* and *porosity,* η. We must first determine the void ratio, which is the ratio of the void volume to the volume of solids:

$$e = V_v / V_s \qquad (28.4)$$

The first step is to determine the ratio of the volume of void spaces to the total volume. We do this by determining the porosity (η) of the soil, which is the ratio of void volume to total volume. Porosity is usually expressed as a percentage.

$$\eta = V_v / V_t \times 100\% \qquad (28.5)$$

Two additional relationships can be developed from the block diagram in Figure 28.4. The terms *moisture content (w)* and *degree of saturation (S)* relate the water content of the soil and the volume of the water in the void space to the total void volume:

$$w = W_w/W_s \times 100\% \qquad (28.6)$$

and

$$S = V_w/V_v \times 100\% \qquad (28.7)$$

Soil Particle Characteristics

Size and shape of particles in the soil, along with density and other characteristics, provide information to the engineer on shear strength, compressibility, and other aspects of soil behavior. These index properties are used to create engineering classifications of soil. Simple classification tests are used to measure index properties (see Table 28.7) in the lab or the field.

Table 28.7.   Index Property of Soils

Soil Type	Index Property
Cohesive (fine-grained)	Water content
	Sensitivity
	Type and amount of clay
	Consistency
	Atterberg limits
Incohesive (coarse-grained)	Relative density
	In-place density
	Particle-size distribution
	Clay content
	Shape of particles

Source: Adaptation from A. E. Kehew (1995).

From the engineering point of view, the separation of the *cohesive* (fine-grained) from the *incohesive* (coarse-grained) soils is an important distinction. Let's take a closer look at these two terms.

A soil's level of cohesion describes soil particles' tendency to stick together. Cohesive soils contain silt and clay, which, along with water content, make these soils hold together through the attractive forces between individual clay and water particles. Because the clay particles so strongly influence cohesion, the index properties of cohesive soils are more complicated than the index properties of cohesionless soils. The soil's *consistency—the arrangement of clay particles—*describes the resistance of soil at various moisture contents to mechanical stresses or manipulations, and is the most important characteristic of cohesive soils.

Sensitivity (the ratio of unconfined compressive strength in the undisturbed state to strength in the remolded state is another important index property of cohesive soils. Soils with high sensitivity are highly unstable.

$$S_t = \frac{strength\ in\ undisturbed\ condition}{strength\ in\ remolded\ condition} \qquad (28.8)$$

As we described earlier, soil water content also influences soil behavior. Water content values of soil (the *Atterburg limits,* a collective designation of the so-called limits of consistency of fine-grained soils determined with simple laboratory tests) are usually presented as the liquid limit (LL), plastic limit (PL), and the plasticity index (PI). Note that plasticity is exhibited over a range of moisture contents referred to as *plasticity limits.* The PL is the lower water level at which soil begins to be malleable in the semisolid state, but while the molded pieces still crumble easily with a little applied pressure. When the volume of the soil becomes nearly constant with further decreases in water content, the soil reaches the shrinkage state.

The upper plasticity limit (or PL) is reached when the water content in a soil–water mixture changes from a liquid to a semifluid or plastic state and tends to flow when jolted. Obviously, a soil that tends to flow when wet presents special problems for both engineering purposes and remediation of contamination.

The range of water content over which the soil is plastic, called the *plasticity index,* provides the difference between the LL and the PL. Soils with the highest plasticity indices are unstable in bearing loads—a key point to remember.

The best known and probably the most useful system of the several systems designed for classifying the stability of soil materials is called the *Unified System of Classification.* This classification gives each soil type (14 classes) a two-letter designation based on particle-size distribution, LL, and PI.

Cohensionless coarse-grained soils are classified by index properties including the size and distribution of particles in the soil. Other index properties (including particle shape, in-place density, and relative density) are important in describing cohesionless soils because they relate how closely particles can be packed together.

Soil Stress

Just as water pressure increases as you go deep into water, the pressure within soil increases as the depth increases. A soil with a unit weight of 75 pounds per cubic feet exerts a pressure of 75 pounds per square inch (psi) at 1-foot depth and

225 psi at 3 feet. Of course, as the pressure on a soil unit increases, soil particles reorient themselves to support the cumulative load. This is critically important information to remember because a soil sample retrieved from beneath the load may not be truly representative once delivered to the surface. Representative samples are essential.

The response of a soil to pressure (*stress*), as when a load is applied to a solid object, is transmitted throughout the material (see Figure 28.5). The load puts the material under pressure, which equals the amount of load divided by the surface area of the external face of the object over which it is applied. The response to this pressure or stress is called *displacement* or *strain*. Stress (like pressure) at any point within the object can be defined as force per unit area.

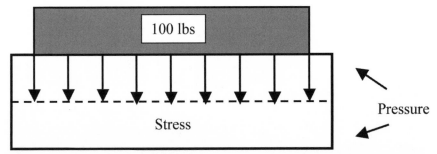

Figure 28.5. A 100-lb weight resting on a block causes pressure on the external surface of the block, and stress on the internal planes in the body.

Soil Compressibility

Compressibility (the tendency of soil to decrease in volume under load) is most significant in clay soils because of inherently high porosity. The actual evaluation process for these properties is accomplished in the *consolidation test*. This test subjects a soil sample to an increasing load. The change in thickness is measured after the application of each load increment.

Soil Compaction

Compaction reduces void ratio and increases the soil density, which affects how materials (including pollutants) travel through soil. Compaction is accomplished by working the soil to reorient the soil grains into a more compact state. Water content sufficient to lubricate particle movement is critical to obtaining efficient compaction.

SOIL FAILURE

Soil structural implications are involved with natural processes such as frost heave (which could damage a septic system and disturb improperly set footings, or shift soils under an improperly seated underground storage tank and its piping) as well as changes applied to soils during remediation efforts (e.g., when excavating to mitigate a hazardous materials spill). When soil cannot support a load, *soil failure* occurs, which can include events as diverse as foundation overload, collapse of the sides of an excavation, or slope failure on the sides of a dike or hill. Because of the safety factors involved, a soil's structural stability is critically important.

Classifying the type of soil to be excavated before an excavation can be accomplished includes determining the soil type. Stable rock, type A, type B, and type C soil are the classifications; finding a combination of soil types at an excavation site is common.

1. Stable rock is generally stable (may lose stability when excavated). It is natural solid mineral material that can be excavated with vertical sides and will remain intact while exposed.
2. Type A soil is the most stable soil. It includes clay, silty clay, sandy clay, clay loam, and sometimes silty clay loam and sandy clay loam.
3. Type B soil is moderately stable soil. It includes silt, silt loam, sandy loam, and sometimes silty clay loam and sandy clay loam.
4. Type C soil is the least stable soil. It includes granular soils like gravel, sand, loamy sand, and submerged soil, as well as soil from which water is freely seeping and unstable submerged rock.

Both visual and manual tests are used to classify soil for excavation.

Visual soil testing concerns soil particle size and type. In a mixture of soils, if the soil clumps when dug, it could be clay or silt. The presence of cracks in walls and spalling (the soil breaks up in chips or fragments) may indicate type B or C soil. Standing water or water seeping through trench walls classifies the soil as type C automatically.

Manual soil testing includes the sedimentation test, wet shaking test, thread test, and ribbon test. A sample taken from soil should be tested as soon as possible to preserve its natural moisture. Soil can be tested either on-site or off-site.

A *sedimentation test* determines how much silt and clay are in sandy soil. Saturated sandy soil is placed in a straight-sided jar with about 5 inches of water. After the sample is thoroughly mixed (by shaking it) and allowed to settle, the percentage of sand is visible. A sample containing 80% sand, for example, is classified as type C.

The *wet shaking test* is another way to determine the amount of sand versus clay and silt in a soil sample. This test is accomplished by shaking a saturated sample by hand to gauge soil permeability based on the following facts: (1) shaken clay resists water movement through it, and (2) water flows freely through sand and less freely through silt.

The *thread test* is used to determine cohesion (remember, cohesion relates to stability—how well the grains hold together). A representative soil sample is rolled between the palms of the hands to about 1/8 inch diameter and several inches in length. The rolled piece is placed on a flat surface, then picked up. If the sample holds together for 2 inches, it's considered cohesive.

The *ribbon test* is used as a backup for the thread test. It also determines cohesion. A representative soil sample is rolled out (using the palms of your hands) to 3/4 inch in diameter, and several inches in length. The sample is then squeezed between thumb and forefinger into a flat unbroken ribbon 1/8- to 1/4-inch thick, which is allowed to fall freely over the fingers. If the ribbon does not break off before several inches are squeezed out, the soil is considered cohesive.

Once soil has been properly classified, the necessary measures for safe excavation can be chosen, based on both soil classification and site restrictions. The two standard protective systems include sloping (or benching) and shoring (or shielding).

SOIL PHYSICS

As a dynamic, heterogeneous body, soil is nonisotropic—it does not have the same properties in all directions. Because soil properties vary directionally, various physical processes are always active in soil, as Windgardner (1996) makes clear: "all of the factors acting on a particular soil, in an established environment, at a specified time, are working from some state of imbalance to achieve a balance" (p. 63).

Soil practitioners must understand the factors involved in the physical processes that are active in soil. These include physical interactions related to soil water, soil grains, organic matter, soil gases, and soil temperature.

WATER AND SOIL

Water is not only a vital component of every living being, it is also essential to plant growth and to the microorganisms that live in the soil, and is important in the weathering process, which involves the breakdown of rocks and minerals to form soil and release plant nutrients. In this section we focus on soil water and its importance in soil. But first, we need to take a closer look at water—what it is and its physical properties.

Water exists as a liquid between 0° and 100° C (32°–212° F); as a solid at or below 0° C (32° F); and as a gas at or above 100° C (212° F). One gallon of water weighs 8.33 pounds (3.778 kg). One gallon of water equals 3.785 liters. One cubic foot of water equals 7.5 gallons (28.35 liters). One ton of water equals 240 gallons. One acre of water equals 43,560 cubic feet (325,900 gallons). Earth's rate of rainfall equals 340 cubic miles per day (16 million tons per second). Finally, water is dynamic (constantly in motion), evaporating from sea, lakes, and the soil; being transported through the atmosphere; falling to Earth; running across the land; and filtering downward into and through the soil to flow along rock strata.

Water: What Is It?*

Water is often assumed to be one of the simplest compounds known on Earth. But water is not simple—nowhere in nature is absolutely simple (pure) water to be found. Here on Earth, with a geologic origin dating back over 3 to 5 billion years, water found in even its purest form is composed of many constituents. Along with H_2O molecules, hydrogen (H^+), hydroxyl (OH^-), sodium, potassium and magnesium, other ions and elements are present.

Water contains additional dissolved compounds, including various carbonates, sulfates, silicates, and chlorides. Rain water (often assumed to be the equivalent of distilled water) is not immune to contamination, which it collects as it descends through the atmosphere. The movement of water across the face of land contributes to its contamination, taking up dissolved gases such as carbon dioxide and oxygen, and a multitude of organic substances and minerals leached from the soil.

SOIL PHYSICAL PROPERTIES

In soil physics, the physical properties (which are also a function of water's chemical structure) of water that concern us are density, viscosity, surface tension, and capillary action. Let's take a closer look at each of these physical properties.

Density is a measure of the mass per unit volume. The number of water molecules occupying the space of a unit volume determines the magnitude of the density. As temperature (which measures internal energy) increases or decreases, the molecules vibrate more or less strongly and frequently (which

---

*Much of the following information is adapted from Spellman, F. R. (2008). *The science of water* (2nd ed.). Boca Raton, FL: CRC.

changes the distance between them), expanding or diminishing the volume occupied by the molecules.

As discussed previously, liquid water reaches its maximum density at 4° C, and its minimum at 100° C. In soil science work, the density may be considered to be a unit weight (62.4 pounds per cubic foot or 1 gram/cubic centimeter).

Viscosity is the measure of the internal flow resistance of a liquid or gas. Stated differently, viscosity is ease of flow of a liquid, or the capacity of a fluid to convert energy of motion (kinetic energy) into heat energy. Viscosity is the result of the cohesion between fluid particles and interchange of molecules between layers of different viscosities. High viscosity fluids flow slowly, while low viscosity fluids flow freely. Viscosity decreases as temperature rises for liquids.

Have you ever wondered why a needle can float on water? Or why some insects can stand on water? The reason is *surface tension.* Surface tension (or cohesion) is the property that causes the surface of a liquid to behave as if it were covered with a weak elastic skin. It is caused by the exposed surface's tendency to contract to the smallest possible area because of unequal cohesive forces between molecules at the surface.

What does surface tension have to do with soil? The surface tension property of water markedly influences the behavior of water in soils. Consider an example you might be familiar with, which will help you understand surface tension and the other important physical properties of the water-soil interface. Water commonly rises in clays, fine silts, and other soils, and surface tension plays a major role. The rise of water through clays, silts, and other soils is termed *capillarity* or *capillary action* (the property of the interaction of the water with a solid), and the two primary factors of capillary rise are surface tension (cohesion) and adhesion (the attraction of water for the solid walls of channels through which it moves).

Why does the water rise? Because the water molecules are attracted to the sides of the tube (or soil pores) and start moving up the tube in response to this attraction. The cohesive force between individual water molecules ensures that water not directly in contact with the side walls is also pulled up the tube (or soil pores). This action continues until the weight of water in the tube counterbalances the cohesive and adhesive forces.

Keep in mind that for water in soil, the rate of movement and the rise in height of soil water are less than one might expect on the basis of soil pore size, because soil pores are not straight like glass tubes, nor are the openings uniform. Also, many soil pores are filled with air, which might prevent or slow down the movement of water by capillarity.

A final word on capillarity—keep in mind that *capillarity* means movement in any direction, not just upward. Since the attractions between soil pores and water are as effective with horizontal pores as with vertical ones, water movement in any direction occurs.

SOIL WATER

Have you ever wondered what happens to water after it enters the soil? The average person probably has not, but the subject is one about which soil scientists must have a full and complete understanding. Water that enters the soil has (in simple terms) four ways it might go:

1. It might move on through the soil and percolate out of the root zone, where it might eventually reach the water table.
2. It might be drawn back to the surface and evaporate.
3. It might be taken up (transpired, or used) by plants.
4. Finally, it might be "saved" in storage in the water profile.

What determines how much water ends up in each of these categories? It depends. Climate and the properties of the particular soil and the requirements of the plants growing in that soil all have an effect on how much water ends up in each of the categories. But don't forget the influence of anthropogenic actions (what we like to call "the heavy hand of humans")—people alter the movement of water, not only by irrigation and stream diversion practices and by building, but also by choosing which crops to plant and the types of tillage practices employed.

References and Recommended Reading

American Society for Testing and Materials (ASTM). (1969). *Manual on water.* Philadelphia: ASTM.

Beazley, J. D. 1992. *The way nature works.* New York: Macmillan.

Blumberg, L., & Gottlieg, R. (1989). *War on waste: Can America win its battle with garbage?* Washington, DC: Island Press.

Brady, N. C., & Weil, R. R. (1996). *The nature and properties of soils* (11th ed.). Upper Saddle River, NJ: Prentice-Hall.

Carson, R. (1962). *Silent spring,* Boston: Houghton Mifflin Company.

Ciardi, J. (1997). In the Stoneworks. In E. M. Cifelli (Ed.). *The Collected Poems of John Ciardi,* Fayetteville, AR: University of Arkansas Press.

Davis, G. H., & Pollock, G. L. (2003). Geology of Bryce Canyon National Park, Utah. In D. A. Sprinkel, et al (Eds.). *Geology of Utah's parks and monuments* (2nd ed.). Salt Lake City: Utah Geological Association.

Environmental justice. (1994). *Christian Science Monitor,* March 15.

Environmental Protection Agency (EPA) (1988). *The solid waste dilemma: An agenda for action—background document,* Washington, DC: EPA/530-SW-88-054A.

Environmental Protection Agency (EPA) (1989). *Decision-makers guide to solid waste management,* Washington, DC: EPA/530-SW89-072.

Environmental Protection Agency (EPA) (1993). *Characterization of municipal solid waste in U.S.: 1992 update*, Washington, DC: EPA/530-5-92-019.

Environmental Protection Agency (EPA). (2007). *Municipal solid waste generation, recycling, and disposal in the United States: Facts and figures for 2007.* Accessed 03/20/09 from www.epa.gov/osw.Eswaran, H. (1993). Assessment of global resources: Current status and future needs, *Pedologie 43*, pp. 19–39.

Foth, H. D. (1992). *Fundamentals of soil science* (6th ed.). New York: John Wiley and Sons.

Franck, I., & Brownstone, D. (1978). *The green encyclopedia.* New York: Prentice-Hall.

Kemmer, F. N. (n.d.). *Water: The universal solvent.* Oak Ridge, IL: NALCO Chemical Company.

Konigsburg, E. L. (1996). *The view from Saturday,* New York: Scholastic.

GLOBE Program (2003). Teacher's guide: Soil investigation chapter. Accessed 03/20/09 from http://archive.globe.gov/tctg/globetg.jsp.

MacKay, D., Shiu, W. Y., & Ma, K-C. (1997). *Illustrated handbook of physical-chemical properties and environmental fate for organic chemicals,* Boca Raton, FL: CRC Press/Lewis Publishers.

Morris, D. (1991). As if materials mattered. *The Amicus Journal 13*(4), 17–21.

Mowet, F. (1957). *The dog who wouldn't be.* New York: Willow Books.

National Park Service (NPS) (2008). *The hoodoo.* Washington, DC: NPS.

Oregon State University (2003). *Soil sampling for home gardens and small acreages.* Extension Service Bulletin. Reprinted April 2003. Accessed 03/20/09 from http://www.agcomm.ads.orst.edu.

O'Reilly, J. T. (1994). *State and local government solid waste management,* Deerfield, IL: Clark, Boardman, Callahan.

Peterson, C. (1987). Mounting garbage problem, *The Washington Post,* April 5.

Richards, B. (1988). Burning issue, *The Wall Street Journal,* June 16.

Spellman, F. R. (1998). *The science of environmental pollution.* Boca Raton: CRC Press.

Spellman, F. R. (2008). *The science of water.* Boca Raton, FL: CRC Press.

Spellman, F. R., & Whiting, N. E. (2006). *Environmental science and technology* (2nd ed.). Rockville, MD: Government Institutes.

Tchobanoglous, G., Theisen, H., & Vigil, S. (1993). *Integrated solid waste management: Engineering principles and management issues.* New York: McGraw-Hill.

Tomera, A. N. (1989). *Understanding basic ecological concepts,* Portland, ME: J. Weston Walch.

Tonge, P. (1987). All that trash, *Christian Science Monitor,* July 6.

U.S. Department of Agriculture (USDA) Soil Survey Staff (1975). *Soil classification: A comprehensive system.* Washington, DC: USDA Natural Resources Conservation Service.

U.S. Department of Agriculture (USDA) Soil Survey Staff (1994). *Keys to soil taxonomy,* Washington, DC: USDA Natural Resources Conservation Service.

U.S. Department of Agriculture (USDA) (1999). *Soil taxonomy: A basic system of soil classification for making and interpreting soil surveys* (2nd ed.). Washington, DC: USDA Natural Resources Conservation Service.

U.S. Department of Agriculture (USDA) (2009). *Urban soil primer.* Washington, DC: USDA.

Wolf, N., & Feldman, E. (1990). *Plastics: America's packaging dilemma,* Washington, DC: Island Press.

Part VI

BIOSPHERE

CHAPTER 29

Earth's Life-Forms and Interrelationships

We do not inherit the earth from our ancestors,
we borrow it from our children.

—Navajo Proverb

Earlier we mentioned that Earth is composed of shells beginning with the inner core and outer core and progressing outward to the mantle and finally the crust. We can also say that there is an additional shell, the absolute outer shell, called the *biosphere*. This absolute outer shell includes land, surface rocks, water, air, and the atmosphere. This is the shell where life occurs, and which biotic processes in turn alter or transform. Moreover, it is the shell that consists of the global ecological system. Within the ecological system, all living beings and their relationships, including their interaction with the elements of the lithosphere, hydrosphere, and atmosphere, are integrated.

Thanks to the life-giving qualities of air and water, Earth is populated by countless species of plants (plants make up over nine-tenths of the total biomass) and animals. Estimates indicate that the entire Earth contains over 75 billion tons of biomass (life—of which 0.6% is accounted for by humans), which lives within various environments comprising the biosphere (U.S. Census Bureau, 2006). Most of the planet's life is found from three meters below the ground to thirty meters above it and in the top 200 meters of the oceans and seas.

In regard to the life-forms that make up the biosphere, life as we know it is only known to exist on Earth. The origin of life is still a poorly understood process, but it is thought to have occurred about 3.9 to 3.5 billion years ago. Once life had appeared, the process of evolution by natural selection resulted in the development of ever-more diverse life-forms. Species that were unable to adjust to the changing environment and competition from other life-forms become extinct.

What Is Life?

Have you ever asked what life is? What does it mean to be alive? Have you ever tried to define life? If so, how did you define it? If these questions strike you as odd, consider them for a moment (they are almost as difficult to answer as to define the origin of life). Of course we all have an intuitive sense of what life is, but if you had difficulty, as is probably the case, with answering these questions, you are not alone. These questions are open to debate and have been from the beginning of time. One thing is certain: life is not a simple concept and it is difficult if not impossible to define.

Along with the impossibility of definitively classifying *life*, it is not always an easy thing to tell the difference between living, dead, and nonliving things. Prior to the 17th century, many people believed that nonliving things could spontaneously turn into living things. For example, it was believed that piles of straw could turn into mice. Obviously, that is not the case. There are some very general rules to follow when trying to decide if something is living, dead, or nonliving. Scientists have identified seven basic characteristics of life. Keep in mind that for something to be described as living, that something must display *all seven* of these characteristics (i.e., "characteristic" is plural). Although many of us have many different opinions about what *living* means, the following characteristics were designated as "characteristics of living things" with the consensus of the scientific community.

- **Living things are composed of cells:** Living things exhibit a high level of organization, with multicellular organisms being subdivided into cells, cells into organelles, organelles into molecules, and so on.
- **Living things reproduce:** All living organisms reproduce, either by sexual or asexual means.
- **Living things respond to stimuli:** All living things respond to stimuli in their environment.
- **Living things maintain homeostasis:** All living things maintain a state of internal balance in terms of temperature, pH, water concentrations, and other factors.
- **Living things require energy:** Some view life as a struggle to acquire energy (from sunlight, inorganic chemicals, or another organism), and release it in the process of forming adenosine triphosphate (ATP). The conventional view is that living organisms require energy, usually in the form ATP. They use this energy to carry out energy-requiring activities such as metabolism and locomotion.
- **Living things display heredity:** Living organisms inherit traits from the parent organisms that created them.

• **Living things evolve and adapt:** All organisms have the ability to adapt or adjust to their surroundings. An example of this is they might be adapting to environmental change, resulting in an increased ability to reproduce.

Again, if something follows one or just a few of the characteristics listed previously, it does not necessarily mean that it is living. To be considered alive, an object must exhibit *all* of the characteristics of living things. A good example of a nonliving object that displays at least one characteristic for living is sugar crystals growing on the bottom of a syrup dispenser. On the other hand, there is a stark exception to the characteristics listed previously. For example, mules cannot reproduce because they are sterile. Another nonliving object that exhibits many of the characteristics of life is a flame. Think about it. A flame:

• Respires
• Requires nutrition
• Reproduces
• Excretes
• Grows
• Moves
• Is irritable
• Is organized

We all know that a flame is not alive, but how do we prove that to the skeptic? The best argument we can make is:

1. Nonliving materials never replicate using deoxyribonucleic acid (DNA) and ribonucleic acid (RNA; hereditable materials).
2. Nonliving material cannot carry out anabolic metabolism.

LIFE: AN AFTER-THOUGHT

The reader, based on the previous section, should have little difficulty in understanding our dilemma in attempting to describe that which is indescribable. In our nebulous attempts and amorphous uses of symphonic prose to describe what life is, we often forget that life is nothing more than living. Thus, the analysis shifts from trying to define *life* to trying to define *living*. Maybe the title of the previous section should not have been "What Is Life?" but "What Is Living?"

What is *living*? The simple answer is that living is a state of mind. The complex answer is What? Maybe a few examples will help us clear the air:

Life is the chirp of a bird.
Life is an ill child who, by accumulated knowledge and skill, has been cured.

Life is the indescribable beauty of a flower.
Life is a high alpine meadow full of wildflowers in full bloom.
Life is that look-around at nature that generates a smile broader than an ocean.
Life is living—the taking of it all in.

Did You Know?

All four divisions of Earth can be and often are present in a single location. For example, a piece of soil will of course have mineral material from the lithosphere. Additionally, there will be elements of the hydrosphere present as moisture within the soil, the biosphere as insects and plants, and even the atmosphere as pockets of air between soil pieces.

Levels of Organization

All living things are organized into several basic levels of organization. The levels of organization from the smallest to the largest are:

- **Atoms:** basic units of matter having a specific chemical property
- **Molecules:** smallest multiatom-containing units having a specific chemical property
- **Macromolecule:** two or more molecules are bonded to form large molecules
- **Organelles:** include many subcellular structures, usually found only in eukaryotes
- **Cell:** basic unit of life
- **Tissue:** group of similar cells acting as a structural or functional unit
- **Organ:** group of dissimilar tissues that act together and have a specific function
- **Organ system:** group of dissimilar organs that together have a specific function
- **Organism:** contains cells and is capable of growing and replicating itself

In addition to the levels of organization for living things, there are also levels of organization in the biosphere. From the smallest to largest group these are:

- **Species:** a group of organisms that resemble one another closely
- **Population:** all organisms of the same species in a specific space and time
- **Community:** all populations in a specific space and time
- **Ecosystem:** community plus abiotic environment in a specific space and time
- **Biome:** a major regional or global biotic community, such as a grassland or desert, characterized chiefly by the dominant forms of plant life and the prevailing climate

- **Biosphere:** part of a planet's outer shell—including air, land, surface rocks, and water—within which life occurs, and which biotic processes in turn alter or transform

Biological Diversity

Biological diversity, or *biodiversity,* refers to the variety of genes, species, and ecosystems found on Earth.

The National Institutes of Health (NIH, 2006) makes the point that perhaps the best way to illustrate the importance of biodiversity is by analogy to the diversity of human knowledge stored in books (an argument made by Tom Lovejoy and others). When the library in Alexandria was consumed by fire in 391 CE, when Constantinople was sacked in 1453, or when Mayan codices were burned in the 16th century, thousands of works of literature were destroyed. Hundreds of works of genius are now known to us only by their titles or from quoted fragments. Thousands more will never be known; several millennia of collective human memory have been irretrievably lost.

Like books, living species represent a kind of memory, the cumulative record of several *million* millennia of evolution. Every species has encountered and survived countless biological problems in its evolutionary history; molecules, cells, and tissues record their solutions. Because we are biological beings ourselves, nature offers a vast library of solutions to many of our current health, environmental, and economic problems. Unfortunately, that precious and irretrievable information is now being destroyed at an unprecedented rate.

Fast-forward to the present. Population pressures and demographic changes threaten biodiversity worldwide. Our cities produce a growing stream of industrial poisons and human wastes. In the countryside, overproduction depletes the soil, and pesticides contaminate water supplies; deforestation for farming, pasture, and building material leads to erosion and heavy flooding. The resulting contraction of natural habitats—including the destruction of species-rich tropical forests—will have profound consequences for the future.

We have identified more than 2 million species of organisms on Earth, but estimate 40 million species inhabit the Earth. Some estimate that there may be millions of species in the tropical rain forest and an unspecified number living in the oceans yet undiscovered. Then there is this: Of the millions of species presently identified how many can we really understand, know, or accurately explain? When studying the various species, it is important to remember the words of Ralph Waldo Emerson: "What is a weed? A plant whose virtues have not yet been discovered."

Classification

For centuries, scientists classified the forms of life visible to the naked eye as either *animal* or *plant*. The Swedish naturalist Carolus Linnaeus organized much of the current knowledge about living things in 1735.

The importance of classifying organisms cannot be overstated. Without a classification scheme, how could we establish criteria for identifying organisms and arranging similar organisms into groups? The most important reason for classification is that a standardized system allows us to handle information efficiently—it makes the vastly diverse and abundant natural world less confusing.

Linnaeus's classification system was extraordinarily innovative. His *binomial system of nomenclature* is still with us today. Under the binomial system, all organisms are generally described by a two-word scientific name, the *genus* and *species*. Genus and species are groups that are part of a hierarchy of groups of increasing size, based on their nomenclature (taxonomy). This hierarchy is:

Kingdom
Phylum
Class
Order
Family
Genus
Species

Using this hierarchy and Linnaeus's binomial system of nomenclature, the scientific name of any organism (as stated previously) includes both the genus and the species name. The genus name is always capitalized, while the species name begins with a lowercase letter. On occasion, when little chance or confusion is present, the genus name is abbreviated to a single capital letter. The names are always in Latin, so they are usually printed in italics or underlined. Some organisms also have English common names. Some microbe names of interest, for example, are listed as follows:

* *Salmonella typhi*—the typhoid bacillus
* *Escherichia coli*—a coliform bacterium
* *Giardia lamblia*—a protozoan

✓ **Interesting Point:** *Escherichia coli* is commonly known as simply *E. coli*, while *Giardia lamblia* is usually referred to by only its genus name, *Giardia*.

Kingdoms of Life

Linnaeus classified all then-known (1700s) organisms into two large groups: the kingdoms Plantae and Animalia. In 1969, Robert Whittaker proposed five kingdoms: Monera, Protista, Fungi, Plantae, and Animalia. Other schemes involving an even greater number of kingdoms have lately been proposed; however, this text employs Whittaker's five kingdoms. Moreover, recent studies suggest that three domains (super-kingdoms) be employed: Archaea, Bacteria, and Eukarya.

The basic characteristics of each kingdom are summarized in the following:

1. **Kingdom Monera** (10,000 species) encompasses unicellular and colonial species, including *archaebacteria* (from the Greek meaning "ancient") and *eubacteria* (meaning "true")

   Archaebacteria include methanogens (producers of methane), halophiles (live in bodies of concentrated salt water), and thermocidophiles (live in the hot acidic waters of sulfur springs). Eubacteria include heterotrophs (decomposers), autotrophs (make food from photosynthesis), and proteobacteria (one of largest phyla of bacteria). All prokaryotic cells (without nuclei and membrane-bound organelles) are in this kingdom. Both reproduce by binary fission, but they do have some ways to recombine genes, allowing change (evolution) to occur.

2. **Kingdom Protista** (250,000 species) includes unicellular protozoans and unicellular and multicellular (macroscopic) algae with cilia and flagella. Kingdom Protista contains all eukaryotes that are not plants, animal, or fungi. Includes Amoebas and *Euglena*.

3. **Kingdom Fungi** (100,000 species) includes eukaryotic, multicellular, and heterotrophic species, having multinucleated cells enclosed in cells with cell walls. Fungi act either as decomposers or as parasites in nature. Includes molds, mildews, mushrooms, and yeast.

4. **Kingdom Plantae** (250,000 species) includes immobile, eukaryotic, multicellular species, and carry out photosynthesis (autotrophs) and have cells encased in cellulose cell walls. Plants are important sources oxygen, food, and clothing and construction materials, as well as pigments, spices, drugs, and dyes.

5. **Kingdom Animalia** (1,000,000) includes multicellular, eukaryotic, heterotrophic species, without photosynthetic pigment; mostly move from place to place. Animal cells have no cell walls.

✓ **Important Point:** Note that recent practice is to place archaebacteria in a separate kingdom, Kingdom Archaebacteria. This is the case because data from DNA and RNA comparisons indicate that archaebacteria are so different that they should not even be classified with bacteria. Thus, a separate and distinct classification scheme higher than kingdom has been devised to accommodate the archaebacteria, called *domain*. In this new system, these organisms are now placed in the domain *Archaea*—the chemosynthetic bacteria. Other prokaryotes, including eubacteria, are placed in the domain *Bacteria*—the disease-causing bacteria. All the kingdoms of eukaryotes, including Protista, Fungi, Plantae, and Animalia, are placed in the domain *Eukarya*.

A brief overview of selected organisms from each of the five major kingdoms is presented in the following sections.

Bacteria

Of all organisms, bacteria are the most widely distributed, the smallest in size, the simplest in morphology, the most difficult to classify, and the hardest to identify. Because of their considerable diversity, even providing a descriptive definition of a bacterial organism can be difficult. About the only generalization that can be made for the entire group is that they are single-celled, prokaryotic, and seldom photosynthetic, and they reproduce by binary fission.

Bacteria cells are usually measured in microns (μ) or micrometers (μm); 1 μm = 0.001 or 1/1000 of a millimeter (mm). A typical coliform bacterial cell that is rod-shaped is about 2 μm long and about 0.7 microns wide. The size of each cell changes with time during growth and death.

The arrangement of bacterial cells, when viewed under the microscope, may be seen as separate (individual) cells or as cells in groupings. Within their species, cells may appear in pairs (diplo), chains, groups of four (tetrads), cubes (Sarcinae), and in clumps. Long chains of cocci result when cells adhere after repeated divisions in one plane; this pattern is seen in the genera *Enterococcus* and *Lactococcus*. In the genus *Sarcina*, cocci divide in three planes, producing cubical packets of eight cells. The form of rod-shaped cells varies, especially the rod's end, which might be flat, cigar-shaped, rounded, or bifurcated. While many rods do occur singly, they might remain together after division to form pairs or chains. These characteristic arrangements are frequently useful in bacterial identification.

Bacteria are found everywhere in our environment. They are present in soil, in water, and in the air. Bacteria are also present in and on the bodies of all living creatures—including people. Most bacteria do not cause disease; they are not pathogenic. Many bacteria carry on useful and necessary functions, related to the life of larger organisms.

However, when we think about bacteria in general terms, we usually think of the damage they cause. In water, for example, the form of water pollution that poses the most direct menace to human health is bacteriological contamination, part of the reason that bacteria are of great significance to water and wastewater specialists. For water treatment personnel tasked with providing the public with safe, potable water, disease-causing bacteria pose a constant challenge (see Table 29.1).

Table 29.1.   Disease-Causing Bacterial Organisms Found in Polluted Water

Microorganism	Disease
Salmonella typhi	typhoid fever
Salmonella species	salmonellosis
Shigella species.	shigellosis
Campylobacter jejuni	campylobacter enteritis
Yersinia enterocolitica	yersiniosis
Escherichia coli	E. coli

The conquest of disease has placed bacteria high on the list of microorganisms of great interest to the scientific community. There is more to this interest and accompanying large research effort than just an incessant search for understanding and the eventual conquest of disease-causing bacteria. Not all bacteria are harmful to people. Some, for example, produce substances (antibiotics) that help in the fight against disease. Others are used to control insects that attack crops. Bacteria also have an effect on the natural cycle of matter. Bacteria work to increase soil fertility, which increases the potential for more food production. With the burgeoning world population, increasing future food productivity is no small matter.

We have a lot to learn about bacteria because we are still principally engaged in making observations and collecting facts, trying wherever possible to relate one set of facts to another, but lacking much of a basis for grand unifying theories. Like most learning processes, gaining knowledge about bacteria is a slow and deliberate process. With more knowledge about bacteria, we can minimize their harmful potential and exploit their useful activities.

Bacteria come in three shapes: elongated rods called *bacilli,* rounded or spherical cells called *cocci,* and spirals (helical and curved) called *spirilla* (the less rigid form) and *spirochaete* (those that are flexible). Elongated, rod-shaped bacteria may vary considerably in length; have square, round, or pointed ends; and may be motile (possess the ability to move) or nonmotile. The spherical-shaped bacteria may occur singly, in pairs, in tetrads, in chains, and in irregular masses. The helical and curved spiral-shaped bacteria exist as slender spirochaetes, spirillum, and bent rods.

Viruses

Viruses are parasitic intracellular particles that are the smallest living infectious agents known. They are not cellular—they have no nucleus, cell membrane, or cell wall. Viruses, like cells, carry genetic information encoded in their nucleic acid, and can undergo mutations and reproduce; however, they cannot carry out metabolism, and thus are not considered alive. They multiply only within living cells (hosts) and are totally inert outside of living cells, but can survive in the environment. Just a single virus cell can infect a host. As far as measurable size goes, viruses range from 20–200 millimicrons in diameter, about 1–2 orders of magnitude smaller than bacteria. More than 100 virus types excreted from humans through the enteric tract could find their way in to sources of drinking water. In sewage, these average between 100–500 enteric infectious units/1000 ml. If the viruses are not killed in various treatment processes and become diluted by a receiving stream, for example, to 0.1–1 viral infectious units/100 ml, the low concentrations make it very difficult to determine virus levels in water supplies. Since tests are usually run on samples of less than 1 ml, at least 1,000 samples would have to be analyzed to detect a single virus unit in a liter of water.

Viruses differ from living cells in at least three ways: (1) they are unable to reproduce independently of cells and carry out cell division; (2) they possess only one type of nucleic acid, either DNA or RNA; and (3) they have a simple cellular organization. Some viruses that may be transmitted by water include hepatitis A, adenovirus (a DNA virus that causes colds and "pink eye"), polio, Coxsackie virus, echo, and Norwalk agent. A virus that infects a bacterium is called a *bacteriophage.*

Lewis Thomas, in *The Lives of a Cell,* points out that when humans "catch diphtheria it is a virus infection, but not of us." That is, when humans are infected by the virus causing diphtheria, it is the bacterium that it really infects—humans simply "blundered into someone else's accident" (1974, p. 76). The toxin of diphtheria bacilli is produced when the organism has been infected by a bacteriophage.

✓ **Interesting Point:** The Papillomavirus is a DNA virus that causes warts. These infectious particles are small, about 15 nm in diameter.

A bacteriophage (phage) is any viral organism whose host is a bacterium. Most of the bacteriophage research that has been carried out has been on the bacterial *Escherichia coli,* one of the gram-negative bacteria that environmental specialists such as water and wastewater operators are concerned about because it is a dangerous typical coliform.

A virus does not have a cell-type structure from which it is able to metabolize or reproduce. However, when the *genome* (a complete haploid set of chromosomes) of a virus is able to enter into a viable living cell (a bacterium) it may "take charge" and direct the operation of the cell's internal processes. When this occurs, the genome, through the host's synthesizing process, is able to reproduce copies of itself, move on, and then infect other hosts. Hosts of a phage may involve a single bacterial species or several bacteria genera.

The most important properties used in classifying bacteriophages are nucleic acid properties and phage morphology. That is, viruses are classified by the type of nucleic acid they contain, and the shape of their protein capsule (capsid). Bacterial viruses might contain either DNA or RNA; most phages have double-stranded DNA.

✓ **Interesting Point**: The Influenza virus causes the flu. It has RNA as its genetic material instead of DNA.

Many different basic structures have been recognized among phages. Phages appear to show greater variation in form than any other viral group. The T-2 phage virus has two prominent structural characteristics: the head (a polyhedral capsid) and the tail.

The effect of phage infection depends on the phage and host and to a lesser extent on conditions. Some phages multiply with and *lyse* (destroy) their hosts. When the host lyses (dies and breaks open), phage progeny are released.

✓ **Important Point:** Bacteriophages invade the host cell, take over the cell, and begin replicating viruses, eventually lysing or bursting the host cell, releasing the new viruses to infect additional cells.

Viruses cause a variety of diseases among all groups of living organisms. Viral diseases include the flu, common cold, herpes, measles, chicken pox, small pox, and encephalitis. Antibiotics are not effective against viruses. Vaccination offers protection for uninfected individuals.

Protists

Protists are unicellular and multicellular eukaryotes that exhibit a great deal of variation in their life cycles. The Protists include heterotrophs, autotrophs, and some organisms that can vary their nutritional mode depending on environmental conditions. Protists occur in freshwater, saltwater, soil, and as symbionts within other organisms; they include protozoa, algae, and slime molds. Because of this tremendous diversity, classification of the Protista is difficult.

✓ **Important Point:** Protists are not plants, animals, or fungi, but they act enough like them that scientist believe protists paved the way for the evolution of early plants, animals, and fungi.

PROTOZOA

The *protozoa* ("first animals" or "little animals") are a large group of eukaryotic organisms (more than 50,000 known species that have adapted a form or cell to serve as the entire body). All protozoans are single-celled organisms. Typically they lack cell walls, but have a plasma membrane that is used to take in food and discharge waste. They can exist as solitary or independent organisms (the stalked ciliates such as *Vorticella* species, for example) or they can colonize like the sedentary *Carchesium* species. As the principal hunters and grazers of the microbial world, protozoa play a key role in maintaining the balance of bacterial, algal, and other microbial life. Protozoa are microscopic and get their name because they employ the same type of feeding strategy as animals. The animal-like protozoans differ from animals in that they are unicellular and do not have specialized tissues, organs, or organ systems that carry out life functions. Most are harmless, but some are parasitic. Some forms have two life stages: active *trophozoites* (capable of feeding) and dormant *cysts*.

✓ **Important Point:** Although they are efficient hunters and grazers and they feed on bacteria, they also eat other protozoa and bits of material that has come off of other living things (organic matter). Protozoans also are themselves an important food source for larger creatures and the basis of many food chains.

As mentioned, as unicellular eukaryotes, protozoa cannot be easily defined because they are diverse and, in most cases, only distantly related to each other. Also, again, protozoa are distinguished from bacteria by their eukaryotic nature and by their usually larger size. Protozoa are distinguished from algae because protozoa obtain energy and nutrients by taking in organic molecules, detritus, or other protists rather than from photosynthesis. Each protozoan is a complete organism and contains the facilities for performing all the body functions for which vertebrates have many organ systems.

Like bacteria, protozoa depend on environmental conditions (the protozoan quickly responds to changing physical and chemical characteristics of the environment), reproduction, and availability of food for their existence. Relatively large microorganisms, protozoans range in size from 4 microns to about 500 microns. They can both consume bacteria (limit growth) and feed on organic matter (degrade waste).

Interest in types of protozoa is high among water treatment practitioners because certain types of protozoans can cause disease. In the United States, the most important of the pathogenic parasitic protozoans is *Giardia lamblia,* which causes a disease known as *giardiasis.* Two other parasitic protozoans that carry waterborne disease are *Entamoeba histolytica* (amoebic dysentery) and *Cryptosporidium* (cryptosporidiosis).

Protozoa are divided into four groups based on their method of motility as shown in Table 29.2.

Table 29.2. Classification of Protozoans

Group	Common Name	Movement	Reproduction
Mastigophora	Flagellates	Flagella	Asexual
Ciliophora	Ciliates	Cili	Asexual by transverse fission
			Sexual by conjugation
Sarcodina	Amoebas	Pseudopodia	Asexual and sexual
Sporozoa	Sporozoans	nonmotile	Asexual and sexual

- **Mastigophora:** These protozoans are mostly unicellular, lack specific shape (have an extremely flexible plasma membrane that allows for the following movement of cytoplasm), and possess whip-like structures called *flagella.* The flagella, which can move in whip-like motion (to move in a relatively straight path, or create currents that spin them through fluids), are used for locomotion, as sense receptors, and to attract food.

  These organisms are common in both fresh and marine waters. The group is subdivided into the *phytomastigophora,* most of which contain chlorophyll and are thus plant-like. A characteristic species of *phytomastigophora* is the *Euglena* species, often associated with high or increasing levels of nitrogen and phosphate in the wastewater treatment process.

✓ **Interesting Point:** The mastigophora trypanosomes require two hosts, one a mammal, to complete their life cycle, and cause African sleeping sickness, Chagas disease, and leishmaniasis. Trichonymphs are symbionts inside the intestines of termites.

- **Ciliophora** (Ciliates): The ciliates are the most advanced and structurally complex of all protozoans. They are heterotrophic and use multiple small cilia for locomotion. The *Paramecium* is probably the most commonly studied ciliate in basic biology classes. Movement and food-getting is accomplished with short, hair-like structures called *cilia* that are present in at least one stage of the organism's life cycle. Three groups of ciliates exist: free-swimmers, crawlers, and stalked. The majority are free-living. They are usually solitary, but some are colonial and others are sessile. They are unique among protozoa in having two kinds of nuclei: a micronucleus and a macronucleus. The micronucleus is concerned with sexual reproduction. The macronucleus is involved with metabolism and the production of RNA for cell growth and function.

✓ **Interesting Point:** To increase strength of the cell boundary, ciliates have a *pellicle,* a sort of tougher membrane that still allows them to change shape.
    The ciliate pellicle may also act as thick armor. In other species, the pellicle may be very thin. The cilia are short and usually arranged in rows. Their structure is comparable to flagella except that cilia are shorter. Cilia may cover the surface of the animal or may be restricted to banded regions.

Like many freshwater protozoans, ciliates are hypotonic; however, removal of water crossing the cell membrane by osmosis is a problem. Therefore, one commonly employed mechanism is a contractile vacuole. Water is collected into the central ring of the vacuole and actively transported from the cell.

- **Sarcodina** (pseudopods, "false feet"): Members of this group have fewer organelles and are simpler in structure than the ciliates and flagellates. Sarcodina move about by the formation of flowing protoplasmic projections called *pseudopodia.* The formation of pseudopodia is commonly referred to as *amoeboid movement.* The *amoebae* are well known for this mode of action (see Figure 6.4). The pseudopodia not only provide a means of locomotion, but also serve as a means of feeding; this is accomplished when the organism puts out the pseudopodium to enclose the food. Most amebas feed on algae, bacteria, protozoa, and rotifers. Several species in the Sarcodina group, including some species of amoebas, cover themselves with protective, shell-like coverings call *tests.* These tests (made of silica) are stippled with many small and large openings through which water can flow in and out and through which the pseudopods protrude.

✓ **Interesting Point:** Pseudopodia are used by many cells, and are not fixed structures like flagella but rather are associated with action near the moving edge of the cytoplasm.

- **Sporozoans:** These protozoans are obligatory intracellular parasites: they must spend at least part if not all of their life cycle in a host animal. They have no special structures used for locomotion. The life cycle often involves more than one host, such as when plasmodium infects both mosquitoes and humans, causing human malaria.

✓ **Interesting Point:** The plasmodium includes the malaria parasites transmitted by Anopheles mosquitoes.

ALGAE

The protists that perform photosynthesis are called *algae.* Algae can be both a nuisance and an ally. Many ponds, lakes, rivers, streams, and bays (e.g., Chesapeake Bay) in the United States and elsewhere are undergoing *eutrophication,* the enrichment of an environment with inorganic substances (phosphorous and nitrogen). When eutrophication occurs, when filamentous algae like *Cladophora* break loose in a pond, lake, stream, or river and washes ashore, algae makes it stinking, noxious presence known. Algae are allies in many wastewater treatment operations. They can be valuable in long-term oxidation ponds where they aid in the purification process by producing oxygen.

Before discussing the specifics and different types of algae, it is important to be familiar with algal terminology.

Algal Terminology

- **Algae:** Large and diverse assemblages of eukaryotic organisms that lack roots, stems, and leaves but have chlorophyll and other pigments for carrying out oxygen-producing photosynthesis.
- **Algology** or **phycology:** The study of algae.
- **Antheridium:** Special male reproductive structures where sperm are produced.
- **Aplanospore:** Nonmotile spores produced by sporangia.
- **Benthic:** Algae attached and living on the bottom of a body of water.
- **Binary fission:** Nuclear division followed by division of the cytoplasm.
- **Chloroplasts:** Packets that contain *chlorophyll a* and other pigments.
- **Chrysolaminarin:** The carbohydrates reserve in organisms of division *Chrysophyta.*
- **Diatoms:** Photosynthetic, circular, or oblong chrysophyte cells.
- **Dinoflagellates:** Unicellular, photosynthetic protistan algae.
- **Epitheca:** The larger part of the frustule (diatoms).

- **Euglenoids:** Contain chlorophylls a and b in their chloroplasts; representative genus is *Euglena.*
- **Fragmentation:** A type of asexual algal reproduction in which the thallus breaks up and each fragmented part grows to form a new thallus.
- **Frustule:** The distinctive two-piece wall of silica in diatoms.
- **Hypotheca:** The small part of the frustule (diatoms).
- **Neustonic:** Algae that live at the water–atmosphere interface.
- **Oogonia:** Vegetative cells that function as female sexual structures in the algal reproductive system.
- **Pellicle:** A *Euglena* structure that allows for turning and flexing of the cell.
- **Phytoplankton:** Made up of algae and small plants.
- **Plankton:** Free-floating, mostly microscopic aquatic organisms.
- **Planktonic:** Algae suspended in water as opposed to attached and living on the bottom (benthic).
- **Protothecosis:** A disease in humans and animals caused by the green algae, *Prototheca moriformis.*
- **Thallus:** The vegetative body of algae.

Algae are autotrophic, contain the green pigment chlorophyll, and are a form of aquatic plant. Algae differ from bacteria and fungi in their ability to carry out photosynthesis—the biochemical process requiring sunlight, carbon dioxide, and raw mineral nutrients. Photosynthesis takes place in the chloroplasts. The chloroplasts are usually distinct and visible. They vary in size, shape, distribution, and numbers. In some algal types, the chloroplast may occupy most of the cell space. They usually grow near the surface of water because light cannot penetrate very far through water. Although in mass (multicellular forms like marine kelp) the unaided eye easily sees them, many are microscopic. Algal cells may be nonmotile, motile by one or more flagella, or exhibit gliding motility as in diatoms. They occur most commonly in water (fresh and polluted water, as well as in salt water), in which they may be suspended (Planktonic) phytoplanktons or attached and living on the bottom (benthic). A few algae live at the water–atmosphere interface and are termed *neustonic.* Within the fresh and saltwater environments, they are important primary producers (the start of the food chain for other organisms). During their growth phase, they are important oxygen-generating organisms and constitute a significant portion of the plankton in water.

According to the five-kingdom system of Whittaker, the algae belong to seven divisions distributed between two different kingdoms. Although seven divisions of algae occur, only five divisions are discussed in this text:

- **Chlorophyta:** Green algae
- **Euglenophyta:** Euglenids
- **Chrysophyta:** Golden-brown algae, diatoms

- **Phaeophyta:** Brown algae
- **Pyrrophyta:** Dinoflagellates

The primary classification of algae is based on cellular properties. Several characteristics are used to classify algae, including: (1) cellular organization and cell wall structure; (2) the nature of the chlorophyll or chlorophylls present; (3) the type of motility, if any; (4) the carbon polymers that are produced and stored; and (5) the reproductive structures and methods.

Algae show considerable diversity in the chemistry and structure of their cells. Some algal cell walls are thin, rigid structures usually composed of cellulose modified by the addition of other polysaccharides. In other algae, the cell wall is strengthened by the deposition of calcium carbonate. Other forms have chitin present in the cell wall. Complicating the classification of algal organisms are the euglenids, which lack cell walls. In diatoms, the cell wall is composed of silica. The frustules (shells) of diatoms have extreme resistance to decay and remain intact for long periods, as the fossil records indicate.

The principal feature used to distinguish algae from other microorganisms (e.g., fungi) is the presence of chlorophyll and other photosynthetic pigments in algae. All algae contain chlorophyll a. Some, however, contain other types of chlorophylls. The presence of these additional chlorophylls is characteristic of a particular algal group. In addition to chlorophyll, other pigments encountered in algae include fucoxanthin (brown), xanthophylls (yellow), carotenes (orange), phycocyanin (blue), and phycoerythrin (red).

Many algae have flagella (a threadlike appendage). As mentioned, the flagella are locomotor organelles that may be the single polar or multiple polar types. The *Euglena* is a simple flagellate form with a single polar flagellum. Chlorophyta have either two or four polar flagella. Dinoflagellates have two flagella of different lengths. In some cases, algae are nonmotile until they form motile gametes (a haploid cell or nucleus) during sexual reproduction. Diatoms do not have flagella, but have gliding motility.

Algae can be either autotrophic or heterotrophic. Most are photoautotrophic; they require only carbon dioxide and light as their principal source of energy and carbon. In the presence of light, algae carry out oxygen-evolving photosynthesis; in the absence of light, algae use oxygen. Chlorophyll and other pigments are used to absorb light energy for photosynthetic cell maintenance and reproduction. One of the key characteristics used in the classification of algal groups is the nature of the reserve polymer synthesized as a result of utilizing carbon dioxide present in water.

Algae may reproduce either asexually or sexually. Three types of asexual reproduction occur: binary fission, spores, and fragmentation. In some unicellular algae, binary fission occurs where the division of the cytoplasm forms new individuals like the parent cell following nuclear division. Some algae reproduce through spores. These spores are unicellular and germinate without fusing with other cells. In fragmentation, the thallus breaks up and each fragment grows to form a new thallus.

Sexual reproduction can involve union of cells where eggs are formed within vegetative cells called *Oogonia* (which function as female structures) and sperm are produced in a male reproductive organ called *antheridia*. Algal reproduction can also occur through a reduction of chromosome number or the union of nuclei.

Characteristics of Algal Divisions

- **Chlorophyta** (green algae): The majority of algae found in ponds belong to this group; they also can be found in salt water and soil. Several thousand species of green algae are known today. Many are unicellular; others are multicellular filaments or aggregated colonies. The green algae have chlorophylls a and b, along with specific carotenoids, and they store carbohydrates as starch. Few green algae are found at depths greater than 7–10 meters, largely because sunlight does not penetrate to that depth. Some species have a holdfast structure that anchors them to the bottom of the pond and to other submerged inanimate objects. Green algae reproduce by both sexual and asexual means. Multicellular green algae have some division of labor, producing various reproductive cells and structures.
- **Euglenophyta** (euglenoids): Euglenophyta are a small group of unicellular microorganisms that have a combination of animal and plant properties. Euglenoids lack a cell wall, possess a gullet, have the ability to ingest food, have the ability to assimilate organic substances, and, in some species, are absent of chloroplasts. They occur in fresh, brackish, and salt waters, and on moist soils. A typical *Euglena* cell is elongated and bound by a plasma membrane; the absence of a cell wall makes them very flexible in movement. Inside the plasma membrane is a structure called the *pellicle* that gives the organisms a definite form and allows the cell to turn and flex. Euglenoids that are photosynthetic contain chlorophylls a and b, and they always have a red eyespot (*stigma*) that is sensitive to light (*photoreceptive*). Some euglenoids move about by means of flagellum; others move about by means of contracting and expanding motions. The characteristic food supply for euglenoids is a lipopolysaccharide. Reproduction in euglenoids is by simple cell division.

✓ **Interesting Point**: Some autotrophic species of *Euglena* become heterotrophic when light levels are low.

- **Chrysophyta** (golden brown algae): The chrysophycophyta group is quite large; it has several thousand diversified members. They differ from green algae and euglenoids in that (1) chlorophylls a and c are present; (2) fuco-xanthin, a brownish pigment, is present; and (3) they store food in the form of oils and leucosin, a polysaccharide. The combination of yellow pigments, fucoxanthin, and chlorophylls causes most of these algae to appear golden-brown. The chrysophycophyta is also diversified in cell wall chemistry and flagellation. The division is divided into three major classes: golden-brown algae, yellow-brown algae, and diatom.

Some chrysophyta lack cell walls; others have intricately patterned cover-ings external to the plasma membrane, such as walls, plates, and scales. The diatoms are the only group that has hard cell walls of pectin, cellulose, or silicon, constructed in two halves (the epitheca and the hypotheca) called a *frustule*. Two anteriorly attached flagella are common among chrysophyta; others have no flagella.

Most chrysophyta are unicellular or colonial. Asexual cell division is the usual method of reproduction in diatoms; other forms of chrysophyta can reproduce sexually.

Diatoms have direct significance for humans. Because they make up most of the phytoplankton of the cooler ocean parts, they are the ultimate source of food for fish.

Water and wastewater operators understand the importance of their abil-ity to function as indicators of industrial water pollution. As water quality in-dicators, their specific tolerances to environmental parameters such as pH, nu-trients, nitrogen, concentration of salts, and temperature have been compiled.

✓ **Interesting Point:** Diatoms secrete a silicon dioxide shell (frustule) that forms the fossil deposits known as diatomaceous Earth, which is used in filters and as abrasives in polishing compounds.

- **Phaeophyta** (brown algae): With the exception of a few freshwater species, all algal species of this division exist in marine environments as seaweed. They are a highly specialized group, consisting of multicellular organisms that are sessile (attached and not free-moving). These algae contain essentially the same pig-ments seen in the golden-brown algae, but they appear brown because of the predominance of and the masking effect of a greater amount of fucoxanthin. Brown algal cells store food as the carbohydrate laminarin and some lipids. Brown algae reproduce asexually. Brown algae are used in foods, animal feeds, and fertilizers, and as source for alginate, a chemical emulsifier added to ice cream, salad dressing, and candy.
- **Pyrrophyta** (dinoflagellates): The principal members of this division are the dinoflagellates. The dinoflagellates comprise a diverse group of biflagellated and nonflagellated unicellular, eukaryotic organisms. The dinoflagellates occupy a

variety of aquatic environments with the majority living in marine habitats. Most of these organisms have a heavy cell wall composed of cellulose-containing plates. They store food as starch, fats, and oils. These algae have chlorophylls a and c and several xanthophylls. The most common form of reproduction in dinoflagellates is by cell division, but sexual reproduction has also been observed.

✓ **Interesting Point:** Cell division in dinoflagellates differs from most protistans, with chromosomes attaching to thze nuclear envelope and being pulled apart as the nuclear envelope stretches. During cell division in most other eukaryotes, the nuclear envelope dissolves.

Fungi

Fungi (singular: *fungus*) constitute an extremely important and interesting group of eukaryotic, aerobic microbes ranging from the unicellular yeasts to the extensively mycelial molds. Fungi first evolved in water but made the transition to land through the development of specialized structures that prevented their drying out. Not considered plants, they are a distinctive life-form of great practical and ecological importance. Fungi are important because, like bacteria, they metabolize dissolved organic matter; they are the principal organisms responsible for the decomposition of carbon in the biosphere. Fungi, unlike bacteria, can grow in low-moisture areas and in low-pH solutions, which aids them in the breakdown of organic matter.

Before discussing specifics and the different types of fungi, it is important to be familiar with fungal terminology:

FUNGAL TERMINOLOGY

- **Hypha** (plural: *hyphae*) is a tubular cell that grows from the tip and might form many branches.
- **Mycelium** consists of many-branched hypha and can become large enough to be seen with the naked eye.
- **Spore** is the reproductive stage of the fungi.
- **Septate hyphae** occurs when a filament has crosswalls.
- **Nonseptate** or **aseptate** occurs when cross walls are not present.
- **Sporangiospores** are spores that form within a sac called a *sporangium*. The sporangia are attached to stalks called *sporangiophore*.
- **Conidia** are asexual spores that form on specialized hyphae called *conidiophores*. Large conidia are called *macroconidia* and small conidia are called *microconidia*.

- **Sexual spores** are used as the basis for four subdivisions in the fungi division mastigomycota: (1) Subdivision *zygomycota* consists of nonseptate hyphae and zygospores. Zygospores are formed by the union of nuclear material from the hyphae of two different strains. (2) Fungi in the subdivision *ascomycotina* are commonly referred to as the *ascomycetes*. They are also called *sac fungi*. They all have septate hyphae. *Ascospores* are the characteristic sexual reproductive spores and are produced in sacs called *asci* (singular: ascus). The mildews and *penicillium* with asci in long fruiting bodies belong to this group. (3) Subdivision *basidiomycotina* consists of mushrooms, puffballs, smuts, rust, and shelf fungi (found on dead trees). The sexual spores of this class are known as *basidiospores,* which are produced on the club-shaped *basidia.* (4) Subdivision *deutermycotina* consists of only one group, the *deuteromycetes.* Members of this class are referred to as the *fungi imperfecti* and include all the fungi that lack sexual means of reproduction.
- **Budding** is the process by which yeasts reproduce.
- **Blastospore** or bud refers to spores formed by budding.

Fungi comprise a large group of organisms that include such diverse forms as molds, mushrooms, puffballs, and yeasts. Because they lack chlorophyll (and thus are not considered plants), they must get nutrition from organic substances. They are either *parasites,* existing in or on animals or plants, or more commonly are *saprophytes,* obtaining their food from dead organic matter. Fungi also are important crop parasites, causing loss of food plants, spoilage of food, and some infectious diseases. Fungi are classified in their own kingdom but the main groups are called *divisions* rather than *phyla.* The study of fungi is called *mycology.*

✓ **Interesting Point:** Fungi range in size from the single-celled organism we know as yeast to the largest known living organism on Earth—a 3.5 mile mushroom dubbed "the humongous fungus," which covers more than 2,000 acres in Oregon's Malheur National Forest.

Fungi may be unicellular or filamentous. They are large, 5–10 microns wide, and can be identified by a microscope. The distinguishing characteristics of the group as a whole include the following characteristics: (1) they are non-photosynthetic, (2) they lack tissue differentiation, (3) they have cell walls of polysaccharides (chitin), and (4) they propagate by spores (sexual or asexual).

Fungi are divided into five classes:

- Myxomycetes, or slime fungi
- Phycomycetes, or aquatic fungi (algae)

- Ascomycetes, or sac fungi
- Basidiomycetes, or rusts, smuts, and mushrooms
- Fungi imperfecti, or miscellaneous fungi

✓ **Interesting Point:** Although fungi are limited to only five classes, more than 80,000 known species exist.

Fungi differ from bacteria in several ways, including in their size, structural development, methods of reproduction, and cellular organization. They differ from bacteria in another significant way as well: their biochemical reactions (unlike the bacteria) are not important for classification; instead, their structure is used to identify them. Fungi can be examined directly, or suspended in liquid, stained, dried, and observed under microscopic examination where they can be identified by the appearance (color, texture, and diffusion of pigment) or their mycelia.

One of the tools available to environmental science students and specialists for use in the fungal identification process is the distinctive terminology used in mycology. Fungi go through several phases in their life-cycle; their structural characteristics change with each new phase. Become familiar with the following listed and defined terms. As a further aid in learning how to identify fungi, relate the defined terms to their diagrammatic representation (see Figure 6.5).

Fungi can be grown and studied by cultural methods. However, when culturing fungi, use culture media that limits the growth of other microbial types—controlling bacterial growth is of particular importance. This can be accomplished by using special agar (culture media) that depresses pH of the culture medium (usually Sabouraud glucose or maltose agar) to prevent the growth of bacteria. Antibiotics can also be added to the agar that will prevent bacterial growth.

✓ **Interesting Point:** Fungi can be found in rising bread, moldy bread, and old food in the refrigerator, and on forest floors.

As part of their reproductive cycle, fungi produce very small spores that are easily suspended in air and widely dispersed by the wind. Insects and other animals also spread fungal spores. The color, shape, and size of spores are useful in the identification of fungal species.

Reproduction in fungi can be either sexual or asexual. The union of compatible nuclei accomplishes sexual reproduction. Most fungi form specialized asexual or sexual spore–bearing structures (fruiting bodies). Some fungal species are self-fertilizing and other species require outcrossing between different but compatible vegetative thalluses (mycelia).

Most fungi are asexual. Asexual spores are often brightly pigmented and give their colony a characteristic color (green, red, brown, back, blue—the blue spores of *Penicillium roquefort* are found in blue or Roquefort cheese).

✓ **Interesting Point:** Fungi usually reproduce without sex. Single-celled yeasts reproduce asexually by budding. A single yeast cell can produce up to 24 offspring.

Asexual reproduction is accomplished in several ways:

- Vegetative cells might bud to produce new organisms. This is very common in the yeasts.
- A parent cell can divide into two daughter cells.
- The most common method of asexual reproduction is the production of spores.

Several types of asexual spores are common:

- A hypha may separate to form cells (*arthrospores*) that behave as spores.
- If a thick wall before separation encloses the cells, they are called *chlamydospores*.
- If budding produces the spore, they are called *blastospores*.
- If the spores develop within sporangia (sac), they are called *sporangiospores*.
- If the spores are produced at the sides or tips of the hypha, they are called *conidiospores*.

Fungi are found wherever organic material is available. They prefer moist habitats and grow best in the dark. Most fungi can best be described as grazers, but a few are active hunters. Most fungi are saprophytes, acquiring their nutrients from dead organic matter, gained when the fungi secrete hydrolytic enzymes, which digest external substrates. They are able to use dead organic matter as a source of carbon and energy. Most fungi use glucose and maltose (carbohydrates) and nitrogenous compounds to synthesize their own proteins and other needed materials. Knowing from what materials fungi synthesize their own protein and other needed materials in comparison to what bacteria are able to synthesize is important to those who work in the environmental disciplines for understanding the growth requirements of the different microorganisms.

✓ **Interesting Point:** Some fungi produce a sticky substance on their hyphae, which then act like flypaper, trapping passing prey.

Plants

The plant kingdom ranks second in importance only to the animal kingdom (at least from the human point of view). The importance of plants and plant communities to humans and their environment cannot be overstated. Some of the important things plants provide are listed:

- **Aesthetics:** Plants add to the beauty of the places we live.
- **Medicine:** Eighty percent of all medicinal drugs originate in wild plants.
- **Food:** Ninety percent of the world's food comes from only 20 plant species.
- **Industrial products:** Plants are very important for the goods they provide. For example, plant fibers provide clothing; wood is used to build homes; and some important fuel chemicals come from plants, such as ethanol from corn and soy diesel from soybeans.
- **Recreation:** Plants form the basis for many important recreational activities, including fishing, nature observation, hiking, and hunting.
- **Air quality:** The oxygen in the air we breathe comes from the photosynthesis of plants.
- **Water quality:** Plants aid in maintaining healthy watersheds, streams, and lakes by holding soil in place, controlling stream flows, and filtering sediments from water.
- **Erosion control:** Plant cover helps to prevent wind or water erosion of the top layer of soil that we depend on.
- **Climate:** Regional climates are affected by the amount and type of plant cover.
- **Fish and wildlife habitat:** Plants provide the necessary habitat for wildlife and fish populations.
- **Ecosystem:** Every plant species serves an important role or purpose in its community.

Though both are important kingdoms of living things, plants and animals differ in many important aspects. Some of these differences are summarized in Table 29.3.

Before discussing the basic specifics of plants, it is important to first define a few key plant terms.

Plant Terminology

- **Apical meristem** consists of meristematic cells located at the tip (apex) of a root or shoot.
- **Cambium** is the lateral meristem in plants.

Table 29.3.

Plants	Animals
Plants contain chlorophyll and can make their own food.	Animals cannot make their own food and depend on plants and other animals for food.
Plants give off oxygen and take in carbon dioxide given off by animals.	Animals give off carbon dioxide, which plants need to make food, and take in oxygen, which they need to breathe.
Plants generally are rooted in one place and do not move on their own.	Most animals have the ability to move fairly freely.
Plants have either no or very basic ability to sense.	Animals have a much more highly developed sensory and nervous system.

- **Chloroplasts** are disk-like organelles with a double membrane found in eukaryotic plant cells.
- **Companion cells** are specialized cells in the phloem that load sugars into the sieve elements.
- **Cotyledons** are leaf-like structure (sometimes referred to as *seed leaf*) that is present in the seeds of flowering plants.
- **Dicot** is one of the two main types of flowering plants characterized by having two cotyledons.
- **Diploid** refers to having two of each kind of chromosome (2n).
- **Guard cells** are specialized epidermal cells that flank stomata and whose opening and closing regulate gas exchange and water loss.
- **Haploid** refers to having only a single set of chromosomes (n).
- **Meristem** is a group of plant cells that can divide indefinitely and provides new cells for the plant.
- **Monocots** are one of two main types of flowering plants characterized by having a single cotyledon.
- **Periderm** is a layer of plant tissue derived from the cork cambium, and then secondary tissue, replacing the epidermis.
- **Phloem** is complex vascular tissue that transports carbohydrates throughout the plant.
- **Sieve cells** conduct cells in the phloem of vascular plants.
- **Stomata** are pores on the underside of leaves that can be opened or closed to control gas exchange and water loss.
- **Thallus** is the main plant body, not differentiated into a stem or leaves.
- **Tropism** refers to plant behavior; controlling the direction of plant growth.

- **Vascular tissue** is tissue found in the bodies of vascular plants that transport water, nutrients, and carbohydrates. The two major kinds are xylem and phloem.
- **Xylem** is vascular tissue of plants that transports water and dissolved minerals from the roots upward to other parts of plant. Xylem often also provides mechanical support against gravity.

Although not typically acknowledged, plants are as intricate and complicated as animals. Plants evolved from photosynthetic protists and are characterized by photosynthetic nutrition, cell walls made from cellulose and other polysaccharides, lack of mobility, and a characteristic life cycle involving an alternation of generations. The phyla and division of plants and examples are listed in Table 29.4.

Table 29.4.   The Main Phyla/Division of Plants

Phlyum/Division	Examples
Bryophyta	Mosses, liverworts, and hornworts
Coniferophyta	Conifers such as redwoods, pines, and firs
Cycadophyta	Cycads, sago palms
Gnetophyta	Shrub trees and vines
Ginkophyta	Gingko is the only genus
Lycophyta	Lycopods (look like mosses)
Pterophyta	Ferns and tree-ferns
Anthophta	Flowering plants including oak, corn, maize, and herbs

THE PLANT CELL

The cell was covered earlier, but a brief summary of plant cells is provided here.

- *Plants have all the organelles animal cells have* (i.e., nucleus, ribosomes, mitochondria, endoplasmic reticulum, Golgi apparatus, etc.).
- *Plants have chloroplasts.* Chloroplasts are special organelles that contain chlorophyll and allow plants to carry out photosynthesis.
- *Plant cells can sometimes have large vacuoles for storage.*
- *Plant cells are surrounded by a rigid cell wall made of cellulose,* in addition to the cell membrane that surrounds animal cells. Those walls provide support.

VASCULAR PLANTS

Vascular plants, also called *tracheophytes,* have special vascular tissue for transport of necessary liquids and minerals over long distances. Vascular tissues are composed of specialized cells that create "tubes" through which materials can flow throughout the plant body. These vessels are continuous throughout the plant, allowing for the efficient and controlled distribution of water and nutrients. In addition to this transport function, vascular tissues also support the plant. The two types of vascular tissue are *xylem* and *phloem.*

- **Xylem** consists of a tube or a tunnel (pipeline) in which water and minerals are transported throughout the plant to leaves for photosynthesis. In addition to distributing nutrients, xylem (wood) provides structural support. After a time, the xylem at the center of older trees ceases to function in transport and takes on a supportive role only.
- **Phloem** tissue consists of cells called *sieve tubes* and *companion cells.* Phloem tissue moves dissolved sugars (carbohydrates), amino acids, and other producers of photosynthesis from the leaves to other regions of the plant.

The two most important tracheophytes are gymnosperms (gymno = naked; sperma = seed) and angiosperms (angio = vessel, receptacle, container).

- **Gymnosperms** represent the sporophyte generation of plants (i.e., the spore-producing phase in the life cycle of a plant that exhibits alternation of generation). Gymnosperms were the first tracheophytes to use seeds for reproduction. The seeds develop in protective structures called *cones.* A gymnosperm contains some cones that are female and some that are male. Female cones produce spores that, after fertilization, become eggs enclosed in seeds that fall to the ground. Male cones produce pollen, which is taken by the wind and fertilizes female eggs by that means. Unlike flowering plants, the gymnosperm does not form true flowers or fruits. Coniferous trees such as firs and pines are good examples of gymnosperms.
- **Angiosperms**, the flowering plants, are the most highly evolved plants and the most dominant in the present. They have stems, roots, and leaves. Unlike gymnosperms such as conifers and cycads, angiosperms' seeds are found in a flower. Angiosperm eggs are fertilized and develop into a seed in an ovary that is usually in a flower.

There are two types of angiosperms: monocots and dicots.

- **Monocots** are angiosperms that start with one seed-leaf (cotyledon); thus, their name is derived from the presence of a single cotyledon during embryonic development. Monocots include grasses, grains, and other narrow-leaved angiosperms. The main veins of their leaves are usually parallel and unbranched, the flower parts occur in multiples of three, and a fibrous root system is present. Monocots include orchids, lilies, irises, palms, grasses, wheat, corn, and oats.

- **Dicots** are angiosperms that grow two seed-leaves (two cotyledons). Most plants are dicots and include maples, oaks, elms, sunflowers, and roses. Their leaves usually have a single main vein or three or more branched veins that spread out from the base of the leaf.

LEAVES

The principal function of leaves is to absorb sunlight for the manufacturing of plant sugars in photosynthesis. The leaves are broad, flattened surfaces that gather energy from sunlight, while apertures on their undersides bring in carbon dioxide and release oxygen. Leaves develop as a flattened surface to present a large area for efficient absorption of light energy. On its two exteriors, the leaf has layers of epidermal cells that secrete a waxy, nearly impermeable cuticle (chitin) to protect against water loss (dehydration) and fungal or bacterial attack. Gases diffuse in or out of the leaf through *stomata,* small openings on the underside of the leaf. The opening or closing of the stomata occurs through the swelling or relaxing of *guard cells.* If the plant wants to limit the diffusion of gases and the transpiration of water, the guard cells swell together and close the stomata. Leaf thickness is kept to a minimum so that gases that enter the leaf can diffuse easily throughout the leaf cells.

CHLOROPHYLL AND CHLOROPLAST

The green pigment in leaves is *chlorophyll.* Chlorophyll absorbs red and blue light from the sunlight that falls on leaves. Therefore, the light reflected by the leaves is diminished in red and blue and appears green. The molecules of chlorophyll are large. They are not soluble in the aqueous solution that fills plant cells. Instead, they are attached to the membranes of disc-like structures, called *chloroplasts,* inside the cells. Chloroplasts are the site of photosynthesis, the process in which light energy is converted to chemical energy. In chloroplasts, the light absorbed by chlorophyll supplies the energy used by plants to transform carbon dioxide and water into oxygen and carbohydrates.

Chlorophyll is not a very stable compound; bright sunlight causes it to decompose. To maintain the amount of chlorophyll in their leaves, plants continuously synthesize it. The synthesis of chlorophyll in plants requires sunlight and warm temperatures. Therefore, during summer, chlorophyll is continuously broken down and regenerated in the leaves of trees.

PHOTOSYNTHESIS

Because our quality of life, and indeed our very existence, depends on photosynthesis, it is essential to understand it. In photosynthesis, plants (and other photosynthetic autotrophs) use the energy from sunlight to create the carbohydrates necessary for cell respiration. More specifically, plants take water and carbon dioxide and transform them into glucose and oxygen:

$$6CO_2 + 6H_2O + \text{light energy} \rightarrow C_6H_{12}O_6 + 6O_2$$

This general equation of photosynthesis represents the combined effects of two different stages. The first stage is called the *light reaction* and the second stage is called the *dark reaction*. The *light reaction* is the photosynthesis process in which solar energy is harvested and transferred into the chemical bonds of ATP; it can only occur in light. The *dark reaction* is the process in which food (sugar) molecules are formed from carbon dioxide from the atmosphere with the use of ATP; the process can occur in the dark as long as ATP is present.

✓ **Interesting Point:** Charles Darwin was the first to discuss how plants respond to light. He found that the new shoot of grasses bend toward the light because the cells on the dark side grow faster than the lighted side.

ROOTS

Roots absorb nutrients and water, anchor the plant in the soil, provide support for the stem, and store food. They are usually below ground and lack nodes, shoots, and leaves. There are two major types of root systems in plants. Taproot systems have a stout main root with a limited number of side-branching roots. Examples of taproot system plants are nut trees, carrots, radishes, parsnips, and dandelions. Taproots make transplanting difficult. The second type of root system, fibrous, has many branched roots. Examples of fibrous root plants are most grasses, marigolds, and beans. Radiating from the roots is a system of root hairs, which vastly increase the absorptive surface area of the roots. Roots also anchor the plant in the soil.

GROWTH IN VASCULAR PLANTS

Vascular plants undergo two kinds of growth (growth is primarily restricted to meristems), primary and secondary growth. *Primary growth* occurs relatively close to the tips of roots and stems. It is initiated by apical meristems and it is primarily involved in the extension of the plant body. The tissues that arise during primary growth are called *primary tissues* and the plant body composed of these tissues is called the *primary plant body*. Most primitive vascular plants are entirely made up of primary tissues. *Secondary growth* occurs in some plants; secondary growth thickens the stems and roots. Secondary growth results from the activity of lateral meristems. Lateral meristems are called **cambia** (cambium) and there are two types:

1. **Vascular cambium** gives rise to secondary vascular tissues (secondary xylem and phloem). The vascular cambium gives rise to xylem to the inside and phloem to the outside.
2. **Cork cambium** forms the *periderm* (bark). The periderm replaces the epidermis in woody plants.

PLANT HORMONES

Plant growth is controlled by plant hormones, which influence cell differentiation, elongation, and division. Some plant hormones also affect the timing of reproduction and germination.

- **Auxins** affect cell elongation (tropism), apical dominance, and fruit drop or retention. Auxins are also responsible for root development, secondary growth in the vascular cambium, inhibition of lateral branching, and fruit development. Auxin is involved in absorption of vital minerals and fall color. As a leaf reaches its maximum growth, auxin production declines. In deciduous plants this triggers a series of metabolic steps, which causes the reabsorption of valuable materials (such as chlorophyll) and their transport into the branch or stem for storage during the winter months. Once chlorophyll is gone, the other pigments typical of fall color become visible.
- **Kinins** promote cell division and tissue growth in leaf, stem, and root. Kinins are also involved in the development of chloroplasts, fruits, and flowers. In addition, they have been shown to delay senescence (aging), especially in leaves, which is one reason that florists use cytokinins on freshly cut flowers—when treated with cytokinins they remain green, protein synthesis continues, and carbohydrates do not breakdown.
- **Gibberellins** are produced in the root growing tips and acts as a messenger to stimulate growth, especially elongation of the stem, and can also end the dormancy period of seeds and buds by encouraging germination. Additionally, gibberellins play a role in root growth and differentiation.

- **Ethylene** controls the ripening of fruits. Ethylene may ensure that flows are carpellate (female) while gibberellin confers maleness on flowers. It also contributes to the senescence of plants by promoting leaf loss and other changes.
- **Inhibitors** restrain growth and maintain the period of dormancy in seeds and buds.

TROPISMS: PLANT BEHAVIOR

Tropism is the movement (and growth in plants) of an organism in response to an external stimulus. For example, tropisms, which are controlled by hormones, are a unique characteristic of sessile organisms such as plants that enable them to adapt to different features of their environment—gravity, light, water, and touch—so that they can flourish. There are three main tropisms:

- **Phototropism** is the tendency of plants growing or bending (moving) in response to light. Phototropism results from the rapid elongation of cells on the dark side of the plant, which causes the plant to bend in the opposite direction. For example, the stems and leaves of a geranium plant growing on the windowsill always turn toward the light.
- **Gravitropism** refers to a plant's tendency to grow toward or against gravity. A plant that displays positive gravitropism (plant roots) will grow downward, toward the center of the Earth. That is, gravity causes the roots of plants to grow down so that the plant is anchored in the ground and has enough water to grow and thrive. Plants that display negative gravitropism (plant stems) will grow upward, away from the Earth. Most plants are negatively gravitropic. Gravitropism is also controlled by auxin. In a horizontal root or stem, auxin is concentrated in the lower half, pulled by gravity. In a positively gravitropic plant, this auxin concentration will inhibit cell growth on the lower side, causing the stem to bend downward. In a negatively gravitropic plant, this auxin concentration will inspire cell growth on that lower side, causing the stem to bend upward.
- **Thigmotropism** refers to the circumstance noticed by some indoor gardeners of houseplants that grow better when they touch the plants and pay attention to them. Touch causes parts of the plant to thicken or coil as they touch or are touched by environmental entities. For instance, tree trunks grow thicker when exposed to strong winds and vines tend to grow straight until they encounter a substrate to wrap around.

PHOTOPERIODISM

Photoperiodism is the response of an organism (e.g., plants) to naturally occurring changes in light during a 24-hour period. The site of perception of photoperiod in plants is leaves. For instance, sunflowers are known for their photoperiodism, or their ability to open and close in response to the changing position of the sun throughout the day.

All flowering plants have been placed in one of three categories with respect to photoperiodism: short-day plants, long-day plants, and day-neutral plants.

- **Short-day plants:** Flowering is promoted by day lengths shorter than a certain critical day length; includes poinsettias, chrysanthemums, goldenrod, and asters.
- **Long-day plants:** Flowering is promoted by day lengths longer than a certain critical day length; includes spinach, lettuce, and most grains.
- **Day-neutral plants:** Flowering is response-insensitive to day length; includes tomatoes, sunflowers, dandelions, rice, and corn.

PLANT REPRODUCTION

Plants can reproduce both sexually and asexually. Each type of reproduction has its benefits and disadvantages. A comparison of sexual and asexual plant reproduction is provided in the following.

- Sexual Reproduction:
  - Sexual reproduction occurs when a sperm nucleus from the pollen grain fuses with the egg cell from an ovary of a pistil (*pistil* is the female reproductive structures in flowers, consisting of the stigma, style, and ovary).
  - Each brings a complete set of genes and produces genetically unique organisms.
  - The resulting plant embryo develops inside the seed and grows when the seed is germinated.
- Asexual reproduction
  - Occurs when a vegetative part of a plant, root, stem, or leaf gives rise to a new offspring plant whose genetic content is identical to the "parent plant." An example is a plant reproducing by root suckers, shoots that come from the root system. The breadfruit tree is an example.
  - Asexual reproduction is also called *vegetative propagation.* It is an important way for plant growers to get many identical plants from one very quickly.
  - By sexual reproduction, plants can spread and colonize an area quickly (e.g., crab grass).

Animals

All animals are members of the Kingdom Animalia. With over 2 million species, Kingdom Animalia is the largest of the kingdoms in terms of its species diversity.

Not surprisingly, in regard to diversity among different animal species, it's difficult to imagine what they all might have in common. First, animals are composed of many cells—they are "multicellular." In most animals, these cells are organized into tissues that make up different organs and organ systems. Second, animals must get their food by eating other organisms, such as plants, fungi, and other animals—they are "heterotrophs." Third, animals are eukaryotic. Fourth, animals develop (or not) an internal cavity called a coelom. In addition, all animals require oxygen for their metabolism and can sense and respond to their environment. Many animals have tissues specialized for specific functions (nerve tissue, muscle), and have the capacity to reproduce sexually (though many reproduce asexually as well). During their development from a fertilized egg to adult, all animals pass through a series of embryonic stages as part of their normal life cycle.

There are two main types of animals, invertebrates and vertebrates. These types are discussed in the following sections.

✓ **Important Point:** The majority of the species in marine ecosystems fall under the Kingdom Animalia.

INVERTEBRATES

Invertebrates—creatures without backbones—are the most abundant creatures on Earth (more than 98% of the known animal species), crawling, flying, floating, or swimming in virtually all of Earth's habitats. Many invertebrates have a fluid-filled, hydrostatic skeleton, like the jelly fish or worm. Others have a hard outer shell, like insect crustaceans. There are many types of invertebrates. The most common invertebrates include the sponges, arachnids, insects, crustaceans, mollusks, and echinoderms.

Mollusks

Mollusks are an amazingly diverse group of animals that live in a wide variety of environments. They can be found inhabiting trees, gardens, freshwater ponds and streams, estuaries, tidal pools, beaches, the continental shelf, and the deep ocean. Some mollusks are excellent swimmers, other crawl or burrow in mud and sand. Others remain stationary by attaching themselves to rocks, other shells, or plants, or by boring into hard surfaces, such as wood or rocks. Adult mollusks can range in size from a few millimeters (0.1 inch) to more than 70 feet in length as documented for some giant deep-sea squids. Their weight can vary from a few milligrams (a fraction of an ounce) to over 227 kilograms (500 pounds) as recorded for the giant South Pacific *Tridacna* clams.

The number of living species of mollusks has been estimated to range from 50,000 to 130,000. Everyone is probably familiar with some type of mollusk. They are the slugs and shelled pests in your backyard garden; the scallops, clams, mussels, or oysters on your dinner plant; the pretty shells you see washed up on the beach; the pearls or other treasures in many jewelry boxes; the octopus or squid at an aquarium.

The word *Mollusca* is translated from Latin as "soft-bodied" but few physical characteristics are unique to all mollusks. The mollusks are invertebrates and therefore lack a backbone; they are unsegmented and most exhibit bilateral symmetry. Most mollusks can be described as free-living, multicellular animals that possess a true heart, and that have a calcareous exterior skeleton that covers at least the back or upper surfaces of the body. This exterior skeleton provides support for a muscular foot and the internal body organs, including the stomach mass. A thin flap of tissue called the *mantle* surrounds the internal organs of most mollusks, and it is this mantle that secretes the animal's shell. The nervous system of mollusks varies greatly from group to group; the clams and tusk shells have very simple nervous systems, while the squids, octopi, and some other mollusks have concentrated complex nerve centers and eyes equivalent to vertebrates.

✓ **Interesting Point:** Because of the many movies in which octopi and squids attack people, boats, and so on, there is a misconception that they are aggressive and dumb creatures. In fact, octopi are probably the most intelligent of all the invertebrates, and there are only two species of octopi that are aggressive (they are located in Australia).

Annelids

Annelids are earthworms, leeches, a large number of mostly marine worms known as *polychaetes* (meaning "many bristles") and other worm-like animals whose bodies are segmented. Segments each contain elements of such body systems as circulatory, nervous, and excretory tracts. Besides being segmented, the body wall of annelids is characterized by being made up of both circular and longitudinal muscle fibers surrounded by a moist, acellular cuticle that is secreted by an epidermal epithelium. All annelids except leeches also have chitinous hair-like structures, called *setae*, projecting from their cuticle. They can reproduce asexually by regeneration, but they usually reproduce sexually. There are about 9,000 species of annelid known today.

✓ **Interesting Point:** Ecologically, annelids range from passive filter feeders to voracious and active predators.

Arthropods

Insects and spiders belong to the group of animals known as *arthropods*. By nearly any measure, they are the most successful animals on the planet. They have conquered land, sea, and air, and make up over three-fourths of all currently known living organisms, or over one million species in all.

Arthropods have segmented bodies with jointed appendages and a chitinous exoskeleton, which must be molted and shed for growth to continue. Insect bodies are divided into three parts: the head, the thorax, and the abdomen. Nearly all insects have wings, and they are the only invertebrate group that can fly. Spiders and their relatives have bodies that are divided into two parts. The head and thorax together are called the cephalothorax, and then comes the abdomen. Most have four pairs of legs.

✓ **Interesting Point:** There are 200 million insects for every person on Earth.

Echinoderms

Echinoderms (from the Greek for "spiny skin") are a phylum of marine animals found at all depths. Along with spiny skin, they are characterized by an endoskeleton, radial symmetry, and a water vascular system. Echinoderms include starfish, sea stars, asteroids, sea daisies, crinoids, feather stars, seal lilies, sand dollars, sea urchins, echinoids, sea cucumbers, brittle stars, and basket stars.

✓ **Interesting Point:** Echinodermata is the largest animal phylum to lack any freshwater or terrestrial representatives.

Chordata

We are most intimately familiar with the *chordata*, because it includes humans and other vertebrates. However, not all chordates are vertebrates. *Chordates* are defined as organisms that posses a notochord, a structure that is present at least during some part of their development. The notochord is a rod that extends most of the length of the body when it is fully developed. Other characteristics shared by chordates include the following (Hickman and Roberts, 1994):

- Bilateral symmetry
- Segmented body, including segmented muscles
- Three germ layers and a well-developed coelom
- Single, dorsal, hollow nerve cord, usually with an enlarged anterior end (brain)
- Tail projecting beyond (posterior to) the anus at some stage of development

- Pharyngeal pouches present at some stage of development
- Ventral heart, with dorsal and ventral blood vessels and a closed blood system
- Complete digestive system
- Bony or cartilaginous endoskeleton usually present

The invertebrate chordates, which do not have a backbone, include the tunicates and lancelets. The adult form of most *tunicates* shows no resemblance to vertebrate animals, but such a resemblance is evident in the larva. The most familiar tunicates are the sea squirts. *Lancelets* are filter feeders with their tails buried in the sand and only their anterior end protruding.

VERTEBRATES

Although *vertebrates* represent only a very small percentage of all animals, their size and mobility often allow them to dominate their environments. Vertebrates include primates, such as humans and monkeys; amphibians; reptiles; birds; and fish. Vertebrates consist of more than 43,000 species of animals with backbones. Vertebrates exhibit all of the chordate characteristics at some point during their lives. The embryonic notochord is replaced by a vertebral column in the adult. The vertebral column is made of individual hard segments (vertebrae) surrounding the dorsal hollow nerve cord. The nerve cord is the one chordate feature present in the adult phase of all vertebrates. The vertebral column, part of a flexible but strong endoskeleton, is evidence that vertebrates are segmented. The vertebrate skeleton is living tissue (either cartilage or bone) that grows as the animal grows. The postanal tail is the only characteristic of chordates that most vertebrates keep throughout their lives.

Human Evolution

Human evolution is the biological and cultural development and change of our hominin (formerly known as *hominid*) ancestors to modern humans. Hominins evolved between 5 to 8 million years ago. To date, fossil records provide evidence of this development and date from about 4.5 million years ago. There were about nine different hominin species. Evidence indicates that *Homo sapiens* made their appearance as early as 300,000 years ago.

Ecosystem

Ecosystem is a term introduced by A. G. Tansley to denote an area that includes all organisms therein and their physical environment. The ecosystem is the major ecological unit in nature. Tansley writes, "There is a constant interchange of the most various kinds within each system, not only between the organisms but between the organic and the inorganic" (Tansley, 1935). Living organisms and their nonliving environment are inseparably interrelated and interact with each other to create a self-regulating and self-maintaining system. To create a self-regulating and self-maintaining system, ecosystems are homeostatic (i.e., they resist any change through natural controls). These natural controls are important in ecology. This is especially the case since it is people through their complex activities who tend to disrupt natural controls.

As stated earlier, the ecosystem encompasses both the living and nonliving factors in a particular environment. The living or biotic part of the ecosystem is formed by two components: autotrophic and heterotrophic. The autotrophic (self-nourishing) component does not require food from its environment but can manufacture food from inorganic substances. For example, some autotrophic components (plants) manufacture needed energy through photosynthesis. Heterotrophic components, on the other hand, depend on autotrophic components for food.

✓ **Important Point:** The nonliving or abiotic part of the ecosystem is formed by three components: inorganic substances, organic compounds (link biotic and abiotic parts), and climate regime.

An ecosystem is a cyclic mechanism in which biotic and abiotic materials are constantly exchanged through biogeochemical cycles. *Biogeochemical cycles* are defined as follows: *bio* refers to living organisms and *geo* to water, air, rocks, or solids. *Chemical* is concerned with the chemical composition of the Earth. Biogeochemical cycles are driven by energy, directly or indirectly from the sun.

In an ecosystem, biotic and abiotic materials are constantly exchanged. Producers construct organic substances through photosynthesis and chemosynthesis. Consumers and decomposers use organic matter as their food and convert it into abiotic components. That is, they dissipate energy fixed by producers through food chains. The abiotic part of a pond, for example, is formed of inorganic and organic compounds dissolved and in sediments such as carbon, oxygen, nitrogen, sulfur, calcium, hydrogen, and humic acids. The biotic part is represented by producers such as rooted plants and phytoplanktons. Fish, crustaceans, and insect larvae make up the consumers. Detritivores, which feed on organic detritus, are represented by mayfly nymphs. Decomposers make up the final abiotic part. They include aquatic bacteria and fungi, which are distributed throughout the pond.

✓ **Important Point:** Again, an ecosystem is a cyclic mechanism. From a functional viewpoint, an ecosystem can be analyzed in terms of several factors. The factors important in this study include biogeochemical cycles, energy, and food chains.

Biomes

When it comes to the terms *ecosystem* and *biome,* there is some confusion. The difference between the two words is slight. An ecosystem is much smaller than a biome. Actually, the larger biome can be considered as many similar ecosystems throughout the world grouped together. An ecosystem can be as large as the Amazon River ecosystem, or as small as a vernal pool. Thus, a biome is a larger region than an ecosystem and is characterized by habitat conditions and by its community structure. The predominant type of plants that grow there characterizes each biome. In biomes there is a strong relationship between climate and life, which suggests that if we know the climate of an area, we can predict what biome will be found there. The distribution of biomes results from interaction of physical features of the Earth. The two key physical features of the Earth are the amount of solar heat and global atmospheric circulation. Together these factors dictate local climate. The two most important climatic factors are precipitation and temperature.

✓ **Important Point:** Each biome contains many ecosystems whose communities have adapted to differences in climate, soil, and other environmental factors.

The major terrestrial biomes of the world include:

• Desert
• Grassland
• Savanna
• Mediterranean shrublands (chaparral)
• Tropical dry forest
• Tropical rainforest
• Temperate deciduous forest
• Taiga or boreal forest
• Tundra

DESERTS

Approximately one-third of the Earth's land surface is desert, arid land with meager rainfall (<25 cm/yr) that supports only sparse vegetation (depends on water conservation) and a limited population of people and animals. Deserts can be cool or cold during parts of the year; temperature can vary greatly during a 24-hour period. Desert organisms have evolved adaptations to help them survive: restricting activity to times of year when water is present; avoiding high temperatures by living in deep, cool, and sometimes moist burrows; emerging only at night when temperatures are lower; and drinking large quantities of water when it is available (camels) and then surviving long, dry periods. The world's great deserts are located in interiors of continents: the Sahara in Africa and the Gobi in Asia.

GRASSLANDS

Grasslands, also known as *temperate grasslands, prairies,* and *steppes,* develop wherever rainfall is not high enough to produce a forest nor light enough to form a desert. Grasslands are widely distributed throughout temperate regions, halfway between equator and poles. Precipitation is approximately 25–75 cm per year. Generally, grasslands tend to be windy with hot summers and cold to mild winters. Grasses make up 60%–90% of vegetation—trees are rare except along water courses due to their need for greater amounts of water. Fire in grasslands is important in preventing invasion of trees and in releasing nutrients from dead plants into the soil, contributing to high fertility of grassland soils. Grasslands are often highly productive when converted to agricultural use. The prairies of the United States and Canada were originally occupied by grasslands. Roots of perennial grasses characteristically penetrate deep into the soil; these soils, therefore, tend to be deep and fertile. Temperate grasslands were often populated by huge herds of grazing mammals. For example, in the United States countless numbers of bison and pronghorns once inhabited prairies—these herds are almost gone now and most prairies have been converted into the richest agricultural region on Earth.

SAVANNA

The term *savanna* is believed to have originally come from an Amerindian word describing "land which is without trees but with much grass either tall or short" (Valdes, 1535). By the late 1800s, it was used to mean "land with both grass and trees." It now refers to land with grass and either scattered trees, or an open canopy of trees. Precipitation is 50 cm to 150 cm/yr but occurs seasonally with periods of heavy rainfall followed by prolonged drought—results in a seasonally structured ecosystem. Predominant plants are grasses with widely spaced, drought-resistant trees. Many animals and plants are active only during rainy season. Fire is common but trees tend to be fire resistant. Savannas are increasingly being converted to agricultural use, causing inhabitants to struggle to survive.

MEDITERRANEAN SHRUBLANDS (CHAPARRAL)

Mediterranean shrublands biome, also known as *chaparral,* is found in a little bit of most of the continents—the west coast of the United States, the west coast of South America, the Cape Town area of South Africa, the western tip of Australia and the coastal areas of the Mediterranean. Precipitation is 40–100 cm/yr with wet, cool winters and hot, dry summers. Vegetation is dominated by woody shrubs adapted to hot, dry summers. Fire is a common feature.

TROPICAL DRY FOREST

The tropical dry forest biome typically experiences an annual hard dry season. Precipitation range is 50–20 cm/yr and many areas exhibit a monsoon climate— several months of heavy rainfall followed by dry periods. Because of highly seasonal rainfall, plants must be drought resistant. Many of the tree species in dry forest are also deciduous. During the driest months these species will drop their leaves. This dry season leaf-drop reduces the water needs of the plant, as there is no evapotranspiration through the leaves.

TROPICAL RAINFOREST

Tropical rainforests are warm (relatively constant) with no frost and are very wet (precipitation is >200 cm/yr). Rainforests, located near the equator, contain many tall trees. They are made up of the most diverse ecosystems on Earth, containing approximately half of all species of terrestrial plants and animals. For example, in one square mile of tropical forest in Brazil there are 1,200 species of butterflies. Rainforest communities are diverse. Each kind of organism is often represented in a given area by only a few individuals. Typically, individuals of the same species are separated by 1 km or more. Most nutrients are tied up in biomass and not in the soil; thus, rainforests do not make good farmlands.

TEMPERATE DECIDUOUS FORESTS

The temperate zones are characterized by having four seasons with summers and winters of roughly equal length. A deciduous forest is primarily composed of deciduous trees—those that lose their leaves once a year. Trees are major producers. A temperate deciduous forest is one that is in a temperate zone. Precipitation ranges from 75 cm/yr to 100 cm/yr and is evenly distributed. Temperate zones have warm summers and mild winters. Plants grow actively for approximately half the year. These zones are often populated by deer, beaver, bear, and raccoon. They generally have a lower number of species but a higher number of individuals per species.

TAIGA

The taiga or boreal forest, one of the largest ecosystems on Earth, exists as a nearly continuous belt of coniferous trees (spruce, fir, larch, etc.) across North America and Eurasia. Coniferous trees with their needle-shaped leaves limit moisture loss and their pyramid shape accommodates snowfall. Mean annual rainfall is 25 cm/yr to 100 cm/yr. Climate is humid due to low evaporation resulting from generally low temperatures. Winters are long and cold—soil freezes in winter. Few people live within these regions because of the very short growing season for farming.

TUNDRA

The tundra is a treeless area between the icecap and the tree line of Arctic regions. It is an area having permanently frozen subsoil (permafrost) and supporting low-growing vegetation such as lichens, mosses, and stunted shrubs. Annual precipitation is approximately 25 cm. Spring and summer temperatures are usually less that 10° C (50° F). Tundra is populated by large grazing mammals such as musk-oxen, caribou, and reindeer; and carnivores such as wolves, foxes, and lynx.

Community Ecology

In ecology, *community ecology* deals with a group of interacting populations in time and space. Communities can be recognized and studied at any number of levels, scales, and sizes. Sometimes a particular subgroup may be specified, such as the fish community in a lake or the soil arthropod community in a forest. Modern community ecology examines patterns such as variation in species, richness, equitability, productivity, and food web structure.

In terms of trophic (feeding) relationships, species comprising communities "function" as:

- **Photosynthesizers**—producers
- **Herbivores**—primary consumers

- **Carnivores**—secondary, tertiary consumers
- **Decomposers**—organisms, such as bacteria, mushrooms, and fungi, that obtain nutrients by breaking down complex matter in the wastes and dead bodies of other organisms
- **Omnivores**—obtain food from more than one trophic level

Emergent properties of communities include:

- Species diversity
- Limits of similarity of competing species
- Food web structure
- Community biomass and productivity

✓ **Important Point:** A central goal in the field of community ecology is to understand the processes that explain the structure (pattern) of a community, that is, the composition of the species, and their abundances and distributions.

Population Ecology

The basics of *population* is explained in the *Webster's Third New International Dictionary* definition:

- "The total number of amount of things especially within a given area."
- "The organisms inhabiting a particular area or biotype."
- "A group of interbreeding biotypes that represents the level of organization at which speciation begins."

A *population system* or life system is a population with its effective environment (Berryman, 1981).

The major components of a population system include:

1. **Population itself:** Organisms in the population can be subdivided into groups according to their age, sex, and other characteristics.
2. **Resources:** Organisms can be identified by their food, shelters, nesting places, space, and other resources.
3. **Enemies:** Organisms can be identified by their predators, parasites, pathogens, and other enemies.
4. **Environment:** Organisms can be identified by the characteristics of their environment, including air, water, soil, temperature composition, and variability of these characteristics in time and space (Sharov, 1997).

Population ecology is the branch of ecology that studies the structure and dynamics of populations. The term *population* is interpreted differently in various sciences. For example, in human demography a population is a set of humans in a given area. In genetics, a population is a group of interbreeding individuals of the same species, which is isolated from other groups. In population ecology, a population is a group of individuals of the same species inhabiting the same area (Sharov, 1996).

✓ **Important Point:** The primary axiom of population ecology is that organisms in a population are ecologically equivalent. Ecological equivalency means:

1. Organisms undergo the same life cycle.
2. Organisms in a particular stage of the life cycle are involved in the same set of ecological processes.
3. The rates of these processes (or the probabilities of ecological events) are basically the same if organisms are put into the same environment (however, some individual variation may be allowed).

When measuring populations, the level of species or density must be determined. Density (*D*) can be calculated by counting the number of individuals in the population (*N*) and dividing this number by the total units of space (*S*) the counted population occupies. Thus, the formula for calculating density becomes:

$$D = N/S \qquad\qquad (29.1)$$

When studying aquatic populations, the occupied space (*S*) is determined by using length, width, and depth measurements. The volumetric space is then measured in cubic units.

Population density may change dramatically. For example, if a dam is closed off in a river midway through spawning season, with no provision allowed for fish movement upstream (a fish ladder), it would drastically decrease the density of spawning salmon upstream. Along with the swift and sometimes unpredictable consequences of change, it can be difficult to draw exact boundaries between various populations. Pianka (1988) makes this point in his comparison of European starlings that were introduced into Australia with starlings that were introduced into North America. He points out that these starlings are no longer exchanging genes with each other; thus, they are separate and distinct populations.

The population density or level of a species depends on natality, mortality, immigration, and emigration. Changes in population density are the result of both births and deaths. The birth rate of a population is called *natality* and the death rate *mortality*. In aquatic populations, two factors besides natality and mortality can affect density. For example, in a run of returning salmon to their spawning grounds, the density could vary as more salmon migrated in or as others left the run for their own spawning grounds. The arrival of new salmon to a population from other places is termed *immigration* (*ingress*). The departure of salmon from a population is called *emigration* (*egress*). Thus, natality and immigration increase population density, whereas mortality and emigration decrease it. The net increase in population is the difference between these two sets of factors.

DISTRIBUTION (DISPERSION)

In ecology, *distribution* relates to where organisms are found on Earth—determined by biotic and abiotic factors. For example, large trees (pines, oaks, hickories) are not found in dry habitats like grasslands and deserts because there is not enough rainfall to sustain their growth—evolution has adapted trees to moist habitats. Moreover, organisms might be absent from a habitat due to the presence of predator species or competing species.

Where a particular species of organism does occur, the spatial relationships of individual organism to one another might take several different forms—this is called *dispersion*. Dispersion is defined as the spatial distribution of individuals of a population. Each organism occupies only those areas that can provide for its requirements, resulting in an irregular dispersion. How a particular population is dispersed within a given area has considerable influence on density. As shown in the following list, organisms in nature may be dispersed in three ways.

1. In *random dispersion,* there is an equal probability of an organism occupying any point in space, and "each individual is independent of the others" (Smith, 1974). A randomly dispersed habitat is relatively uniform so individuals are neither repelled nor attracted to one another.
2. In a *regular* or *uniform distribution,* in turn, organisms are spaced more evenly; they are not distributed by chance; they are all about the same distance from one another. Animals compete with each other and effectively defend a specific territory, excluding other individuals of the same species. Another example is seen in desert shrubs that compete for water and often display a regular dispersion pattern. In regular or uniform distribution, the competition between individuals can be quite severe and antagonistic to the point where spacing generated is quite even (Odum, 1983).
3. The most common distribution is the *contagious, clumped,* or *aggregated dispersion* where organisms are found in groups; this may reflect the heterogeneity of the habitat. Smith points out that contagious or clumped dispersions "produce

aggregations, the result of response by plants and animals to habitat differences" (Smith, 1974). Organisms that exhibit a contagious, clumped, or aggregated dispersion may develop social hierarchies in order to live together more effectively. Resources and suitable habitat may be patched in dispersion, which causes organisms to form clumps (e.g., pillbugs under rocks). Some organisms may come together in clumps for breeding purposes (e.g., seals).

✓ **Important Point:** Animals within the same species have evolved many symbolic aggressive displays that carry meanings that are not only mutually understood but also prevent injury or death within the same species. For example, in some mountainous regions, dominant male bighorn sheep force the juvenile and subordinate males out the territory during breeding season. In this way, the dominant male gains control over the females and need not compete with other males.

POPULATION GROWTH

The size of animal populations is constantly changing due to natality, mortality, emigration, and immigration. As mentioned, the population size will increase if the natality and immigration rates are high. On the other hand, it will decrease if the mortality and emigration rates are high. Each population has an upper limit on size, often called the *carrying capacity*. Carrying capacity can be defined as being the "optimum number of species' individuals that can survive in a specific area over time" (Enger et al, 1989). Stated differently, the carrying capacity is the maximum number of species that can be supported in a bioregion. A pond may be able to support only a dozen frogs depending on the food resources for the frogs in the pond. If there were thirty frogs in the same pond, at least half of them would probably die because the pond environment wouldn't have enough food for them to live. Carrying capacity is based on the quantity of food supplies, the physical space available, the degree of predation, and several other environmental factors.

The carrying capacity is of two types: ultimate and environmental. Ultimate carrying capacity is the theoretical maximum density; that is, it is the maximum number of individuals of a species in a place that can support itself without rendering the place uninhabitable. The environmental carrying capacity is the actual maximum population density that a species maintains in an area. Ultimate carrying capacity is always higher than environmental.

✓ **Important Point:** The population growth for a certain species may exhibit several types of growth.

Density-dependent factors are those that increase in importance as the size of the population increases. For example, as the size of a population grows, food and space may become limited. The population has reached the carrying capacity. When food and space become limited, growth is suppressed by competition. Odum describes density-dependent factors as acting "like governors on an engine and for this reason are considered one of the chief agents in preventing overpopulation" (Odum, 1983).

✓ **Important Point:** Density-independent factors are those that have the same effect on population regardless of size. Typical examples of density-independent factors are devastating forest fires, streambeds drying up, or the destruction of the organism's entire food supply by disease. Thus, population growth is influenced by multiple factors. Some of these factors are generated within the population, others from without. Even so, usually no single factor can account fully for the curbing of growth in a given population. It should be noted, however, that humans are, by far, the most important factor; their activities can increase or exterminate whole populations.

POPULATION RESPONSE TO STRESS

As mentioned earlier, population growth is influenced by multiple factors. When a population reaches its apex of growth (its carrying capacity), certain forces work to maintain population at a certain level. On the other hand, populations are exposed to small or moderate environmental stresses. These stresses work to affect the stability or persistence of the population. Ecologists have concluded that a major factor that affects population stability or persistence is species diversity. Species diversity is a measure of the number of species and their relative abundance.

Species diversity is related to several important ecological principles. For example, under normal conditions, high species diversity, with a large variety of different species, tends to spread risk. This is to say that ecosystems in a fairly constant or stable environment, such as a tropical rain forest, usually have higher species diversity. However, as Odum (1983) points out, "diversity tends to be reduced in stressed biotic communities."

If the stress on an ecosystem is small, the ecosystem can usually adapt quite easily. Moreover, even when severe stress occurs, ecosystems have a way of adapting. Severe environmental change to an ecosystem can result from such natural occurrences as fires, earthquakes, and floods; and from people-induced changes such as land clearing, surface mining, and pollution.

Ecosystems can and do change. For example, if a forest is devastated by a fire, it will grow back, eventually, because of ecological succession. *Ecological succession* is the observed process of change (a normal occurrence in nature) in the species structure of an ecological community over time; that is, a gradual and orderly replacement of plant and animal species takes place in a particular area over time. The result of succession is evident in many places. For example, succession can be seen in an abandoned pasture. It can be seen in any lake and any pond. Succession can even be seen where weeds and grasses grow in the cracks in a tarmac, roadway, or sidewalk.

Additional specific examples of observable succession include the following:

1. Consider a red pine planting area where the growth of hardwood trees (including ash, poplar, and oak) occurs. The consequence of this hardwood tree growth is the increased shading and subsequent mortality of the sun-loving red pines by the shade-tolerant hardwood seedlings. The shaded forest floor conditions generated by the pines prohibit the growth of sun-loving pine seedlings and allow the growth of the hardwoods. The consequence of the growth of the hardwoods is the decline and senescence of the pine forest.
2. Consider raspberry thickets growing in the sunlit forest sections beneath the gaps in the canopy generated by wind-thrown trees. Raspberry plants require sunlight to grow and thrive. Beneath the dense shade canopy particularly of red pines but also dense stands of oak, there is not sufficient sunlight for the raspberry's survival. However, in any place in which there has been a tree fall, the raspberry canes proliferate into dense thickets. Within these raspberry thickets, by the way, are dense growths of hardwood seedlings. The raspberry plants generate a protected "nursery" for these seedlings and prevent a major browser of tree seedlings (the white-tailed deer) from eating and destroying the trees. By providing these trees a shaded haven in which to grow, the raspberry plants are setting up the future tree canopy that will extensively shade the future forest floor and consequently prevent the future growth of more raspberry plants!

Succession usually occurs in an orderly, predictable manner. It involves the entire system. The science of ecology has developed to such a point that ecologists are now able to predict several years in advance what will occur in a given ecosystem. For example, scientists know that if a burned-out forest region receives light, water, nutrients, and an influx or immigration of animals and seeds, it will eventually develop into another forest through a sequence of steps or stages.

Two types of ecological succession are recognized by ecologists: primary and secondary. The particular type that takes place depends on the condition at a particular site at the beginning of the process.

Primary succession, sometimes called *bare-rock succession,* occurs on surfaces such as hardened volcanic lava, bare rock, and sand dunes, where no soil exists, and where nothing has ever grown before. Obviously, in order to grow, plants need soil. Thus, soil must form on the bare rock before succession can begin. Usually this soil formation process results from weathering. Atmospheric exposure—weathering, wind, rain, and frost—forms tiny cracks and holes in rock surfaces. Water collects in the rock fissures and slowly dissolves the minerals out of the rock's surface. A pioneer soil layer is formed from the dissolved minerals and supports such plants as lichens. Lichens gradually cover the rock surface and secrete carbonic acid, which dissolves additional minerals from the rock. Eventually, the lichens are replaced by mosses. Organisms called *decomposers* move in and feed on dead lichen and moss. A few small animals such as mites and spiders arrive next. The result is what is known as a *pioneer community*. The pioneer community is defined as the first successful integration of plants, animals, and decomposers into a bare-rock community.

After several years, the pioneer community builds up enough organic matter in its soil to be able to support rooted plants like herbs and shrubs. Eventually, the pioneer community is crowded out and is replaced by a different environment. This, in turn, works to thicken the upper soil layers. The progression continues through several other stages until a mature or climax ecosystem is developed, several decades later. It is interesting to note that in bare-rock succession, each stage in the complex succession pattern dooms the stage that existed before it. According to Tomera (1989), "mosses provide a habitat most inhospitable to lichens, the herbs will eventually destroy the moss community, and so on until the climax stage is reached" (p. 33).

Secondary succession is the most common type of succession. Secondary succession occurs in an area where the natural vegetation has been removed or destroyed but the soil is not destroyed. For example, succession that occurs in abandoned farm fields, known as *old field succession,* illustrates secondary succession. An example of secondary succession can be seen in the Piedmont region of North Carolina. Early settlers of the area cleared away the native oak-hickory forests and cultivated the land. In the ensuing years, the soil became depleted of nutrients, reducing the soil's fertility. As a result, farming ceased in the region a few generations later, and the fields were abandoned. Some 150 to 200 years after abandonment, the climax oak-hickory forest was restored.

Flow of Energy in an Ecosystem

Simply defined, energy is the ability or capacity to do work. For an ecosystem to exist, it must have energy. All activities of living organisms involve work, which is the expenditure of energy. This means the degradation of a higher state of energy to a lower state. The flow of energy through an ecosystem is governed by two laws: the first and second laws of thermodynamics.

The first law, sometimes called the *conservation law,* states that energy may not be created or destroyed. The second law states that no energy transformation is 100% efficient. That is, in every energy transformation, some energy is dissipated as heat. The term *entropy* is used as a measure of the nonavailability of energy to a system. Entropy increases with an increase in dissipation. Because of entropy, input of energy in any system is higher than the output or work done; thus, the resultant efficiency is less than 100%.

Odum (1983) explains that "the interaction of energy and materials in the ecosystem is of primary concern of ecologists" (p. 316). In biogeochemical nutrient cycles, it is the flow of energy that drives these cycles. Moreover, it should be noted that energy does not cycle as nutrients do in biogeochemical cycles. For example, when food passes from one organism to another, energy contained in the food is reduced step by step until all the energy in the system is dissipated as heat. Price refers to this process as "a unidirectional flow of energy through the system, with no possibility for recycling of energy" (Price, 1984, p. 66). When water or nutrients are recycled, energy is required. The energy expended in this recycling is not recyclable. And, as Odum points out, this is a "fact not understood by those who think that artificial recycling of man's resources is somehow an instant and free solution to shortages" (Odum, 1975, p. 118).

As pointed out earlier, the principal source of energy for any ecosystem is sunlight. Green plants, through the process of photosynthesis, transform the sun's energy into carbohydrates, which are consumed by animals. This transfer of energy, as stated previously, is unidirectional from producers to consumers. Often this transfer of energy to different organisms is called a *food chain.*

All organisms, alive or dead, are potential sources of food for other organisms. All organisms that share the same general type of food in a food chain are said to be at the same trophic level (nourishment or feeding level). Since green plants use sunlight to produce food for animals, they are called the *producers,* or the *first trophic level.* The *herbivores,* which eat plants directly, are called the *second trophic level* or the *primary consumers.* The *carnivores* are flesh-eating consumers; they include several trophic levels from the third on up. At each transfer, a large amount of energy (about 80% to 90%) is lost as heat and wastes. Thus, nature normally limits food chains to four or five links. However, in aquatic ecosystems, "food chains are commonly longer than those on land" (Dasmann, 1984, p. 202). The aquatic food chain is longer because several predatory fish may be feeding on the plant consumers. Even so, the built-in inefficiency of the energy-transfer process prevents development of extremely long food chains.

Only a few simple food chains are found in nature. Most simple food chains are interlocked. This interlocking of food chains forms a food web. A food web can be characterized as a map that shows what eats what (Miller, 1988). Most ecosystems support a complex food web. A food web involves animals that do not feed on one trophic level. For example, humans feed on both plants and animals. The point is an organism in a food web may occupy one or more trophic levels. Trophic level is determined by an organism's role in its particular community, not by its species. Food chains and webs help to explain how energy moves through an ecosystem.

An important trophic level of the food web that has not been discussed thus far is composed of the decomposers. The decomposers feed on dead plants or animals and play an important role in recycling nutrients in the ecosystem. As Miller (1988) points out, "there is no waste in ecosystems. All organisms, dead or alive, are potential sources of food for other organisms" (p. 312).

FOOD CHAIN EFFICIENCY

Earlier, we pointed out that energy from the sun is captured (via photosynthesis) by green plants and used to make food. Most of this energy is used to carry on the plant's life activities. The rest of the energy is passed on as food to the next level of the food chain.

✓ **Important Point:** A food chain is the path of food from a given final consumer back to a producer.

It is important to note that nature limits the amount of energy that is accessible to organisms within each food chain. Not all food energy is transferred from one trophic level to the next. For ease of calculation, ecologists often assume an ecological efficiency of 10% (10% rule) to estimate the amount of energy transferred through a food chain. For example, if we apply the 10% rule to the diatoms–copepods–minnows–medium fish–large fish food chain, we can predict that 1,000 grams of diatoms produce 100 grams of copepods, which will produce 10 grams of minnows, which will produce 1 gram of medium fish, which, in turn, will produce 0.1 gram of large fish. Thus, only about 10% of the chemical energy available at each trophic level is transferred and stored in usable form at the next level. What happens to the other 90%? The other 90% is lost to the environment as low-quality heat in accordance with the second law of thermodynamics.

✓ **Important Point:** When an organism loses heat, it represents one-way flow of energy out of the ecosystem. Plants only absorb a small part of energy from the sun. Plants store half of the energy and lose the other half. The energy plants lose is metabolic heat. Energy from a primary source will flow in one direction through two different types of food chains. In a grazing food chain, the energy will flow from plants (producers) to herbivores, and then through some carnivores. In detritus-based food chains, energy will flow from plants through detrivores and decomposers. In terms of the weight (or biomass) of animals in many ecosystems, more of their body mass can be traced back to detritus than to living producers. Most of the time the two food webs intersect one another.

ECOLOGICAL PYRAMIDS

As we proceed in the food chain from the producer to the final consumer, it becomes clear that a particular community in nature often consists of several small organisms associated with a smaller and smaller number of larger organisms. A grassy field, for example, has a larger number of grass and other small plants, a smaller number of herbivores like rabbits, and an even smaller number of carnivores like fox. The practical significance of this is that we must have several more producers than consumers.

This pound-for-pound relationship, which takes more producers than consumers, can be demonstrated graphically by building an ecological pyramid. In an ecological pyramid, the number of organisms at various trophic levels in a food chain is represented by separate levels or bars placed one above the other with a base formed by producers and the apex formed by the final consumer. The pyramid shape is formed due to a great amount of energy loss at each trophic level. The same is true if numbers are substituted by the corresponding biomass or energy. Ecologists generally use three types of ecological pyramids: pyramids of number, biomass, and energy. Obviously, there will be differences among them. Some generalizations follow:

1. **Energy pyramids** must always be larger at the base than at the top (because of the second law of thermodynamics relating to the dissipation of energy as it moves from one trophic level to another).
2. Likewise, **biomass pyramids** (in which biomass is used as an indicator of production) are usually pyramid-shaped. This is particularly true of terrestrial systems and aquatic ones dominated by large plants (marshes), in which consumption by heterotrophs is low and organic matter accumulates with time. It is important to point out, however, that biomass pyramids can sometimes be inverted. This is especially common in aquatic ecosystems, in which the primary producers are microscopic planktonic organisms that multiply very

rapidly, have very short life spans, and are heavily grazed by herbivores. At any single point in time, the amount of biomass in primary producers is less than that in larger, long-lived animals that consume primary producers.

3. **Numbers pyramids** can have various shapes (and not be pyramids at all, actually) depending on the sizes of the organisms that make up the trophic levels. In forests, the primary producers are large trees and the herbivore level usually consists of insects, so the base of the pyramid is smaller than the herbivore level above it. In grasslands, the number of primary producers (grasses) is much larger than that of the herbivores above (large grazing animals).

To get a better idea of how an ecological pyramid looks and how it provides its information, we need to look at an example. The example used here is the energy pyramid. According to Odum (1983), the energy pyramid is a fitting example because among the "three types of ecological pyramids, the energy pyramid gives by far the best overall picture of the functional nature of communities" (p. 183).

In an experiment conducted in Silver Springs, Florida, Odum measured the energy for each trophic level in terms of kilocalories. A kilocalorie is the amount of energy needed to raise 1 cubic centimeter of water 1° C. When an energy pyramid is constructed to show Odum's findings, it takes on the typical upright form (as it must because of the second law of thermodynamics).

Simply put, according to the second law of thermodynamics, no energy transformation process is 100% efficient. This fact is demonstrated, for example, when a horse eats hay. The horse cannot obtain, for his own body, 100% of the energy available in the hay. For this reason, the energy productivity of the producers must be greater than the energy production of the primary consumers. When human beings are substituted for the horse, it is interesting to note that, according to the second law of thermodynamics, only a small population could be supported. But this is not the case. Humans also feed on plant matter, which allows a larger population. Therefore, if meat supplies become scarce, we must eat more plant matter. This is the situation we see today in countries where meat is scarce. Consider that if we all ate soybean, there would be at least enough food for 10 times as many of us as compared with a world where we all eat beef (or pork, fish, chicken, etc.).

PRODUCTIVITY

As mentioned previously, the flow of energy through an ecosystem starts with the fixation of sunlight by plants through photosynthesis. In evaluating an ecosystem, the measurement of photosynthesis is important. Ecosystems may be classified into highly productive or less productive. Therefore, the study of ecosystems must involve some measure of the productivity of that ecosystem.

Smith defines *production* (or, more specifically, *primary production,* because it is the basic form of energy storage in an ecosystem) as being "the energy accumulated by plants" (1974, p. 411). Stated differently, primary production is the rate at which the ecosystem's primary producers capture and store a given amount of energy, in a specified time interval. In even simpler terms, primary productivity is a measure of the rate at which photosynthesis occurs. Odum (1971) lists four successive steps in the production process as follows:

1. Gross primary productivity: the total rate of photosynthesis in an ecosystem during a specified interval
2. Net primary productivity: the rate of energy storage in plant tissues in excess of the rate of aerobic respiration by primary producers
3. Net community productivity: the rate of storage of organic matter not used
4. Secondary productivity: the rate of energy storage at consumer levels

When attempting to comprehend the significance of the term *productivity* as it relates to ecosystems, it is wise to consider an example. Consider the productivity of an agricultural ecosystem such as a wheat field. Often its productivity is expressed as the number of bushels produced per acre. This is an example of the harvest method for measuring productivity. For a natural ecosystem, several 1-square-meter plots are marked off, and the entire area is harvested and weighed to give an estimate of productivity as grams of biomass per square meter per given time interval. From this method, a measure of net primary production (net yield) can be measured.

Productivity, both in the natural and cultured ecosystem, may vary considerably, not only between type of ecosystems, but also within the same ecosystem. Several factors influence year-to-year productivity within an ecosystem. Such factors as temperature, availability of nutrients, fire, animal grazing, and human cultivation activities are directly or indirectly related to the productivity of a particular ecosystem.

The ecosystem that is of greatest importance in this particular study is the aquatic ecosystem. Productivity can be measured in several different ways in the aquatic ecosystem. For example, the production of oxygen may be used to determine productivity. Oxygen content may be measured in several ways. One way is to measure it in the water every few hours for a period of 24 hours. During daylight, when photosynthesis is occurring, the oxygen concentration should rise. At night the oxygen level should drop. The oxygen level can be measured by using a simple x-y graph. The oxygen level can be plotted on the y-axis with time plotted on the x-axis.

Another method of measuring oxygen production in aquatic ecosystems is to use light and dark bottles. Biochemical oxygen demand bottles (300 ml) are filled with water to a particular height. One of the bottles is tested for the initial dissolved oxygen (DO), and then the other two bottles (one clear, one dark) are suspended in the water at the depth they were taken from. After a 12-hour period, the bottles are collected and the DO values for each bottle are recorded. Once the oxygen production is known, the productivity in terms of grams per meter per day can be calculated.

References and Recommended Reading

Abrahamson, D. E. (Ed.). (1988). *The challenge of global warming.* Washington, DC: Island Press.

Adams, V. D. (1990). *Water and wastewater examination manual.* Chelsea, MI: Lewis Publishers.

Allen, J. D. (1996). *Stream ecology: Structure and function of running waters.* London: Chapman & Hall, p. 23.

American Heritage Dictionary of the English Language (4th ed.). (2000). Boston: Houghton Mifflin Company.

Barbour, M. T., Gerritsen, J., Snyder, B. D., & Stribling, J. B. (1997). *Revision to rapid bioassessment protocols for use in streams and rivers, periphyton, benthic macroinvertebrates and fish.* Washington, DC: United States Environmental Protection Agency. EPA 841-D-97-002, 1997.

Berryman, A. A. (1981). *Population systems: A general introduction.* New York: Plenum Press, p. 89.

Bradbury, I. (1991). *The biosphere.* New York: Belhaven Press.

Brown, L. R. (1994). *Facing food insecurity.* In Brown L. R., et al. (Eds.). *State of the world.* New York: W. W. Norton, pp. 179–187.

Camann, M. (1996). *Freshwater aquatic invertebrates: Biomonitoring.* Retrieved from www.humboldt.edu., pp. 1–4.

Campbell, N. A. (2004). *Biology: Concepts and connections* (4th CD-Rom ed). New York: Benjamin-Cummings.

Carson, R. (1962). *Silent spring.* Boston: Houghton Mifflin Company.

Cave, C. (1998). *Ecology.* Retrieved from http://web.netcom.com/cristiindex.htm, p. 1.

Cave, C. (2000). *How a river flows.* http://web.netcome/cristiindex.htm/cristi@ix.netcom.com, p. 3.

Clark, L. R., Gerier, P. W., Hughes, R. D., & Harris, R. F. (1967). *The ecology of insect populations.* London: Methuen, p. 73.

Cummins, K. W. (1994). Structure and function of stream ecosystems. *Bioscience* 24:631–641.

Cummins, K. W., & Klug, M. J. (1979). Feeding ecology of stream invertebrates. *Annual Review of Ecology and Systematics* 10:631–641.

Darwin, C. (1998). *The origin of species,* Suriano, G. (Ed.). New York: Grammercy, p. 112.

Davis, M. L., & Cornwell, D. A. (1991). *Introduction to environmental engineering*. New York: McGraw-Hill, Inc.

Dasmann, R. F. (1984). *Environmental conservation*. New York: John Wiley & Sons.

Enger, E., Kormelink, J. R., Smith, B. F., & Smith, R.J. (1989). *Environmental science: The study of interrelationships*. Dubuque, IA: Wm. C. Brown.

Environmental Protection Agency (EPA). (2006). *What is the scientific method?* Accessed 11/01/10 from http://www.epa.gov/maia/html/scientific.html.

Evans, R. (1965). Industrial wastes and water supplies. *Journal of the American Water Works Association* 57:625–628.

Evans, E. D., & Neunzig, H. H. (1996). *Megaloptera and aquatic nueroptera*. In R. W. Merritt & K. W. Cummins (Eds.). *Aquatic insects of North America* (3rd ed.). Dubuque, IA: Kendall/Hunt Publishing Company, pp. 298–308.

Freedman, B. (1989). *Environmental ecology*. New York: Academic Press.

Hamburg, M. (1987). *Statistical analysis for decision making*. New York: Harcourt Brace Jovanovich.

Hewitt, C. N., & Allott, R. (1992). *Understanding our environment: An introduction to environmental chemistry and pollution*, R. M. Harrison (Ed.). Cambridge, Great Britain: The Royal Society of Chemistry.

Hickman, C. P., Roberts, L. S., & Hickman, F. M. (1998). *Integrated principles of zoology*. St. Louis: Times Mirror/Mosby College Publishing.

Hickman, C. P., Roberts, L. S., & Hickman, F. M. (1990). *Biology of animals*. St. Louis: Times Mirror/Mosby College Publishing.

Huxley, T. H. (1876). *Science & Education, Volume III, Collected Essays*. New York: D. Appleton & Company.

Goshorn, D. (2006). Proceedings—DELMARVA Coastal Bays Conference III: Tri-State Approaches to Preserving Aquatic Resources. US Environmental Protection Agency.

Jeffries, M., & Mills, D. (1990). *Freshwater ecology: Principles and applications*. London: Belhaven Press.

Jones, A. M. (1997). *Environmental biology*. New York: Routledge.

Karr, J. R., & Dudley, D. R. (1981). Ecological perspective on water quality goals. *Environmental Management* 5:55–68.

Karr, C. (1991). Biological integrity and the goal of environmental legislations: Lessons for conservation biology. *Conservation Biology* 4:66–84.

Keeton, W. T. (1996). *Biological science*. Norwell, MA: R. S. Means Company.

Kimmel, W. G. (1983). The impact of acid mine drainage on the stream ecosystem. In S. K. Majumdar & W. W. Miller (Eds.). *Pennsylvania coal: Resources, technology and utilization*. Pennsylvania Academy of Science, pp. 424–437.

King, R. M. (2003). *Biology made simple*. New York: Broadway Books.

Koch, R. (1882). "Uber die Atiologie der Tuberkulose." In *Verhandlungen des Knogresses fur Innere Medizin*. Erster Kongress, Wiesbaden.

Koch, R. (1884). *Mitt Kaiser Gesundh* 2:1–88.

Koch R. (1893). *J Hyg Inf* 14:319–333.

Lafferty, P., & Rowe, J. (1993). *The dictionary of science*. New York: Simon & Schuster.

Larsson, K. A. (1993). Prediction of the pollen season with a cumulated activity method. *Grana*, 32:111–114.

Leopold, L. B. (1970). *A view of the river.* Cambridge, MA: Harvard University Press, p. 36.

Mackie, G. I. (1998). *Applied aquatic ecosystem concepts.* Ontario, Canada: University of Guelph.

Madsen, J. (1985). *Up on the river.* New York: Lyons Press, pp. 8–15.

Mason, C. F. (1990). *Biological aspects of freshwater pollution. Pollution: Causes, effects, and control,* R. M. Harrison (Ed.). Cambridge, Great Britain: The Royal Society of Chemistry.

Masters, G. M. (1991). *Introduction to environmental engineering & science.* Englewood Cliffs, NJ: Prentice Hall.

McCafferty, P. W. (1981). *Aquatic entomology, the fishermen's and ecologists' illustrated guide to insects and their relatives.* Boston: Jones and Bartlett.

McGhee, T. J. (1991). *Water supply and sewerage.* New York: McGraw-Hill.

Med Net. (2006). *Definition of Koch's postulates.* Medicine Net.com.

Michaud, J. P. (1994). *A citizen's guide to understanding and monitoring lakes and streams. Publications #94-149.* Olympia, Washington State Dept. of Biology, pp. 1–13.

Miller, G. T. (1988). *Environmental science: An introduction.* Belmont, CA: Wadsworth.

Moran, J. M., Morgan, M. D., & Wiersma, J. H. (1986). *Introduction to environmental science.* New York: W. H. Freeman and Company.

Naar, J. (1990). *Design for a livable planet.* New York: Harper & Row.

O'Connor, D., & Dobbins, W. (1975). The mechanism of reaeration in natural streams. *Journal of Hydraulics Division: Proceedings of American Society of Civil Engineers* 101:1315.

Odum, E. P. (1971). *Fundamentals of ecology.* Philadelphia: Saunders College.

Odum, E. P. (1975). *Ecology: The link between the natural and the social sciences.* New York: Holt, Rinehart and Winston.

Odum, E. P. (1983). *Basic ecology.* Philadelphia: Saunders College.

Overcash, M. R., & Davedson, J. M. (1981). *Environmental impact of nonpoint source pollution.* Ann Arbor, MI: Ann Arbor Science.

Penneck, R. W. (1989). *Freshwater invertebrates of the United States* (3rd ed.), Hoboken, NJ: Wiley, p. 189.

Pianka, E. R. (1988). *Evolutionary ecology.* New York: Harper Collins.

Porteous, A. (1992). *Dictionary of environmental science and technology.* New York: John Wiley & Sons.

Price, P. W. (1984). *Insect ecology.* New York: John Wiley & Sons.

SGM. (2006). *The scientific method, fish health and pfiesteria.* College Park: University of Maryland, National Oceanic and Atmospheric Administration.

Sharov, A. A. (1992). Life-system approach: A system paradigm in population ecology. *Oikos* 63:485–494.

Sharov, A. (1997). *Population ecology.* http://www.gypsymoth.ent.vt.edu/sharov/population/welcome.html, p. 1.

Sharov, A. (1996). *What is population ecology?* Blacksburg, VA: Department of Entomology Virginia Tech. University, pp. 1–2.

Smith, R. A., Alexander, R. B., & Wolman, M. G. (1987). Water-quality trends in the Nation's rivers. *Science* 235(4796):1607–1615.

Smith, R. L. (1974). *Ecology and field biology.* New York: Harper & Row.

Spellman, F. R. (2009). *Biology for non-biologists.* Latham, MD: Government Institutes.

Spellman, F. R., & Price-Bayer, J. (2011). *In defense of science.* Latham, MD: Government Institutes.

Spellman, F. R., & Whiting, N. E. (2006). *Environmental science and technology* (2nd ed.). Rockville, MD: Government Institutes.

Spieksma, F. T. (1991). Aerobiology in the nineties: Aerobiology and pollinosis, *International Aerobiology Newsletter* 34:1–5.

Stewart, K. W., & Stark, B. P. (1998). *Nymphs of North American stonefly general (Plecoptera),* Volume 12. Denton, TX: Thomas Say Foundation.

Tansley, A. G. (1935). The use and abuse of vegetational concepts and terms. *Ecology* 16:284–307.

Tomera, A. N. (1989).*Understanding basic ecological concepts.* Portland, ME: J. Weston Walch.

U.S. Census Bureau (2006). World population information. Accessed 11/01/10 from http://www.census.gov/ipc/www/world.html.

Wackernagel, M. (1997). *Framing the sustainability crisis: Getting from concerns to action.* Accessed from http://www.sdri.ubc.ca/publications/wacherna.html

Welch, P. S. (1963). *Limnology.* New York: McGraw-Hill.

Westman, W. E. (1985). *Ecology, impact assessment, and environmental planning.* New York: John Wiley & Sons.

Wetzel, R. G. (1983). *Limnology.* New York: Harcourt Brace Jovanovich College.

Wooton, A. (1984). *Insects of the world.* New York: Facts on File.

World Resources 1986. New York: World Resources Institute (WRI) and International Institute of Environment and Development (IIED).

World resources 1988–89. New York: World Resource Institute (WRI) and International Institute of Environment and Development (IIED), p. 133.

Part VII
GEODESY

CHAPTER 30

Measuring Earth

> In the next century, planet earth will don an electronic skin.
> It will use the Internet as a scaffold to support and transmit
> its sensations. This skin is already being stitched together.
> It consist of millions of embedded electronic measuring de-
> vices: thermostats, pressure gauges, pollution detectors, cam-
> eras, microphones, glucose sensors, EKGs, electroencephalo-
> graphs. These will probe and monitor cities and endangered
> species, the atmosphere, our ships, highways and fleets of
> trucks, our conservations, our bodies—even our dreams.
>
> —Neil Gross, 1999

Based on experience, we have found that it is not uncommon to lump the geoscience disciplines of geography and geodesy into the same category as one entity. Many think they have the same meaning. In the past, geography, the study of physical and human landscapes, and the processes that affect them was the main discipline and geodesy was one of its subdisciplines. Currently, as a stand-alone discipline, geodesy is the science of measuring the size, shape, and gravity field of the Earth, which requires exact measurements.

To gain insight into the actual, applied function of geodesy, we have included a brief introduction to the parts of geography that are closely related to it; then we provide a complete description of geodesy and its separate importance in the practice of geoscience.

Geography

This section has been reproduced from F. R. Spellman (2010). *Geography for Non-Geographers*. Boca Raton, FL: CRC.

For many of us, about the only thing we remember about the geography we were taught in school were a few of those capitals we had to memorize. Also, there were those maps, maps everywhere: maps pinned to every wall, in textbooks, and on that globe of the world that stood in the corner of the classroom. Those maps and globes caused many of us to gaze at them and to let our minds escape in wonderment to so many mysterious, far-off places.

While the word geography is derived from Greek (*geo*-referring to Earth, and *–graphy* meaning picture or writing) and literally means "to write about the Earth," the subject of geography, as mentioned, is much more than the names of capitals and maps. Geography is a science, an all encompassing science that seeks to bridge and understand, through an understanding of location and place, the world's physical and human features. Geography teaches us where things are and how they got there; it looks at the spatial connection between people, places, and the Earth.

Some have called geography the "mother of all sciences." This title is well-suited and fitting because all other disciplines took root from its existence. Geography is not only an analytic tool, but also a spatial science (dealing with people, landscapes, money, and other infinite uses) that deals with many aspects of social science (e.g., history, psychology, and anthropology), physical science (e.g., geology, weather, and climate), and technical science (e.g., geographic information systems [GIS], geodesy, and remote sensing) (Bell, 2007).

Did You Know?

Most people have an idea what a geologist does but don't have a clue of what a geographer does. While geography is commonly divided into human geography and physical geography, the difference between physical geography and geology is often confusing. We distinguish the difference by pointing out that geographers study the surface of the Earth, its features, and why they are where they are. Geologists look deeper into the Earth and study its rocks, as well as every child's favorite, the dinosaurs.

Another view of geography states, "The more I work in the social-studies field the more convinced I become that Geography is the foundation of all. When I call it the 'queenly science', I do not visualize a bright-eyed young woman recently a princess but rather an elderly, somewhat beat-up dowager, knowing in the way of power." The author of this statement, James A. Michener (1970), who is also the author of such best-selling historical novels as *The Source, Hawaii, Centennial, Alaska, Texas, Caribbean, Chesapeake, Caravans, The Drifters*, and several others, knew a few things about geography and several other subjects. Using his wide-ranging geographic knowledge of Earth, Michener simply makes the point that geography is a parent to biology, geology, chemistry, physics, history, and economics. In light of this, geographers are the sons and daughters of the science geographers who as scientists study the relationship between people and their environments.

Michener (1970) made another important point about geography, one that is germane to the purpose of this text. Consider the following: "With growing emphasis on ecology and related problems of the environment, geography will undoubtedly grow in importance and relevance. I wish that the teaching of it were going to improve commensurately; most of the geography courses I have known were rather poorly taught and repelled the general student like me" (p. 188). Based on personal experience, I have found that Michener's sentiments are all too true, even at the college level.

Looking back, the state of training and the relevance of geography have not always been problematic. Consider the following by John Rennie Short (2004):

> There was a time when "formal geography texts constitute[d] most of the body of schoolbooks in the eighteenth and early nineteenth century in the United States. Texts on history, for example, did not appear until the 1880s. These early geography texts were compendiums of knowledge, widely used in schools and the more literate homes. The early geography texts were the encyclopedia of their day.

Mother of All Sciences

We need to be more careful when, in public discourse and popular culture, we use the stock phrase *the mother of all* (fill in the blank). Remember Saddam Hussein promised in a speech "the mother of all battles" if the U.S.-led coalition forces attempted to evict his army of occupation from Kuwait in 1991. The truth be told, the coalition-led invasion turned out for Hussein to be the mother of all slaughters, the mother of all debacles, the mother of all embarrassments. General Patton, who said in World War II that U.S. forces would run through the enemy "like crap through a goose," would have been proud. Hussein's army was our goose. Today, it is not uncommon to see or hear the current financial crisis gripping the globe as the mother of all bailouts. Who can argue against that? One thing is certain: If you are going to label something "as the mother of" anything, then you better be able to back it up.

The point is that when we say that geography is the mother of all sciences, we have to be careful with such usage. Some might question the validity of this statement; they might be somewhat baffled. However, the professional geographer knows that this statement is true and has little difficulty in defending its accuracy. The professional geographer would simply point out that nearly every other type of scientific profession or pursuit, be it ecology, geology, population study, demographics, agriculture, literacy, or wildfire or environmental management is based on the geographic location of the place under study. Geography plays a role in nearly every decision we make. Choosing sites; targeting market sectors; planning distribution networks; responding to emergencies; or redrawing city, county, state, or country boundaries—all of these problems involve questions in geography. Moreover, with the developing technology of GIS, which integrates software and data for capturing, managing, analyzing, and displaying forms of geographically referenced material, geography has been propelled into the digital age—and for the geographer and the rest of us, this is a good thing.

The professional geographer is a naturalist, an observer of the main facets of nature; he or she recognizes and understands the richness and variety of nature and can recognize the contours of landscapes, correlate satellite images with the area, read maps, and interpret landscape. Most of all, a geographer is a preserver of nature—the ultimate Mother of us all.

The Big Picture

Unlike the important but narrowly focused science of geology—the science that deals with and answers many questions about planet Earth—geography is holistic in its approach; it presents the big picture that displays the intertwining of environmental events and human actions.

All of the disciplines emanate from the core of geographic sciences. No one made this point clearer than the renowned American geologist and geographer, Nevin Fenneman. Dr. Fenneman (1865–1945) was noted for his work on the physiography (i.e., study of processes and patterns in the natural environment) of the United States (Short, 2004). He opined that there is in geography a central core that is pure geography and nothing else, but there is much beyond this core that is nonetheless geography, though it belongs also to overlapping sciences. Figure 30.1 is derived from Fenneman's work and it demonstrates his contention that "the seeds are in the core, and the core is regional geography, and this is why the subject propagates itself and maintains a separate existence" (Bell, 2007).

History of Geography

One of the main foundational factors adding to the credence of geography being classified as the mother of all sciences is its history. Geography has been

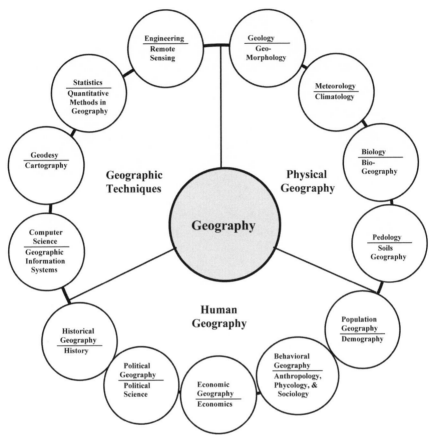

Figure 30.1. Shows the disciplines emanating from the core of geographic sciences.

Adaptation from Bell (2007).

around for more than 2,200 years, dating back to the Greeks when Eratosthenes supposedly coined the term from the words, as pointed out earlier, *geo* (Earth) and *graphein* (to write)—literally meaning *description of the Earth*. As geography developed, many other scientific disciplines developed later and became increasingly specialized.

As mentioned, the name *geography* is attributed to Eratosthenes, but it is Anaximander of Miletus (c. 610–545 BCE) who is considered to be the true founder of geography. Anaximander is credited with the early measurement of latitude and with the prediction of eclipses. From Europe to Africa to Asia, Eratosthenes, the Chinese, and Arabs amassed and synthesized incredible amounts of information about the Earth. Building on the body of knowledge of early geographers, Emmanuel Kant (1724–1804) distinguished between the two ways of classifying things: spatially (having the nature of space) and the category of time (temporal). Modern geography evolved from the works of German geographers Alexander von Humboldt and Carl Ritter. They moved geography away from the study of tomes of data to the study of regions for the ultimate understanding of Earth. Table 30.1 lists notable geographers.

Table 30.1.   Notable Geographers

Geographer	Period	Major Accomplishments
Eratosthenes	276–194 BCE	Calculated the size of the Earth.
Ptolemy	c. 90–168	Compiled Greek and Roman knowledge into the book *Geographica*.
Gerardus Mercator	1512–1594	Produced the Mercator projection.
Alexander von Humboldt	1769–1859	Considered the father of modern geography.
Carl Ritter	1779–1859	Also considered the father of modern geography.
Arnold Henry Guyot	1807–1884	Noted glaciologist.
William Rorris Davis	1850–1934	Considered the father of American geography; developer of the cycle of erosion.
Paul Vidal de la Blanche	1845–1918	Wrote the principles of human geography.
Sir Halford John Mackinder	1861–1947	Founder of the Geographical Association.
Walter Christaller	1893–1969	Human geographer and inventor of central place theory.
Yi-Fu Tuan	1930–	Started humanistic geography.
David Harvey	1935–	Author of theories on spatial and urban geography.
Edward Soja	1941–	Noted for his work on regional development, planning, and governance.
Michael Frank Goodchild	1944–	GIS scholar.
Doreen Massey	1944–	Key scholar in the space and places of globalization and its pluralities.
Nigel Thrift	1949–	Originator of nonrepresentational theory.

Branches of Geography

Earlier, in Figure 30.1, it was pointed out that geography is a merged, mingled, or integrative science and a synthesizer (combiner of parts) of knowledge. Each specialized field in geography, as shown in Figure 30.1, overlaps other related branches of science.

There are two major branches of geography:

- Physical geography is the study of the physical features and changes on the Earth's surface; focuses on geography as an Earth science.
- Cultural geography is the study of humans and their ideas, patterns, and processes that affect and shape human ideas and actions on the Earth.

Did You Know?

In the context of this text, *culture* is defined as the way of life that distinguishes a group of people (i.e., their religions, language, customs, beliefs, knowledge, law, lifestyles, foods, music, etc.).

Other branches of geography include:

- Biophysical geography is the study of the natural environment and the interrelationships of all the living things in that environment.
- Topography studies the shapes of the land and the bodies of water in a given location.
- Political geography studies the political organization of areas.
- Social geography studies groups of people, and the interrelationships among groups and communities.
- Economic geography is the study of resources and resource use.
- Historical geography is the study of the ways in which the relationships between people and their environments have changed over time.
- Urban geography focuses on cities.
- Cartography is the art and science of map-making
- Environmental geography describes the spatial aspects of interactions between humans and the natural world.
- Geomatics involves the use of traditional spatial techniques used in cartography and topography and their application to computers.
- Regional geography studies the regions of all areas across the Earth.

Tools of Geography

Every branch of science requires certain tools. Geography is no different. The most important tools of geography appear in the following list:

- Globes and maps
- GIS, remote sensing
- Observations
- Surveys
- Mathematical models

GLOBES AND MAPS

Because they are the most useful models of the Earth, globes and maps are the two basic tools of geographers. Globes and maps are not reality. Globe and map makers look at the surface of the Earth and then decide what to place on the globe or the map. The art and science of making globes and maps is called *cartography*. Again, globes and maps are not reality; they are subjective and not objective. We can get globes and maps to say or depict anything we want. So to define globes and maps, we can say that they are three-dimensional models (globes) or graphical representation of features (qualitative and quantitative) of the surface of the Earth.

Globes of Earth are often used in geography because they accurately represent the shape of the Earth, the shapes of landmasses and bodies of water. And they represent parallels and meridians, direction, and distance. Globes are the best tools to show the shortest distance between two places. The problem with globes is that they can be big and bulky as well as expensive. Another problem with using a globe is that only one half of the globe can be viewed at one time.

Although maps are flat representations of the Earth, they are easier to use than globes. For example, it is easier to carry a map just about anywhere; they are portable. A map provides an easy-to-use reference and can show the Earth's entire surface or just a particular part. Maps also show more detail such as a wide range of topics including physical and cultural features of Earth. Again, the problem with maps is that they are not reality and can have distortions (inaccuracies) because they are flat and not three-dimensional representations.

Map Key Terms

The following key terms and definitions, taken from the U.S. Geological Survey (USGS, 2006), are useful in any discussion of maps and globes.

- **Azimuth:** Refers to the angle measured in degrees between the base line radiating from a center point and another line radiating from the same point. Normally, the base line points north and degrees are measured clockwise from the base line.
- **Aspect:** Individual azimuthal map projections are divided into three aspects: the polar aspect, which is tangent at the pole; the equatorial aspect, which is tangent at the equator; and the oblique aspect, which is tangent anywhere else.
- **Conformality:** A map projection is conformal when at any point the scale is the same in every direction. Therefore, meridians and parallels intersect at right angles and the shapes of very small areas and angles with very short sides are preserved. The size of most areas, however, is distorted.
- **Developable surface:** Refers to a simple geometric form capable of being flattened without stretching. Many map projections can then be grouped by a particular developable surface: cylinder, cone, or plane.
- **Equal areas:** A map projection is equal in area if every part, as well as the whole, has the same area as the corresponding part on the Earth, at the same reduced scale. No flat map can be both equal in area and conformal.
- **Equidistant:** An equidistant map shows true distances only from the center of the projection or along a special set of lines. For example, an azimuthal equidistant map centered at Washington, DC, shows the correct distance between Washington, DC, and any other point on the projection. It shows the correct distance between Washington, DC, and San Diego and between Washington, DC, and Seattle. But it does not show the correct distance between San Diego and Seattle. No flat map can be both equidistant and equal in area.
- **Graticule:** Refers to the spherical coordinate system based on lines of latitude and longitude.
- **Great circle:** Refers to a circle formed on the surface of a sphere by a plane that passes through the center of the sphere. The equator, each meridian, and each other full circumference of the Earth forms a great circle. The arc of the great circle shows the shortest distance between points on the surface of the Earth.
- **Linear scale:** Refers to the relation between a distance on a map and the corresponding distance on the Earth. Scale varies from place to place on every map. The degree of variation depends on the projection used in making the map.
- **Map projection:** Refers to a systematic representation of a round body such as the Earth or a flat (plane) surface. Each map projection has specific properties that make it useful for specific purposes.
- **Rhumb line:** Refers to a line on the surface of the Earth cutting all meridians at the same angle. A rhumb line shows true direction. Parallels and meridians, which also maintain constant true directions, may be considered special cases of the rhumb line. A rhumb line is a straight line on a Mercator

projection. A straight rhumb line does not show the shorter distance between points on the Equator or on the same meridian.

Parts of Maps

Maps consist of parts that can be used to read the maps to analyze the physical and human landscapes of the world. Map parts are described in the following list.

- **Title** identifies map and contents.
- **Legend (key)** explains the meaning of colors and symbols used on the map and may include the key to elevation (distance above or below sea level).
- **Direction indicator** identifies direction or orientation on a map. Usually direction is provided on a map by a single arrow labeled "N" that points north. Other maps are printed with a compass rose (directional indicator) symbol that indicates direction on a map with arms that point to the cardinal and intermediate directions (see Figure 30.2). The cardinal directions are north, south, east, and west, and the intermediate directions are northeast, southeast, northwest, and southwest (see Figure 30.2).
- **Map scales** provide information used to measure distances on maps. Different scales are necessary on different maps for developing map representations because the size of a map in relation to the size of the real world is different; scale is shown by giving the ratio between distances on the map and actual

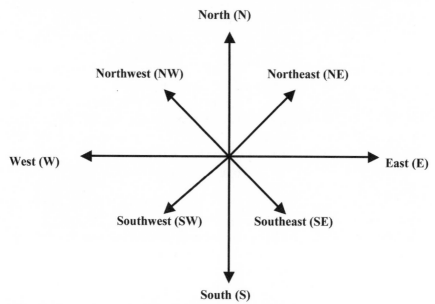

Figure 30.2. Map direction finder; compass rose.

distances on Earth. Areas can be represented using a variety of scales. The amount of detail shown on a map depends on the scale used.

- **Written or statement scale** uses a statement or phrase to relate a distance on the map to the distance it represents on Earth (i.e., 1 inch equals 4 miles; English units are used in the United States).
- **Representative fraction** uses fractions or ratios to relate distance on a map to distance on the Earth (i.e., 1:250,000 or 1/250,000).
- **Graphic or bar scale** is a short line that represents the number of miles or kilometers on the Earth's surface equal to a distance on the map (divided into equal parts labeled with miles or kilometers, as shown in Figure 30.3).

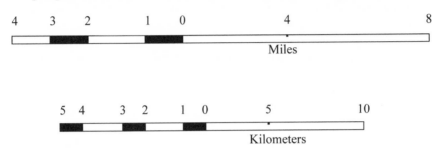

Figure 30.3.

- **Grid** is used to locate places on a map. Global grids are formed by the crossing of parallels (line of latitude) and meridians (lines of longitude). A letter or number coordinate grid system is used to locate places on maps of smaller places like state maps, city maps, and highway maps.

Types of Maps

Maps vary according to their purpose. All the different kinds of maps can be put into two broad categories based on the type of information they show.

- **General map** displays qualitative information, mainly showing a place or area size, or where things are located and their qualities—nominal and cardinal data.
- **Thematic map** is a general-purpose map that shows specific quantitative data or information, often on a single theme or topic. Examples of thematic maps include population, economic resource, language, ethnicity, climate, highway, and vegetation maps.

Map Projections (USGS, 2006)

Globes provide the most accurate depiction of surface features on Earth; they are the only true representation of distance, area, direction, and proximity.

However, you can't fold up a globe and put it into your pocket. Moreover, if you need to locate a particular city street in a specific city, you would need a very large globe. You could, of course, make a huge globe and then just cut the section or sections you need from the globe. I think you get my point. Globes are accurate but not hands-on-practical or easy to transport from place to place. This is the reason flat maps were developed—three-dimensional Earth is projected onto a two-dimensional map. You can fold up a flat map and carry it almost anywhere. However, even though these flat maps are convenient and do display spatial information, they give a distorted view. At present, there is no flat map that does not have some type of distortion. Simply, when a map projection is used to portray all or part of the Earth on a flat surface, it can't be done with some distortion.

Every map projection has its own set of advantages and disadvantages. There is no "best" projection. Some projection types are conical, planar, or cylindrical. Choice of projection depends on the ultimate use of the map.

- **Mercator projection** is a cylindrical projection that preserves shape. It is used for navigation or maps of equatorial regions. Distances are true only along the equator. This projection is not perspective, equal in area, or equidistant.
- **Transverse Mercator projection** is used by the USGS for many quadrangle maps; it is also used for mapping large areas that are mainly north–south in extent. It is mathematically projected on a cylinder tangent to a meridian.
- **Miller cylindrical projection** is used to represent the entire Earth in a rectangular frame. It is popular for world maps. The map is not equal in area, equidistant, conformal, or perspective.
- **Robinson projection** uses tabular coordinates rather than mathematical formulas to make the world "look right." It is not conformal, equal in area, equidistant, or perspective.
- **Orthographic projection** is used for perspective views of the Earth, Moon, and other planets. The Earth appears as it would on a photograph from deep space. The map is perspective but not conformal or equal in area.
- **Gnomonic projection** (meaning "sun dial-like" or "great circles in straight lines") is used along with the Mercator by some navigators to find the shortest path between two points. It is considered to be the oldest projection. It is geometrically projected on a plane. The point of projection is the center of a globe.
- **Lambert conformal conic** is used by the USGS for many 7.5- and 15-minute topographic maps and for the state base map series. It is also used to show a country or region that is mainly east–west in extent. This conic projection is mathematically projected on a cone conceptually secant at two standard parallels.

GEOGRAPHIC INFORMATION SYSTEMS (USGS, 2007)

GIS is the newest development on the geographical frontier. GIS unites databases and maps through the power of the computer. In addition, the computer aides in drawing maps rapidly and without the use of conventional drawing materials (pens and ink). Moreover, GIS can be used for scientific investigations, resource management, and development planning. For example, GIS might be used to find wetlands that need protection from pollution. GIS has revolutionized many of the geographer's tasks, from map making to spatial analysis.

So, exactly what is a GIS and how does it work? A GIS is a computer system that weds mapping, database, procedures, and operating personnel for spatial analysis by capturing, storing, analyzing, and displaying geographically referenced information; that is, data identified according to location. The power of GIS comes from the ability to relate different information in a spatial context and to reach a conclusion about this relationship. Most of the information we have about our world contains a location reference, placing that information at some point on the globe. When rainfall information is collected, it is important to know where the rainfall is located. This is done by using a location reference system, such as longitude and latitude, and perhaps elevation. Comparing the rainfall information with other information, such as the location of marshes across the landscape, may show that certain marshes receive little rainfall. This fact may indicate that these marshes are likely to dry up, and this inference can help us make the most appropriate decisions about how humans should interact with the marsh. A GIS, therefore, can reveal important new information that leads to better decision-making.

Many computer databases that can be directly entered into a GIS are being produced by federal, state, tribal, and local governments; private companies; academia; and nonprofit organizations. Different kinds of data in map form can be entered into a GIS (e.g., digital line graph of rivers, a map-like layer of digital information about vegetative covers, or how much pesticide has been applied to a parcel of land).

REMOTE SENSING

Remote sensing is the process of detecting and monitoring the physical characteristics of an area by measuring its reflected and emitted radiation at a distance from the targeted area. *Remote sensing* is used in this text to refer to methods that are solely or primarily deployed through air or space. Included in this concept are studies of biological populations using remote imaging techniques. Related methods that are used most frequently on the ground (e.g., photography), whether underwater, from airplanes, or on satellites, are not included in the term *remote sensing* (USGS, 2009).

OBSERVATIONS

Observations refers to geographers' use of questionnaires or participant observation to investigate spatial phenomena.

MATHEMATICAL MODELS

Geographers use quantifiable observations, trends, and so on, to predict or understand complex spatial relationships. These models could describe typical or human phenomenon (e.g., pollution models, watershed and flood modes, etc.).

SURVEYS

Geographers conduct several annual or on-going large-scale topographical surveys.

Five Themes of Geography

All sciences evolve with time. Geography and its formal presentation to students of geography evolved from Pattison's (1963) four traditions of geography, which included spatial tradition, area studies tradition, man-land tradition, and Earth science traditions, to Taaffe's (1973) spatial view in context. The spatial view in context held three views: the man-land view, the area study view, and the spatial view—Earth science is considered to be subsumed in the other three traditions. In 1984, Natoli introduced *Guidelines for Geographic Education and the Fundamental Themes in Geography* (popularly known as *The Five Themes of Geography*). The five themes answer five important questions that can help organize information about places. These themes are:

- **Location** (What is the location of a place?)
  ○ Relative location
  ○ Absolute location
- **Place** (What is the character of a place?)
  ○ Human characteristics
  ○ Physical characteristics

- **Human–environmental interactions** (How do people interact with the natural environment of a place?)
  - Humans adapt to the environment
  - Humans modify the environment
  - Humans depend on the environment
- **Movement** (How do people, goods, and ideas move between places?)
  - People
  - Goods
  - Ideas
- **Regions** (How are places similar to, and different from, other places?)
  - Formal
  - Functional
  - Vernacular (perceptual)

The five themes became the framework on which the content of geography could be taught in the K–12 classrooms until the national geography standards were published in 1994. Because the five themes and six standards are wedded, the themes remain a valuable tool for learning a geographic perspective. The standards, discussed later, work to strengthen instructional planning.

Did You Know?

Imaginary lines that mark positions on the surface of the Earth include the equator, which circles the globe halfway between the north and south poles. It divides the world into two hemispheres, northern and southern. Lines of latitude (parallels) are imaginary lines that run parallel to the equator. Latitude is a location's distance from the equator, either north or south. Latitude is measured in degrees from 0 degrees to 90 degrees north, and from 0 degrees to 90 degrees south. Lines of longitude (meridians) are imaginary lines that run north and south from pole to pole. Measuring longitude is the distance a place is east or west of the prime meridian (an imaginary line that runs through Greenwich, England, at 0 degrees longitude; it divides the Earth into two hemispheres: eastern and western).

LOCATION

An important theme in geography is *location*. Every place on the Earth can be described in both absolute and relative terms. An *absolute location* is usually found by using imaginary lines (latitude and longitude—global location) that mark positions on the surface of the Earth or a street location (local location).

The global grid formed by lines of latitude and longitude crossing each other can be used to name the precise or absolute location of any place on Earth. Every place has only one absolute location that never changes.

Relative locations are those where the position of a place is in relation to another place. Every place has many relative locations, which can change over time. Relative locations are described by landmarks, time, direction, or distance from one place to another and may associate place with another.

PLACE

Another important geographic theme is *place*. Each place on the Earth has its own physical features. It can be described in terms of its land, water, weather, soil, altitude, and plant and animal life. Each place also has its human or cultural characteristics such as government, relation, language, ethnic groups, art, literature, tools, laws, customs, lifestyles, population, economy, types of agriculture, and so forth. The image people have of a place is based on their experiences, both intellectual and emotional.

HUMAN–ENVIRONMENTAL INTERACTION

An important theme in geography is human and environmental interaction. How do humans and the environment affect each other? There are three key concepts to human–environmental interaction: Humans adapt to the environment, humans modify the environment, and humans depend on the environment. All places and interactions have pluses and minuses for humans and the environment (e.g., clearing a forest for farming produces food but also destroys trees; burning coal provides energy but also pollutes the air). One person's advantage may be another person's disadvantage. People adapt to their surroundings by using different types of clothing, food, and shelter. People change their physical surroundings. Some of these changes are intentional and others have been accidental (i.e., tunnels, clearing of forests, planting of crops, use of land and natural resources).

MOVEMENT

Another important geographic theme is movement. This theme includes the movement and exchange of people, goods, services, and ideas (types of transportation, written materials, telecommunications, travel, imports, exports, etc.). We interact with each other through travel, trade, cyberspace, and political events. We live in a global village and global economy. Movement may also include natural movements of Earth's physical features: water cycle, ocean currents, wind, volcanic eruptions, mass wasting, animal migration, and so on.

REGIONS

Yet another important theme in geography is regions. Regions, groups of places with at least one common characteristic, are important because they make the study of geography manageable. Geographers often divide the world into regions, or areas (they may be any size), based on certain physical features such as land type, landform, water bodies, climate, or plant and animal life. They also divide the world into regions based on certain human characteristics, such as the way people are governed, religion, industry, agriculture, or the kind of language they speak. The boundaries of regions may overlap. There are three types of regions:

- **Formal regions** are areas in which a certain characteristic is found throughout the area; they are defined by governmental or administrative boundaries (e.g., United States, Canada, Seattle).
- **Functional regions** consist of one central place and the surrounding places affected by it (e.g., the Tennessee Valley Authority, United Airlines landing area, the Andes Mountains, the Amazon drainage basin, or a military route service area). If the function ceases to exist, the region no longer exists.
- **Vernacular (native) regions** are areas loosely defined by people's perception (i.e., the Middle East, the South, the Pacific Northwest, east of the mountains in Washington State, etc.).

Key Points of Geographical Mapping and Location

- Every map projection has some degree of distortion because a curved surface cannot be represented on a flat surface without distorting curvature.
- Regions have in common a relative location, spatial extent, and boundaries.
- A street address is an example of *absolute location*.
- *Spatial interaction* is the movement of people and things between places.
- *Situation* is the relative location of a place or activity.
- *Connectivity* describes the paths and ways in which different places are linked.
- *Dispersion* is the amount of spread of an item over an area.

- *Density* is the quantity of an item within a unit area.
- *Latitude* is a measure of distance north and south of the equator.
- *Latitude lines* are always parallel to each other
- On a globe, lines of latitude intersect meridians of longitude at right angles.
- An isometric map is a type of quantitative thematic map.
- The size and location of a place described by its local physical characteristics is called its *site*.
- A *contour* is an isoline showing points of equal elevation.
- All *meridians* are one-half the length of the equator.
- The scale on the surface of the globe is the same in every direction.
- *Density, dispersion, and pattern* are elements common to all spatial distributions.
- *Perceptual regions* reflect personal or popular impression of territory and spatial divisions.
- An *isoline* on a map connects points of equal value to the mapped item.
- The four main properties of maps are area, shape, distance, and direction.
- The *north and south poles, equator, and prime meridian* are all key reference points in the grid system.
- The *map scale* defines the relationship between the size of an Earth feature and its size on the map.

National Geography Standards (National Aeronautics and Space Administration [NASA], 2009)

It was mentioned earlier that the science of geography is an evolving science, basically a work in progress. Nowhere is this more apparent than in the current trend to blend the 1994 national geography standards with the five themes. In reality, these geography standards are a framework of benchmarks against which the context of various geography courses can be measured in the United States. The 18 standards in Table 30.2 have been categorized into 6 essential elements.

Geodesy

Information in this section is modified from National Oceanic and Atmospheric Administration. (2005). Welcome to geodesy. Accessed 08/30/11 from www.oceanservice.noaa.gov/education.

Table 30.2. National Geography Standards

The World in Spatial Terms

The geographically informed person knows and understands:

Standard 1: How to use maps and other geographic representations, tools, and technologies to acquire, process, and report information from a spatial perspective.

Standard 2: How to use mental maps to organize information about people places, and environments in a spatial context.

Standard 3: How to analyze the spatial organization of people, places, and environments on Earth's surface.

Places and Regions

The geographically informed person knows and understands:

Standard 4: The physical and human characteristics of places.

Standard 5: That people create regions to interpret Earth's complexity.

Standard 6: How culture and experience influence people's perceptions of places and regions.

Physical Systems

The geographically informed person knows and understands:

Standard 7: The physical process that shapes the patterns of Earth's surface.

Standard 8: The characteristics and spatial distribution of ecosystems on Earth's surface.

Human Systems

The geographically informed person knows and understands:

Standard 9: The characteristics, distribution, and migration of human populations on Earth's surface.

Standard 10: The characteristics, distribution, and complexity of Earth's culture mosaics.

Standard 11: The patterns and networks of economic interdependence on Earth's surface.

Standard 12: The processes, patterns, and functions of human settlement.

Standard 13: How the forces of cooperation and conflict among people influence the division and control of Earth's surface.

Environment and Society

The geographically informed person knows and understands:

Standard 14: How human actions modify the physical environment.

Standard 15: How physical systems affect human systems.

Standard 16: The changes that occur in the meaning, use, distribution, and importance of resources.

The Uses of Geography

The geographically informed person knows and understands:

Standard 17: How to apply geography to interpret the past.

Standard 18: How to apply geography to interpret the present and plan for the future.

As mentioned, *geodesy* (from the Greek—"division of the Earth") is a separate branch of geoscience; it deals with the measurement and representation of the Earth, including its gravitational field, in a three-dimensional, time-varying space. Geodesy professionals (geodesists) also study geodynamical phenomena such as crustal motion, tides, and polar motion. The National Oceanic and Atmospheric Administration (NOAA, 2005) points out that geodesists basically assign addresses to points all over the Earth. If you were to stick pins in a model of the Earth and then give each of those pins an address, then you would be doing what a geodesist does. By looking at the height, angles, and distances between these locations, geodesists create a spatial reference system (SRS) that everyone can use.

Building roads and bridges, conducting land surveys, and making maps are some of the important activities that depend on an SRS. For example, if you build a bridge, you need to know where to start on both sides of the river. If you don't, your bridge may not meet in the middle. Some of the SRS systems in existence are:

- Universal transverse Mercator coordinate system
- British national grid reference system
- Hellenic geodetic reference system 1987
- Irish grid reference system
- Irish transverse Mercator
- Israeli transverse Mercator
- Israeli Cassini Soldner
- Jordan transverse Mercator
- Military grid reference system
- United States national grid

Because positioning and navigation have become fundamental to the functions of society, geodesy has become increasingly important.

Geodesy helps the transportation industry ensure safety and reliability, while reducing costs. Without geodesy, planes might land next to—rather than at—airports, and ships could crash onto land. Geodesy also helps shipping companies save time and money by shortening their ships' and airplanes' routes and reducing fuel consumption.

Geologists, oceanographers, meteorologists, and even paleontologists use geodesy to understand physical processes on, above, and within the Earth. Because geodesy makes extremely accurate measurements (to the centimeter level), scientists can use its results to determine exactly how much the Earth's surface has changed over very short and very long periods (Careers in Geodesy, 1986).

The Earth's surface changes for many reasons. For instance, its surface rises and falls about 30 centimeters (about 1 foot) every day due to the gravitational influences of the moon and the sun. The Earth's outermost layer, the crust, is made up of a dozen or more "plates" that ride atop a sea of molten rock, called *magma,* which flows beneath the surface of the Earth.

As mentioned in Chapter 22, *plate tectonics* is the scientific discipline that looks at how these plates shift and interact, especially in relation to earthquakes and volcanoes. Although these phenomena are violent and usually affect large areas of land, even smaller events, such as erosion and storms, have an influence on shaping the Earth's surface. Geodesy helps us determine exactly where and how much the Earth's surface is changing.

HISTORY OF GEODESY

Throughout history, the shape of the Earth has been debated by philosophers and scientists. By 500 BCE most scholars thought the Earth was completely spherical. Aristotle (384–322 BCE) is credited as the first person to try to calculate the size of the Earth by determining its circumference (the length around the equator). He estimated this distance to be 400,000 stades (a *stadia* is a Greek measurement equaling about 600 feet). With 1 mile equal to 5,280 feet, Aristotle calculated the distance around the Earth to be about 45,500 miles (Smith, 1988).

As mentioned in Chapter 1, around 250 BCE, another Greek philosopher, Eratosthenes, measured the circumference of the Earth using the following Equation:

$$(360° ÷ \theta) = (s) \tag{30.1}$$

In this calculation, (s) is the distance between two points that lie north and south of each other on the surface of the Earth. If you were to draw a line from each of the points to the center of the Earth, the angle formed between them would be θ.

Obviously, Eratosthenes could not get to the center of the Earth, so he got the angle measurement using the rays of the sun. At noon on the longest day of the year, the summer solstice, the sun shone directly into a deep well at Syene (which is now Aswan, Egypt), casting no shadow.

At the same time in Alexandria, Egypt, he found that the sun cast a shadow equivalent to about one-fiftieth of a circle or 7.12 degrees. Eratosthenes combined this measurement with the distance between Syene and Alexandria, about 4,400 stades.

If we plug these numbers into the previous equation, we get (360° ÷ 7.12°), which equals 50; and 50 × 4,400 equals 220,000 stades, or about 25,000 miles. The accepted measurement of the Earth's circumference today is about 24,855 miles (Smith, 1988). Given the simple tools and technology that Eratosthenes had at his disposal over 2,000 years ago, his calculations were quite remarkable.

As technology developed, scientists and surveyors began to use different techniques to measure distance. In the 16th and 17th centuries, triangulation started to be used widely. Triangulation is a method of determining the position of a fixed point by measuring the angles to it from two other fixed points that are a known distance apart. Triangulation formed the basis for many national surveys. By the end of the 19th century, major triangulation networks covered the United States, India, Great Britain, and large parts of Europe.

At the end of the 16th century, the Royal Society in London and the L'Academie Royale des Sciences in Paris were founded. Soon they became locked in a battle to determine the shape of the Earth. The French argued that the Earth was *prolate,* or shaped like an egg. The English, using Sir Isaac Newton's universal theory of gravity and the knowledge that the Earth spun around on its axis, thought that the Earth was *oblate,* or flattened at the poles. To prove their idea, the Academy in Paris staged two expeditions, one to Peru (now Ecuador) at the equator, and the other to the border of Sweden and Finland in the northern hemisphere. Their objective was to measure the north–south curvature of the Earth at each location's latitude and determine whose concept of the Earth's shape was correct. The Academy's efforts prove that Newton was right. The Earth is flattened into the shape of an oblate sphere (Smith, 1988).

During the last 100 years, geodesy and its applications have advanced tremendously. The 20th century brought space-based technology, making geodetic measurements extremely precise. Today, NAVSTAR global positioning system (GPS) satellites allow scientists to measure changes in the Earth's surface to the centimeter.

ELEMENTS OF GEODESY: THE FIGURE OF THE EARTH

The Earth's surface is nearly spherical, with a radius of about 3,963 miles (6,378 km), and its surface is very irregular. Mountains and valleys make actually measuring this surface impossible because an infinite amount of data would be needed. For example, if you wanted to find the actual surface area of the Grand Canyon, you would have to cover every inch of land. It would take you many lifetimes to measure every crevice, valley, and rise. You could never complete the project because it would take too long.

To measure the Earth and avoid the problems that places like the Grand Canyon present, geodesists use a theoretical mathematical surface called the *ellipsoid*. Because the ellipsoid exists only in theory and not in real life, it can be completely smooth and does not take any irregularities—such as mountains or valleys—into account. The ellipsoid is created by rotating an ellipse around its shorter axis. This matches the real Earth's shape, because the Earth is slightly flattened at the poles and bulges at the equator.

While the ellipsoid gives a common reference to geodesists, it is still only a mathematical concept. Geodesists often need to account for the reality of the Earth's surface. To meet this need, the geoid, a shape that refers to global mean sea level, was created. If the geoid really existed, the surface of the Earth would be equal to a level in between the high-tide and low-tide marks.

Although a geoid may seem to be a smooth, regular shape, it isn't. The Earth's mass is unevenly distributed, meaning that certain areas of the planet experience more gravitation "pull" than others. Because of these variations in gravitational force, the "height" of different parts of the geoid is always changing, moving up and down in response to gravity. The geoidal surface is an irregular shape with a wavy appearance; there are rises in some areas and dips in others (Geodesy for the Layman, 1984).

ELEMENT OF GEODESY: DATUMS

Datums (sets of data) are the basis for all geodetic survey work. They act as reference points in the same way that starting points do when you give someone directions. For instance, when you want to tell someone how to get to your house, you give the person a known starting point, like a road or a building. Geodesists and surveyors use datums as starting or reference points when they create maps; mark off property boundaries; and plan, design, and build roads, bridges, and other structures.

Another way to think about a datum is as a set of information that acts as a foundation for other data. For example, when a skyscraper is about to be built, the construction team must first pour the foundation. Without this element, the skyscraper would be unstable and unsafe. This is the same concept as a datum. While a datum is a mathematical and geometric concept, it acts like the concrete foundation of a skyscraper. Once the foundation is set, the construction workers can build on top of it, creating the building's structure. After the building is complete, offices or apartments can be created inside the building. If the structure is an apartment building, its tenants can bring in furniture and decorate as they please. Although the foundation of the building probably isn't the first thing on the minds of the tenants, without it, the building would not be a safe place to live.

In geodesy, two main datums create the foundation for navigation and transportation in the United States. These datums—called the *horizontal* and *vertical datums*—make up the National Spatial Reference System (NSRS). Geodesists, surveyors, and people interested in precise positioning use the NSRS as their foundation for reference.

ELEMENT OF GEODESY: HORIZONTAL DATUMS

The horizontal datum is a collection of specific points on the Earth that have been identified according to their precise northerly or southerly location (latitude) and easterly or westerly location (longitude) (National Geodetic Survey, 1986).

To create the horizontal datum, or network of horizontal positions, surveyors marked each of the positions they had identified, typically with a brass, bronze, or aluminum disk or monument (see Figure 30.4). These markers were placed so that surveyors could see one marked position from another. To maxi-

Figure 30.4.   Shows co-author photographing horizontal brass monument datum point at 8,044 foot summit of Little Mount Baldy, Sequoia Kings Canyon, California.

mize the line of sight between monuments, they were usually set on mountain-tops or at high elevations. When monuments were set on flat land, towers were built above them to aid surveyors in locating them.

To "connect" the horizontal monuments into a unified network, or datum, surveyors have used a variety of methods, including triangulation. As technology improved, surveyors now rely almost exclusively on the GPS to identify locations on the Earth and incorporate them into existing datums.

ELEMENT OF GEODESY: VERTICAL DATUMS

The vertical datum is a collection of specific points on the Earth with known heights either above or below mean sea level. Near coastal areas, mean sea level is determined with a tide gauge. In areas far away from the shore, mean sea level is determined by the shape of the geoid.

Similar to the survey markers used to identify known positions in the horizontal datum, round brass plates mark positions in the vertical datum. The traditional method for setting these vertical benchmarks is called *differential leveling*. This method uses a known elevation at one location to determine the elevation at another location. As with horizontal datums, the advanced technology of GPS has almost completely replaced this classical technique of vertical measure.

ELEMENT OF GEODESY: GRAVITY

Gravity is the force that pulls all objects in the universe toward each other. On Earth, gravity pulls all objects "downward" toward the center of the planet. According to Sir Isaac Newton's Universal Law of Gravitation, the gravitational attraction between two bodies is stronger when the masses of the objects are greater and closer together. This rule applies to the Earth's gravitational field as well. Because the Earth rotates and its mass and density vary at different locations on the planet, gravity also varies.

One reason geodesists measure variations in the Earth's gravity is because gravity plays a major role in determining mean sea level. Geodesists calculate the elevation of locations on the Earth's surface based on the mean sea level. So knowing how gravity changes sea level helps geodesists make more accurate measurements. In general, in areas of the planet where gravitational forces are stronger, the mean sea level will be higher. In areas where the Earth's gravitational forces are weaker, the mean sea level will be lower.

To measure the Earth's gravity field, geodesists use instruments in space and on land. In space, satellites gather data on gravitational changes as they pass over points on the Earth's surface. On land, devices called *gravimeters* measure the Earth's gravitational pull on a suspended mass. With this data, geodesists can create detailed maps of gravitation fields and adjust elevations on existing maps. Gravity principally affects the vertical datum because it changes the elevation of the land surface (Geodesy for the Layman, 1984).

NATIONAL SPATIAL REFERENCE SYSTEM

All of the elements of geodesy are joined together in the NSRS. For almost 200 years, the National Geodetic Survey (NGS) and its predecessors have been using geodesy to map the U.S. shoreline, determine land boundaries, and improve transportation and navigation safety. NGS evolved from the Coast Survey, an agency established by Thomas Jefferson in 1807. The creation of the United States's first civilian scientific agency was prompted by the increasing importance of waterborne commerce to the fledging country. As the nation grew westward, NGS's mission began to include surveys of the North American interior.

With numerous surveys being conducted simultaneously across the growing nation, the surveyors needed to establish a common set of reference points. This would ensure that surveyors' maps and charts, which often covered hundreds of miles, would align with each other and not overlap. The common set of reference points they used were the benchmarks from the horizontal and vertical datums. Today, the complete set of vertical and horizontal benchmarks for the United States is known as the *NSRS*. This defined group of reference points acts as the foundation for innumerable activities requiring accurate geodetic information.

Think of it this way: When construction workers begin to build, they have to be sure that the area where they are building is free from dangerous power lines. The construction team will have to find out where the power lines are and make sure they are not building on top of them. To ensure success, the team needs to know the coordinates of the building site and of the local power lines. The NSRS provides a framework for identifying these coordinates. The team can then compare the two sets of coordinates and make sure they do not overlap.

To identify the benchmarks in the NSRS, NGS has traditionally placed markers, or permanent monuments, where the coordinates have been determined. These markers are brass or bronze disks (metals that sustain weathering) and are set in concrete or bedrock. Each marker is about 9 centimeters wide and has information about NGS printed on its surface.

With the advent of the GPS, NGS began to use different kinds of markers. These are made from long steel rods, driven to refusal (pushed into the ground until they won't go any farther). The top of each rod is then covered with a metal plate. This method ensures that the mark won't move and that people can't destroy or remove it. After tying these marks into a specific horizontal or vertical datum, the mark can be included in the NSRS database. Once the coordinates of the mark are entered into this database, they are available for anyone to use.

GLOBAL POSITIONING SYSTEM

Using the GPS, every point on Earth can be given its own unique address—its latitude, longitude, and height. The U.S. Department of Defense developed GPS satellites as a strategic system in 1978. But now anyone can gather data from them. For instance, many new cars have a GPS receiver built into them. These receivers help drivers know exactly where they are, and can help them from getting lost.

GPS uses a constellation of satellites that orbit approximately 11,000 miles above the Earth and transmit radio wave signals to receivers across the planet. By determining the time that it takes for a GPS satellite signal to reach your receiver, you can calculate your distance to the satellite and figure out your exact location on the Earth. Sound easy? In fact it is a very complicated process. For the GPS system to work, you need to have incredibly precise clocks on the satellites and receivers, and you must be able to access and interpret the signals from several orbiting satellites simultaneously. Fortunately, the receivers take care of all the calculations.

Let's tackle the distance calculation first. GPS satellites have very precise clocks that tell time to within 40 nanoseconds or 40 billionths (0.000000040) of a second. There are also clocks in the GPS receivers. Radio wave signals from the satellites travel at 186,000 miles per second. To find the distance from a satellite to a receiver, use the following equation: (186,000 mi/sec) × (signal travel time in seconds) = Distance of the satellite to the receiver in miles.

Knowing the distance of your GPS receiver to a single satellite is useful, but it will not provide you with enough information to determine your exact position on the Earth. For that, you need to simultaneously access the signals from four satellites. By calculating its distance from three satellites simultaneously, a GPS receiver can determine its general position with respect to latitude, longitude, and elevation.

You may wonder why the GPS receiver needs a fourth satellite. Essentially, it provides for even greater precision. To accurately calculate the distance from the GPS receiver to each of the three orbiting satellites, you must know the precise radio signal transmission and reception times. To do this, the clocks in the satellites and the clocks in the receivers must be perfectly synchronized. A mismatch of as little as one-millionth of a second between the clock in the satellite and the clock in the receiver could translate into a positioning error of as much as 900 feet (Herring, 1996). The atomic clocks in the satellites are extremely accurate, but the clocks in the GPS receivers are not, which creates a timing error. The signal from the fourth satellite is used to adjust for the receiver clock error and calculate the receiver's potion. But if the receiver can access more satellite signals, it will calculate a more accurate position (Tinambunan, 2003).

CONTINUOUSLY OPERATING REFERENCE STATIONS AND GEOGRAPHIC INFORMATION SYSTEMS

In a field of study that is thousands of years old, GPS represents a quantum leap in geodesy. As advanced as GPS technology is, most commercially available GPS receivers are only accurate within several meters. Considering that the Earth is almost 25,000 miles in circumference, the difference of a few meters may not seem important. This level of accuracy may be adequate for a hiker in the woods or someone driving a car. But there are many scientific, military, and engineering activities that require much higher levels of positioning accuracy, often to within a few centimeters or less.

To provide measurements at this level of accuracy, NGS developed the continuously operating reference stations (CORS) network. CORS is a network of hundreds of stationary, permanently operating GPS receivers throughout the United States. Working 24 hours a day, 7 days a week, CORS stations continuously receive GPS radio signals and integrate their positional data into the NSRS. This data is then distributed over the Internet. After logging onto the CORS website, users can determine the accuracy of their coordinates to the centimeter. This system has been especially useful in assessing the integrity of buildings and bridges in areas that are geologically active or have been affected by natural disasters such as hurricanes or floods.

Another powerful tool that has evolved along with GPS technology is the GIS. A GIS is composed of three parts: spatial information (location), special software, and a computer. These components work together to provide a digital platform for viewing and processing layers of spatial information.

A GIS assembles information from several sources, including ground surveys, existing maps, aerial photos, and satellite imagery. In a GIS, specific information about a place, such as the locations of utility lines, roads, streams, buildings, and even trees and animal populations, is layered over a set of geodetic data. Using special software, regional planners and scientists can examine the layers individually or in various combinations to improve traffic flow, merge construction with utility systems, develop around environmentally sensitive areas, and protect the public from potential natural disasters. Because a GIS stores data digitally, information can be quickly and economically updated, easily reproduced, and made widely available. In fact, because of its power and speed, GIS technology is doing most of the cartographic (mapmaking) work that, in the past, was laboriously done by hand on paper charts and maps.

The most important element needed to reconstruct geographic reality in a GIS is good spatial information. If the spatial information provided to a GIS is sparse or of poor resolution, then the world created by the computer will be a lifeless digital shell—a sharp contrast to the complexity of our living Earth.

References and Recommended Reading

Bell, K. M. (2007). *What is geography?* Retrieved 05/12/09 from http://74.6.239.67/ search/cache?ei= UTF-8&p=geography+mother+of+all+sciences&icp+l&w+geograph.

Bergman, J. (2005). *Rocks and the rock cycle.* Accessed 05/23/08 from www.windows .ucar.edu/ tour/ling=/earth/geology/rocks_intro.html.

Blue, J. (2007). Descriptor terms *Gazetteer of Planetary Nomenclature.* USGS. Retrieved on 05/20/2008 from http://planetarynames.wr.usgs.gov/jsp/apped5.jsp.

Campbell, N. A. (2004). *Biology: Concepts and connections* (4th CD-Rom ed.). New York: Benjamin-Cummings.

Careers in geodesy. (1986). *The American Geophysical Union Section of Geodesy.* Washington, DC: American Geophysical Union.

Geodesy for the layman. (1984). Washington, DC: Defense Mapping Agency. Accessed from http://www.ngs.noaa.gov/PUBS_LIB/Geodesy4Layman/TR80003A.HTM.

Geodetic glossary. (1986). Rockville, MD: National Geodetic Survey.

Herring, T. (1996, February). The global positioning system. *Scientific American,* 44–50.

Hurn, J. (1989). *GPS: A guide to the next utility.* Sunnyvale, CA: Trimble Navigation.

Huxley, T. H. (1876). *Science and education, Volume III, Collected essays.* New York: D. Appleton & Company.

Larsson, K. A. (1993). Prediction of the pollen season with a cumulated activity method. *Grana,* 32:111–114.

Michener, J. A. (1970). The mature social studies teacher. *Social Education,* November, pp. 760–767.

Natoli, S. J. (1984). *Guidelines for geographic education and the fundamental themes in geography.* New York: Association of American Geographers.

Pattison, W. D. (1963). *The four traditions of geography.* Columbus, OH: Annual Convention for National Council of Geography Education.

Press, R., & Siever, F. (2001). *Earth* (3rd ed.). New York: W. H. Freeman & Co.

Schultz, S. S., & Wallace, R. E. (1997). *The San Andreas fault.* United States Geological Survey. Accessed from http://pubs.usgs.gov/gip/earthg3/safaultgip.html.

Short, J. R. (2004). *Representing the republic: Mapping the United States, 1600–1900.* Chicago: Reaktion.

Smith, J. R. (1988). *Basic geodesy.* Rancho Cordova, CA: Landmark Enterprises.

Spellman, F. R. (1998). *Environmental science and technology: Concepts and applications.* Lancaster, PA: Technomic.

Spellman, F. R. (2008). *The science of air: Concepts and applications.* Boca Raton, FL: CRC.

Spellman, F. R., & Whiting, N. E. (2006). *Environmental science and technology* (2nd ed.). Rockville, MD: Government Institutes.

Taaffe, E. J. (1973). *Geography of transportation* (2nd ed.). Englewood Cliffs, NJ: Prentice-Hall.

Tinambunan, D. (2003). How do FPS devices work? *ScientificAmerican.com.* Accessed from http://www.sciam.com/askexpert_question.cfm?articleID=000349D4-D6FC-1CFC-93F6809ECS880000&catID=3

U.S. Geological Survey (USGS) (2006). *Map projections.* Accessed 05/15/09 from http://egsc.usgs.gov/isb/pubs/MapProjections/projections.html.

U.S. Geological Survey (USGS) (2007). *Geographic information systems.* Accessed 05/16/09 from http://egsc.usgs.gov/isb/pubs/gis_poster/.

U.S. Geological Survey (USGS) (2009). *Remote sensing.* Accessed 05/01/09 from http://usgs.gov/science.php?term=981.

Vanicek, P. (2000). *An online tutorial in geodesy.* Academic Press. Retrieved from http://einsteing.gge.unb.ca/tutorial.htm.

Glossary

This section was partly compiled and adapted from several sources, including *Soil Taxonomy* (1976). Washington, DC: U.S. Department of Agriculture; *Resource Conservation Glossary* (1982). Ankeny, IA: Soil Conservation Society of America; *Glossary of Soil Science Terms* (1987). Madison, WI: Soil Science Society of America.

7.5-minute Quadrangle: A U.S. Geological Survey paper map product at 1:24,000 scale covering 7.5 minutes of latitude and 7.5 minutes of longitude. Features shown include elevation contours, roads, railroads, water bodies, building, urban developments, and wetlands. This is a basic layer of information for many ecological and natural resource applications. An automated version of the 7.5-minute quadrangle is called the digital raster graphic (DRG). Informally known as *7.5-minute quad.*

Ablation Till: A superglacial coarse-grained sediment or till, accumulating as the subadjacent ice melts and drains away, finally deposited on the exhumed subglacial surface.

Absorption: Any process by which one substance penetrates the interior of another substance.

Accuracy: The closeness of results of observations, computations, or estimates to the true values or to values that are accepted as being true (American Society of Photogrammetry, 1984). In the U.S. Geological Survey–National Park Service Vegetation Mapping Program there are two aspects of accuracy: thematic and positional accuracy.

Accuracy Assessment: The process of determining the positional and thematic accuracy of the spatial vegetation community data. This is an independent process performed after the vegetation mapping and classification is complete. See "Producing rigorous and consistent accuracy assessment procedures,"

1996, at http://biology.usgs.gov/npsveg/standards/NPSVI_Accuracy_Assess
ment_Guidelines_ver2.pdf for more information.

Accuracy Assessment Point: A location where accuracy assessment data are collected. See "Producing rigorous and consistent accuracy assessment procedures,"
1996, at http://biology.usgs.gov/npsveg/standards/NPSVI_Accuracy_Assess
ment_Guidelines_ver2.pdf for more information.

Acid: Has a pH of water less than 5.5; pH modifier used in the U.S. Fish and
Wildlife Service wetland classification system; in common usage, acidic water
has a pH less than 7.

Acidic Deposition: The transfer of acidic or acidifying substances from the atmosphere to the surface of the Earth or to objects on its surface. Transfer can
be either by wet-deposition processes (rain, snow, dew, fog, frost, hail) or by
dry deposition (gases, aerosols, or fine to coarse particles).

Acid Rain: Atmospheric precipitation with pH values less than about 5.6, the
acidity being due to inorganic acids such as nitric and sulfuric that is formed
when oxides of nitrogen and sulfur are emitted into the atmosphere.

Acid Soil: A soil with a pH value of less than 7 or neutral. Soils may be naturally
acidic from their origin, by leaching, or may become acidic from decaying
leaves or from soil additives such as aluminum sulfate (alum). Acid soils can
be neutralized by the addition of lime products.

Acre-Foot (acre-ft.): The volume of water needed to cover an acre of land to a
depth of 1 foot; equivalent to 43,560 cubic feet or 32,851 gallons.

Actinomycetes: A group of organisms intermediate between the bacteria and the
true fungi that usually produce a characteristic branched mycelium. Includes
many (but not all) organisms belonging to the order of Actinomycetales.

Activated Carbon: A very porous material that, after being subjected to intense
heat to drive off impurities, can be used to adsorb pollutants from water.

Adhesion: Molecular attraction that holds the surfaces of two substances (e.g.,
water and sand particles) in contact.

Adsorption: The process by which one substance is attracted to and adheres
to the surface of another substance, without actually penetrating its internal
structure.

Aeration: A physical treatment method that promotes biological degradation of
organic matter. The process may be passive (when waste is exposed to air), or
active (when a mixing or bubbling device introduces the air).

Aeration, Soil: The process by which air in the soil is replaced by air from the
atmosphere. In a well-aerated soil, the soil air is similar in composition to the
atmosphere above the soil. Poorly aerated soils usually contain more carbon
dioxide and correspondingly less oxygen than the atmosphere above the soil.

Aerial Photography: Analog imagery taken from an airplane. In this program,
the optical axis is oriented perpendicular to the Earth's surface so that the film

is parallel to the surface being photographed (also *vertical aerial photography*). A sequence of aerial photographs will overlap so the photos can be used in stereoscopic analysis (stereo pairs). The overlap is referred to as *endlap* (top-to-bottom area in common, same flightline) and *sidelap* (side-to-side area in common, different flightlines) (Portions from American Society of Photogrammetry, 1984). Aerial photography used in the program is 9 inch by 9 inch vertical, stereoscopic, color, or color infrared photography.

Aerobic Bacteria: A type of bacteria that requires free oxygen to carry out metabolic function.

Aggregates (Soil): Soil structural units of various shapes, composed of mineral and organic material, formed by natural processes, and having a range of stabilities.

Agronomy: A specialization of agriculture concerned with the theory and practice of field crop production and soil management. The scientific management of land.

Air Capacity: Percentage of soil volume occupied by air spaces or pores.

Air Porosity: The proportion of the bulk volume of soil that is filled with air at any given time or under a given condition, such as a specified moisture potential; usually the large pores.

Algae: Chlorophyll-bearing nonvascular, primarily aquatic species that have no true roots, stems, or leaves; most algae are microscopic, but some species can be as large as vascular plants.

Algal Bloom: The rapid proliferation of passively floating, simple plant life, such as blue-green algae, in and on a body of water.

Alkali: A substance capable of liberating hydroxide ions in water, measured by a pH of more than 7, and possessing caustic properties; it can neutralize hydrogen ions, with which it reacts to form a salt and water, and is an important agent in rock weathering.

Alkaline: Has a pH greater than 7; pH modifier in the U.S. Fish and Wildlife Service wetland classification system; in common usage, a pH of water greater than 7.4.

Alliance: A physiognomically uniform group of associations sharing one or more diagnostic (dominant, differential, indicator, or character) species that, as a rule, are found in the uppermost stratum of the vegetation (Federal Geographic Data Committee). This is the second finest level in the National Vegetation Classification Standard hierarchy.

Alluvial Aquifer: A water-bearing deposit of unconsolidated material (sand and gravel) left behind by a river or other flowing water.

Alluvium: General term for sediments of gravel, sand, silt clay, or other particulate rock material deposited by flowing water, usually in the beds of rivers and streams, on a flood plain, on a delta, or at the base of a mountain.

Alpine Snow Glade: A marshy clearing between slopes above the timberline in mountains.

Amalgamation: The dissolving or blending of a metal (commonly gold and silver) in mercury to separate it from its parent material.

Amendment, Soil: Any substance other than fertilizers (such as compost, sulfur, gypsum, lime, and sawdust) used to alter the chemical or physical properties of a soil, generally to make it more productive.

Ammonia: A compound of nitrogen and hydrogen (NH_3) that is a common byproduct of animal waste. Ammonia readily converts to nitrate in soils and streams.

Ammonification: The production of ammonia and ammonium-nitrogen through the decomposition of organic nitrogen compounds in soil organic matter.

Anadromous Fish: Migratory species that are born in freshwater, live mostly in estuaries and ocean water, and return to freshwater to spawn.

Anaerobic: Pertaining to, taking place in, or caused by the absence of oxygen.

Anderson Classification System: A land cover–land use classification system developed for use with remote sensing systems in the 1970s adopted for the U.S. Geological Survey–National Park Service Vegetation Mapping Program to map cultural land cover (Anderson et al., 1976).

Level I	Level II
1. Urban or built-up land	Residential
	Commercial and services
	Industrial
	Transportation, communications, and utilities
	Industrial and commercial complexes
	Mixed urban or built-up land
	Other urban or built-up land
2. Agricultural land	Cropland and pasture
	Orchard, groves, vineyards, nurseries, and ornamental horticultural areas
	Confined feeding operations
	Other agricultural lands
5. Water (nonvegetated portion)	Streams and canals
	Lakes
	Reservoirs
	Bays and estuaries
7. Barren land	Dry salt flats
	Beaches
	Sandy areas other than beaches

(*continued*)

Level I	Level II
	Strip mines, quarries, and gravel pits
	Transitional areas
	Mixed barren lands
8. Perennial snow or ice	Perennial snowfields
	Glaciers

Note: This is not the complete Anderson Level II Classification. Areas of natural vegetation are classified under the National Vegetation Classification Standards.

Anion: An atom that has gained one or more negatively charged electrons and is thus itself negatively charged.

Anomalies: As related to fish, externally visible skin or subcutaneous disorders, including deformities, eroded fins, lesions, and tumors.

Anthropogenic: Having to do with or caused by humans.

Anticline: A fold in the Earth's crust, convex upward, whose core contains stratigraphically older rocks.

Aquaculture: The science of farming organisms that live in water, such as fish, shellfish, and algae.

Aquatic: Living or growing in or on water.

Aquatic Guidelines: Specific levels of water quality that, if reached, may adversely affect aquatic life. These are nonenforceable guidelines issued by a governmental agency or other institution.

Aquifer: A geologic formation, group of formations, or part of a formation that contains sufficient saturated permeable material to yield significant quantities of water to springs and wells.

Arroyo: A small, deep, flat-floored channel or gully of an ephemeral or intermittent stream, usually with nearly vertical banks cut into unconsolidated material.

Artificial Recharge: Augmentation of natural replenishment of groundwater storage by some method of construction, spreading of water, or by pumping water directly into an aquifer.

Aspect (of Slopes): The direction that a slope faces with respect to the sun.

Assimilation: The taking up of plant nutrients and their transformation into actual plant tissues.

Association: The finest level of the National Vegetation Classification Standard. The association is a physiognomically uniform group of vegetation stands that share one or more diagnostic (dominant, differential, indicator, or character) overstory and understory species. These elements occur as repeatable patterns of assemblages across the landscape, and are generally found under similar habitat conditions (Federal Geographic Data Committee). Within the program, *association* is the preferred term, but it is also synonymous with *community, community type, plant community, type, vegetation community,* and *vegetation type.*

Association for Biodiversity Information (ABI): A nonprofit organization dedicated to developing and providing knowledge about the world's natural diversity. In cooperation with the Natural Heritage Network, ABI collects and develops authoritative information about the plants, animals, and ecological communities of the Western hemisphere. ABI maintains databases to support the United States National Vegetation Classification Standard and the plot data that it is based on. ABI cooperates closely with the program to develop vegetation community classifications.

Atmospheric Deposition: The transfer of substances from the air to the surface of the Earth, either in wet form (rain, fog, snow, dew, frost, hail) or in dry form (gases, aerosols, particles).

Atmospheric Pressure: The pressure exerted by the atmosphere on any surface beneath or within it; equal to 14.7 pounds per square inch at sea level.

Atterburg Limits: Water contents of fine-grained soils at different states of consistency.

Attribute: (Digital data) A numeric, text, or image data field in a relational database table (such as a geographic information system) that describes a spatial feature (point, line, polygon, cell) (Environmental Systems Research Institute, 1994).

Automate: The process of entering data into a computer. Synonymous with *digitize.*

Autotrophs: Plants and microorganisms capable of synthesizing organic compounds from inorganic materials by either photosynthesis or oxidation reactions.

Available Water: The portion of water in a soil that can be readily absorbed by plant roots. The amount of water released between the field capacity and the permanent wilting point.

Average Discharge: As used by the U.S. Geological Survey, the arithmetic average of all complete water years of record of surface water discharge, whether consecutive or not. The term *average* generally is reserved for average of record and *mean* is used for averages of shorter periods, namely, daily, monthly, or annual mean discharges.

Background Concentration: A concentration of a substance in a particular environment that is indicative of minimal influence by human (anthropogenic) sources.

Backwater: A body of water in which the flow is slowed or turned back by an obstruction such as a bridge or dam, an opposing current, or the movement of the tide.

Bacteria: Single-celled microscopic organisms.

Bank: The sloping ground that borders a stream and confines the water in the natural channel when the water level, or flow, is normal.

Bank Storage: The change in the amount of water stored in an aquifer adjacent to a surface-water body resulting from a change in stage of the surface-water body.

Barrier Bar: An elongate offshore ridge, submerged at least at high tide, built up by the action of waves or currents.

Base Flow: The sustained low flow of a stream, usually groundwater inflow to the stream channel.

Base Map: The source or control from which all spatial data are developed and geo-referenced to. Photo-interpreted data are transferred to a base to rectify and register the data. For this program, base maps consist of the U.S. Geological Survey's digital orthophoto quadrangles or specially made orthophotos.

Basic: The opposite of acidic; water that has a pH of greater than 7.

Basin and Range Physiography: A region characterized by a series of generally north-trending mountain ranges separated by alluvial valleys.

Bed Material: Sediment comprising the streambed.

Bed Sediment: The material that is temporarily stationary in the bottom of a stream or other watercourse.

Bedload: Sediment that moves on or near the streambed and is in almost continuous contact with the bed.

Bedrock: A general term used for solid rock that underlies soils or other unconsolidated material.

Benthic Invertebrates: Insects, mollusk, crustaceans, worms, and other organisms without a backbone that live in, on, or near the bottom of lakes, streams, or oceans.

Benthic Organism: A form of aquatic life that lives on or near the bottom of stream, lakes, or oceans.

Bioaccumulation: The biological sequestering of a substance at higher concentrations than that at which it occurs in the surrounding environment or medium. Also, the process whereby a substance enters organisms through the gills, epithelial tissues, dietary, or other sources.

Bioavailability: The capacity of a chemical constituent to be taken up by living organisms either through physical contact or by ingestion.

Biochemical: Refers to chemical processes that occur inside or are mediated by living organisms.

Biochemical Process: A process characterized by, produced by, or involving chemical reactions in living organism.

Biochemical Oxygen Demand (BOD): The amount of oxygen required by bacteria to stabilize decomposable organic matter under aerobic conditions.

Biodegradation: Transformation of a substance into new compounds through biochemical reactions or the actions of microorganisms such as bacteria.

Biological Resources Division (BRD): A division of the U.S. Geological Survey in which the Center for Biological Informatics is located. The BRD mission is to work with others to provide the scientific understanding and technologies needed to support the sound management and conservation of our nation's biological resources. Also known as the *Biological Resources Discipline.*

Biological Treatment: A process that uses living organisms to bring about chemical changes.

Biomass: The amount of living matter, in the form of organisms, present in a particular habitat, usually expressed as weight-per-unit area.

Biota: All living organisms of an area.

Bog: A nutrient-poor, acidic wetland dominated by a waterlogged, spongy mat of sphagnum moss that ultimately forms a thick layer of acidic peat; generally has no inflow or outflow; fed primarily by rain water.

Brackish Water: Water with a salinity intermediate between seawater and freshwater (containing from 1,000 to 10,000 milligrams per liter of dissolved solids).

Breakdown Product: A compound derived by chemical, biological, or physical action on a pesticide. The breakdown is a natural process that may result in a more toxic or a less toxic compound and a more persistent or less persistent compound.

Breakpoint Chlorination: The addition of chlorine to water until the chlorine demand has been satisfied and free chlorine residual is available for disinfection.

Bureau of Reclamation (BOR, USBOR): A U.S. Department of Interior agency, created in 1902, charged with developing irrigation and hydropower projects in 17 western states in an environmentally and economically sound manner in the interest of the American public. The Remote Sensing and Geographic Information Systems Group of the BOR is managing several park projects for the U.S. Geological Survey–National Park Service Vegetation Mapping Program.

Calcareous: A rock or substance formed of calcium carbonate or magnesium carbonate by biological deposition or inorganic precipitation, or containing those minerals in sufficient quantities to effervesce when treated with cold hydrochloric acid.

Calcareous Soil: Containing sufficient calcium carbonate (often with magnesium carbonate) to effervesce visibly when treated with hydrochloric acid.

Caliche: A layer near the surface, more or less cemented by secondary carbonates of calcium or magnesium precipitated from the soil solution. It may occur as a soft, thin soil horizon, as a hard, thick bed just beneath the solum, or as a surface layer exposed by erosion.

Capillary Fringe: The zone above the water table in which water is held by surface tension. Water in the capillary fringe is under a pressure less than atmospheric.

Capillary Water: Held within the capillary pores of soils; mostly available to plants.

Carbonate Rocks: Rocks (such as limestone or dolostone) that are composed primarily of minerals (such as calcite and dolomite) containing a carbonate ion.

Catena: The sequence of soils that occupy a slope transect, from the topographic divide to the bottom of the adjacent valley.

Cation: An atom that has lost one or more negatively charged electrons and is thus itself positively charged.

Center for Biological Informatics (CBI): A U.S. Geological Survey (USGS) Science Center. CBI serves as the operating agent for the National Biological Information Infrastructure (NBII). In addition, CBI manages the USGS–National Park Service Vegetation Mapping Program along with other national data collection programs that complement and strengthen its role within the NBII.

Center Pivot Irrigation: An automated sprinkler system involving a rotating pipe or boom that supplies water to a circular area of an agricultural field through sprinkler heads or nozzles.

Channel Scour: Erosion by flowing water and sediment on a stream channel; results in removal of mud, silt, and sand on the outside curve of a stream bend and the bed material of a stream channel.

Channelization: The straightening and deepening of a stream channel to permit the water to move faster or to drain a wet area for farming.

Chelate (Greek, *claw*): A complex organic compound containing a central metallic ion surrounded by organic chemical groups.

Chemical Treatment: A process that results in the formation of a new substance or substances. The most common chemical water treatment processes include coagulation, disinfection, water softening, and filtration.

Chlordane: Octachlor-4,7-methanotetrahydroindane; an organochlorine insecticide no longer registered for use in the United States. Technical chlordane is a mixture in which the primary components are cis- and trans-chlordane, cis- and trans-nonachlor, and heptachlor.

Chlorinated Solvent: A volatile organic compound containing chlorine. Some common solvents are trichloroethylene, tetrachloroethylene, and carbon tetrachloride.

Chlorofluorocarbons: A class of volatile compounds consisting of carbon, chlorine, and fluorine. Commonly called *freon,* which has been in refrigeration mechanisms, as blowing agents in the fabrication of flexible and rigid foams, and, until banned from use several years ago, as propellants in spray cans.

Chlorination: The process of adding chlorine to water to kill disease-causing organisms or to act as an oxidizing agent.

Cienaga: A marshy area where the ground is wet due to the presence of seepage of springs.

Class: The level in the National Vegetation Classification Standard hierarchy based on the structure of the vegetation and determined by the relative percentage of cover and the height of the dominant, uppermost life forms (Grossman et al., 1998).

Class, Soil: A group of soils having a definite range in a particular property such as acidity, degree of slope, texture, structure, land-use capability, degree of erosion, or drainage.

Classification Accuracy: How closely the map classes match the vegetation communities found on the landscape. This is determined by accuracy assessment protocols. See "Producing rigorous and consistent accuracy assessment procedures" at http://biology.usgs.gov/npsveg/standards/NPSVI_Accuracy_Assessment_Guidelines_ver2.pdf for more information.

Clay: A soil separate consisting of particles smaller than 0.0002 mm in equivalent diameter.

Clean Water Act (CWA): Federal law dating to 1972 (with several amendments) with the objective to restore and maintain the chemical, physical, and biological integrity of the nation's waters. Its long-range goal is to eliminate the discharge of pollutants into navigable waters, and to make national waters fishable and swimmable.

Climate: The sum total of the meteorological elements that characterize the average and extreme conditions of the atmosphere over a long period at any one place or region of the Earth's surface.

Coagulants: Chemicals that cause small particles to stick together to form larger particles.

Coagulation: A chemical water treatment method that causes very small suspended particles to attract one another and form larger particles. This is accomplished by the addition of a coagulant that neutralizes the electrostatic charges that cause particles to repel each other.

Cohesion: Holding together: force holding a solid or liquid together, owing to attraction between like molecules. Decreases with rise in temperature.

Coliform Bacteria: A group of bacteria predominantly inhabiting the intestines of humans or animals, but also occasionally found elsewhere. Presence of the bacteria in water is used as an indication of fecal contamination (contamination by animal or human wastes).

Colloidal: Matter of very fine particle size.

Color: A physical characteristic of water. Color is most commonly tan or brown from oxidized iron, but contaminants may cause other colors, such as green or blue. Color differs from turbidity, which is water's cloudiness.

Combined Sewer Overflow: A discharge of untreated sewage and storm water to a stream when the capacity of a combined storm–sanitary sewer system is exceeded by storm runoff.

Communicable Diseases: Usually cased by *microbes*—microscopic organisms including bacteria, protozoa, and viruses. Most microbes are essential components of our environment and do not cause disease. Those that do are called *pathogenic organisms,* or simply *pathogens.*

Community: An assemblage of species that co-occur in defined areas at certain times and have the potential to interact with one another (Grossman et al., 1998).

Community Element Global (CEGL): The Association for Biodiversity Information's unique identifier code to a vegetation association (community) in their central biodiversity database; also known as *Elcode.*

Community Water System: A public water system that serves at least 15 service connections used by year-round residents, or regularly serves at least 25 year-round residents.

Complex: A group of associations that are not distinguishable from one another on aerial photography and so are grouped into a map unit.

Composite Sample: A series of individual or grab samples taken at different times from the same sampling point and mixed together.

Concentration: The ratio of the quantity of any substance present in a sample of a given volume or a given weight compared with the volume or weight of the sample.

Cone of Depression: The depression of heads around a pumping well caused by withdrawal of water.

Confined Aquifer (Artesian Aquifer): An aquifer that is completely filled with water under pressure and that is overlain by material that restricts the movement of water.

Confining Layer: A body of impermeable or distinctly less permeable material stratigraphically adjacent to one or more aquifers that restricts the movement of water into and out of the aquifers.

Confluence: The flowing together of two or more streams; the place where a tributary joins the main stream.

Conglomerate: A coarse-grained sedimentary rock composed of fragments larger than 2 millimeters in diameter.

Constituent: A chemical or biological substance in water, sediment, or biota that can be measured by an analytical method.

Consumptive Use: The quantity of water that is not available for immediate rescue because it has been evaporated, transpired, or incorporated into products, plant tissue, or animal tissue.

Contact Recreation: Recreational activities, such as swimming and kayaking, in which contact with water is prolonged or intimate, and in which there is a likelihood of ingesting water.

Contaminant: A toxic material found as an unwanted residue in or on a substance.

Contamination: Degradation of water quality compared with original or natural conditions due to human activity.

Contingency Table: A table that compares mapped data with ground data to determine accuracy. The "known" classes derived from accuracy assessment plots are compared to the classes derived from photo interpretation. The results are then tabulated in the form of a contingency table to determine the degree of misclassification that has occurred between classes. Also referred to as *error matrix, confusion matrix,* or *misclassification matrix.* For an example of a contingency table see http://biology.usgs.gov/npsveg/ftp/vegmapping/scbl/aa_matrix.xls.

Contributing Area: The area in a drainage basin that contributes water to streamflow or recharge to an aquifer.

Convection: A process of heat transfer in a fluid involving the movement of substantial volumes of the fluid concerned. Convection is very important in the atmosphere, and to a lesser extent, in the oceans.

Coordinate System: A reference system to represent horizontal and vertical locations and distances on a map. The geographic coordinate system is the latitude and longitude with respect to a reference spheroid. A local coordinate system is one that is not aligned with the Earth's surface. Most coordinate systems are based on projections of the Earth's surface to a plane. All spatial data in the program are represented in the Universal Transverse Mercator coordinate system.

Core sample: A sample of rock, soil, or other material obtained by driving a hollow tube into the undisturbed medium and withdrawing it with its contained sample.

Cover: The area of ground covered by the vertical projection of the aerial parts of plants of one or more species (Federal Geographic Data Committee, 1997).

Cover Type: A designation based on the plant species forming a plurality of composition within a given area (e.g., Oak-Hickory) (Federal Geographic Data Committee, 1997). Also refers to an alliance or group of alliances in the U.S. National Vegetation Classification.

Coverage: A file format used by Arc/Info software for vector spatial data.

Cowardin Classification: Wetland classification system commonly referred to as the *Cowardin classification,* after the first author. This is the Federal Geographic Data Committee standard for wetland classification (Cowardin, 1979).

Criterion: A standard rule or test on which a judgment or decision can be based.

Cross Connection: Any connection between safe drinking water and a nonpotable water or fluid.

Crosswalk: Relationship between the elements of two classification systems. For example, there is a crosswalk between map classes and units of the National Vegetation Classification Standard. This relationship is often shown in a look-up table.

Cultural Vegetation: Vegetation planted or actively maintained by humans such as annual croplands, orchards, and vineyards. Contrast with natural vegetation (Grossman et al., 1998).

C×T Value: The product of the residual disinfectant concentration *C*, in milligrams per liter, and the corresponding disinfectant contact time *T*, in minutes. Minimum C×T values are specified by the surface water treatment rule as a means of ensuring adequate kill or inactivation of pathogenic microorganisms in water.

Datum: A mathematical model that describes the size and shape of the ellipsoid (the Earth is not a sphere but an ellipsoid distorted by rotation about its axis, with the globe bulging at the equator and flattened at the poles). The flattening is not uniform around the Earth due to the influence of the continents location (Snyder, 1982). Using the wrong datum in relation to geographic coordinates can result in errors of hundreds of meters in position. This text uses the North American Datum (NAD) of 1983 or NAD83.

Denitrification: The biochemical reduction of nitrate or nitrite to gaseous nitrogen, either as molecular nitrogen or as an oxide of nitrogen.

Density: Density is the relationship between the area covered by the overstory of a vegetation community and the total area of a polygon in which the community is found. One of the physiognomic modifiers classified in the U.S. Geological Survey–National Park Service Vegetation Mapping Program. Density in map units is classified as Closed/Continuous > 60 %, Discontinuous 40%–60%, Dispersed 25%–40%, Sparse 10%–25%, Rare 2%–10%.

Detritus: Debris from dead plants and animals.

Diagnostic Species: Used to evaluate (i.e., diagnose) an area, or site, for some characteristic. For example, the presence and relative density of a *Vaccinium stamineum var. stamineum* (gooseberry) understory existing beneath a canopy of chestnut oak, black oak, and Virginia pine indicates that the site is xeric (or dry). The oaks and pines can inhabit a wide range of sites, wet to dry. But the gooseberry understory is the indicator of a drier habitat (which is probably due to a combination of factors, including soil type, slope, aspect, elevation, and site history). Also known as *indicator species* (Federal Geographic Data Committee, 1997).

Dichotomous Field Key: A document that identifies vegetation communities on the basis of exclusive characteristics. An example of exclusive characteristics is forested versus nonforested. Also known as *vegetation field key* and *vegetation key*. This key is an important product of each vegetation mapping project. For an example of a dichotomous field key, visit http://biology.usgs.gov/npsveg/agfo/agforpt.pdf.

Diffusion: The movement of atoms in a gaseous mixture, or ions in a solution, primarily as a result of their own random motion.

Digital Orthophoto Quadrangle (DOQ): The U.S. Geological Survey digital product derived from high-altitude aerial photography. These digital images are rectified and registered to locations on the Earth and cover approximately one quarter of a 7.5-minute quadrangle. Also called digital orthophoto quarter quadrangle, DOQ, and 3.75-minute DOQ. DOQs are often used as base maps to register the photo interpreted data in this program.

Digital Raster Graphic (DRG): A scanned image of a paper U.S. Geological Survey topographic map. The geographic information is georeferenced in the universal transverse Mercator projection with the accuracy and datum of the original map. The minimum scanning resolution is 250 dots per inch. DRGs are useful layers in a geographic information system

Digitize: The process of entering data into a computer. There are several methods of entering spatial data into a computer, including manual digitizing, scan digitizing, and soft copy photogrammetric methods. Synonymous with *automate.*

Division: The highest level in the National Vegetation Classification Standard separating Earth cover into either vegetated or nonvegetated categories (Federal Geographic Data Committee, 1997).

Dominance: The extent to which a given species or life-form predominates in a community because of its size, abundance, or cover, and affects the fitness of associated species (Federal Geographic Data Committee, 1997).

Dominant Life Form: An organism, group of organisms, or taxon that by its size, abundance, or coverage exerts considerable influence on an association's biotic (such as structure and function) and abiotic (such as shade and relative humidity) conditions (Federal Geographic Data Committee, 1997).

Drainage: The removal of excess water, both surface and subsurface, from plants. All plants (except aquatics) will die if exposed to an excess of water.

Duff: The matted, partly decomposed organic surface layer of forest soils.

Ecological Groups: Classification of vegetation communities based on plant assemblages, physical environments, and dynamic processes useful for conservation planning. These groups are classified on total floristic composition, physiognomy (vertical structure), spatial pattern (horizontal structure), physical environment (landscape position and soil), chemical variables (e.g., soil pH), and disturbance regimes. Some factors are difficult to measure directly, and must be inferred from knowledge of species ecology, spatial patterns, and ecological processes. These groups often occur between the floristic and physiognomic levels of the National Vegetation Classification Standard.

Ecological Society of America (ESA): A nonpartisan, nonprofit organization of scientists founded in 1915 to promote ecological science and ensure the appropriate use of ecological science in environmental decision making. The ESA Panel on Vegetation Classification was constituted to support and facilitate the creation of standardized, scientifically credible North American vegetation classification.

Ecoregion: An area of similar climate, landform, soil, potential natural vegetation, hydrology, or other ecologically relevant variables.

Ecosystem: A community of organisms considered together with the nonliving factors of its environment.

Effluent: Outflow from a particular source, such as a stream that flows from a lake or liquid waste that flows from a factory or sewage-treatment plant.

Effluent Limitations: Standards developed by the Environmental Protection Agency to define the levels of pollutants that could be discharged into surface waters.

Electrodialysis: The process of separating substances in a solution by dialysis, using an electric field as the driving force.

Electronegativity: The tendency for atoms that do not have a complete octet of electrons in their outer shell to become negatively charged.

Emergent Plants: Erect, rooted, herbaceous plants that may be temporarily or permanently flooded at the base but do not tolerate prolonged inundation of the entire plant.

Enhanced Surface Water Treatment Rule (ESWTR): A revision of the original Surface Water Treatment Rule that includes new technology and requirements to deal with newly identified problems.

Environment: The sum of all conditions and influences affecting the life of organisms.

Environmental Sample: A water sample collected from an aquifer or stream for the purpose of chemical, physical, or biological characterization of the sampled resource.

Environmental Setting: Land area characterized by a unique combination of natural and human-related factors, such as row-crop cultivation or glacial-till soils.

Ephemeral Stream: A stream or part of a stream that flows only in direct response to precipitation; it receives little or no water from springs, melting snow, or other sources; its channel is at all times above the water table.

EPT Richness Index: An index based on the sum of the number of taxa in three insect orders, *Ephemeroptera* (mayflies), *Plecoptera* (stoneflies), and *Trichoptera* (caddis flies), that are composed primarily of species considered to be relatively intolerant to environmental alterations.

Erosion: The process whereby materials of the Earth's crust are loosened, dissolved, or worn away and simultaneously moved from one place to another.

Error: The distance of results of observations, computations, or estimates from the true values or to values that are accepted as being true. Also refers to the misclassification of thematic data.

Eutrophication: The process by which water becomes enriched with plant nutrients, most commonly phosphorus and nitrogen.

Evaporite Minerals (Deposits): Minerals or deposits of minerals formed by evaporation of water-containing salts. These deposits are common in arid climates.

Evaporites: A class of sedimentary rocks composed primarily of minerals precipitated from a saline solution as a result of extensive or total evaporation of water.

Evapotranspiration: The process by which water is discharged to the atmosphere as a result of evaporation from the soil and surface-water bodies and transpiration by plants.

Exfoliation: Mechanical or physical weathering that involves the disintegration and removal of successive layers of rock mass.

Existing Vegetation: The plant species existing at a location at the present time. The U.S. Geological Survey–National Park Service Vegetation Mapping Program classifies and maps existing vegetation (Ecological Society of America, 1999).

Facultative Bacteria: A type of anaerobic bacteria that can metabolize its food either aerobically or anaerobically.

Fall Line: Imaginary line marking the boundary between the ancient, resistant crystalline rocks of the Piedmont province of the Appalachian Mountains, and the younger, softer sediments of the Atlantic Coastal Plain province in the eastern United States. Along rivers, this line commonly is reflected by waterfalls.

Fecal Bacteria: Microscopic single-celled organisms (primarily fecal coliforms and fecal streptococci) found in the wastes of warm-blooded animals. Their presence in water is used to assess the sanitary quality of water for body-contact recreation or for consumption. Their presence indicates contamination by the wastes of warm-blooded animals and the possible presence of pathogenic (disease-producing) organisms.

Federal Geographic Data Committee (FGDC): Coordinates the development of the National Spatial Data Infrastructure (NSDI). The NSDI encompasses policies, standards, and procedures for organizations to cooperatively produce and share geographic data. The 17 federal agencies that make up the FGDC are developing the NSDI in cooperation with organizations from state, local, and tribal governments; the academic community; and the private sector. The program complies with FGDC standards for vegetation classification, metadata, spatial data transfer, and positional accuracy.

Federal Water Pollution Control Act (1972): The Act outlines the objective "to restore and maintain the chemical, physical, and biological integrity of the nation's waters." This 1972 act and the subsequent Clean Water Act amendments are the most far-reaching water pollution control legislation ever enacted. They provided for comprehensive programs for water pollution control, uniform laws, and interstate cooperation. They provided grants for research, investigations, training, and information on national programs on surveillance, the effects of pollutants, pollution control, and the identification and measurement of pollutants. Additionally, they allot grants and loans for the construction of treatment works. The Act established national discharge standards with enforcement provisions. The Federal Water Pollution Control Act established several milestone achievement dates. It required secondary treatment of domestic waste by publicly owned treatment works (POTWs), and application of "best practicable" water pollution control technology by industry by 1977. Virtually all industrial sources have achieved compliance (because of economic difficulties and cumbersome federal requirements, certain POTWs obtained an extension to July 1, 1988, for compliance). The Act also called for new levels of technology to be imposed during the 1980s and 1990s, particularly for controlling toxic pollutants. The Act mandates a strong pretreatment program to control toxic pollutants discharged by industry into POTWs. The 1987 amendments require that storm water from industrial activity must be regulated.

Fertility, Soil: The quality of a soil that enables it to provide essential chemical elements in quantities and proportions for the growth of specified plants.

Fertilizer: Any of a large number of natural or synthetic materials, including manure and nitrogen, phosphorus, and potassium compound, spread on or worked into soil to increase its fertility.

Field Reconnaissance: Preliminary field visits by photo interpreters and vegetation classification experts to gain an overview of the vegetation of the project area and how it relates to the National Vegetation Classification Standard. Communication between photo interpreters and vegetation classification experts during this fieldwork is key to developing an accurate classification system. Observation point data are collected during this reconnaissance.

Field Verification: Field visits by photo interpreters after photo interpretation is complete to check for correctness of photo interpretation. At this point, changes may be made to the photo interpretation. This occurs prior to accuracy assessment.

Filtrate: Liquid that has been passed through a filter.

Filtration: A physical treatment method for removing solid (particulate) matter from water by passing the water through porous media such as sand or an artificial filter.

Fixation: The transformation in soil of a plant nutrient from an available to an unavailable state.

Flight Line: Refers to a line or strip of aerial photography. Usually designated on the film as "flightline number–photo number." Technical: A line connecting the principal points of sequential vertical aerial photographs (Portions from American Society of Photogrammetry, 1984).

Flocculation: The water treatment process following coagulation. It uses gentle stirring to bring suspended particles together so that they will form larger, more settleable clumps called *floc*.

Flood: Any relatively high streamflow that overflows the natural or artificial banks of a stream.

Flood Attenuation: A weakening or reduction in the force or intensity of a flood.

Flood Irrigation: The application of irrigation water whereby the entire surface of the soil is covered by ponded water.

Flood Plain: A strip of relatively flat land bordering a stream channel that is inundated at times of high water.

Floristics: The kinds and number of plant species in particular areas and their distribution (Ecological Society of America, 1999).

Flow Line: The idealized path followed by particles of water.

Flowpath: An underground route for ground-water movement, extending from a recharge (intake) zone to a discharge (output) zone such as a shallow stream.

Fluvial: Deposits of parent materials laid down by rivers or streams.

Formation: A level in the National Vegetation Classification Standard hierarchy below subgroup which represents vegetation types that share a definite physiognomy or structure within broadly defined environmental factors, relative landscape positions, or hydrologic regimes (Grossman et al., 1998).

Freshwater: Water that contains less than 1,000 milligrams per liter of dissolved solids.

Freshwater Chronic Criteria: The highest concentration of a contaminant that freshwater aquatic organisms can be exposed to for an extended period (4 days) without adverse effects.

Friable: A soil consistency term pertaining to the ease of crumbling of soils.

Geographic Information System (GIS): An organized collection of geographically (spatially) referenced information (Portions from Environmental Systems Research Institute, 1994).

Georeference: The process of converting a map or image into real-world coordinates. A nongeoreferenced map or image is said to be in "digitizer-inches" or "scanner-inches" (i.e., it has no real-world coordinates).

Global Positioning System (GPS): A system of satellites, ground receiving stations, and handheld receivers that allow accurate measurement of feature coordinates on the face of the Earth. GPS receivers are used to measure the location of field plots, reconnaissance points, and accuracy assessment points.

Grab Sample: A single water sample collected at one time from a single point.

Gradsect: Gradient-directed transect sampling. This approach is based on the distribution of patterns along environmental gradients. The gradsect sampling design is intended to provide a description of the full range of biotic variability (e.g., vegetation) in a region by sampling along the full range of environmental variability. Transects that contain the strongest environmental gradients in a region are selected to optimize the amount of information gained in proportion to the time and effort spent during the vegetation survey (Grossman & Goodin, et al., 1999).

Ground Photograph: An image recorded with the photographer standing on the ground (as opposed to an aerial photograph).

Ground Truth: The process of taking aerial photographs into the field to verify the ground condition compared with how that condition appears in the photograph.

Groundwater: The fresh water found under the Earth's surface, usually in aquifers. Groundwater is a major source of drinking water, and a source of a growing concern in areas where leaching agricultural or industrial pollutants, or substances from leaking underground storage tanks are contaminating groundwater.

Group: The level in the National Vegetation Classification Standard hierarchy below subclass based on leaf characteristics and identified and named in conjunction with broadly defined macroclimatic types to provide a structural-geographic orientation (Grossman et al., 1998).

Habitat: The combination of environmental or site conditions and ecological processes influencing a plant community (Ecological Society of America, 1999).

Habitat Type: (1) A collective term for all parts of the land surface supporting, or capable of supporting, the same kind of climax plant association (Daubenmire, 1978); (2) an aggregation of land areas having a narrow range of environmental variation and capable of supporting a given plant association (Gabriel & Talbot, 1984).

Hardness: A characteristic of water caused primarily by the salts of calcium and magnesium. It causes deposition of scale in boilers, damage in some industrial processes, and sometimes objectionable taste. It may also decrease soap's effectiveness.

Headwaters: The source and upper part of a stream.

Heaving: The partial lifting of plants, buildings, roadways, fence posts, and other objects out of the ground as a result of freezing and thawing of the surface soil during the winter.

Hectare: A metric unit of measure equal to 10,000 m^2 or approximately 2.471 acres.

Height: Height of the overstory of a vegetation community. One of the physiognomic modifiers classified in the U.S. Geological Survey–National Park Service Vegetation Mapping Program. Height in map units is classified as less than 0.5 meters, 0.5– 2 meters, 2–5 meters, 5–15 meters, 15–35 meters, 35–50 meters, and more than 50 meters.

Heterotroph: An organism capable of deriving energy for life processes only from the decomposition of organic compounds, and incapable of using inorganic compounds as sole sources of energy or for organic synthesis.

Horizon, Soil: A layer of soil, approximately parallel to the soil surface, differing in properties and characteristics from adjacent layers below or above it.

Humus: More or less stable fraction of the soil organic matter (usually dark in color) remaining after the major portions of added plant and animal residues have decomposed.

Hydration: The incorporation of water into the chemical composition of a mineral, converting it from an anhydrous to a hydrous form; the term is also applied to a form of weathering in which hydration swelling creates tensile stress within a rock mass.

Hydraulic Conductivity: The capacity of a rock to transmit water. It is expressed as the volume of water at the existing kinematic viscosity that will move in unit time under a unit hydraulic gradient through a unit area measured at right angles to the direction of flow.

Hydraulic Gradient: The change of hydraulic head per unit of distance in a given direction.

Hydrogen Bonding: The term used to describe the weak but effective attraction that occurs between polar covalent molecules.

Hydrograph: Graph showing variation of water elevation, velocity, streamflow, or other property of water with respect to time.

Hydrologic Cycle: Literally the water–Earth cycle. The movement of water in all three physical forms through the various environmental mediums (air, water, biota, and soil).

Hydrology: The science that deals with water as it occurs in the atmosphere, on the surface of the ground, and underground.

Hydrolysis: The reaction between water and a compound (commonly a salt). The hydroxyl from the water combines with the anion from the compound undergoing hydrolysis to form a base; the hydrogen ion from the water combines with the cation from the compound to form an acid.

Hydrostatic pressure: The pressure exerted by the water at any given point in a body of water at rest.

Hygroscopic: A substance that readily absorbs moisture.

Hygroscopic Coefficient: The amount of moisture in a dry soil when it is in equilibrium with some standard relative humidity near a saturated atmosphere (about 98%), expressed in terms of percentage on the basis of oven-dry soil.

Impermeability: The incapacity of a rock to transmit a fluid.

Index of Biotic Integrity (IBI): An aggregated number, or index, based on several attributes or metrics of a fish community that provides an assessment of biological conditions.

Indicator Sites: Stream sampling sites located at outlets of drainage basins with relatively homogeneous land use and physiographic conditions; most indicator-site basins have drainage areas ranging from 20 to 200 square miles.

Infiltration: The downward movement of water from the atmosphere into soil or porous rock.

Inventory and Monitoring (I&M) Program: A National Park Service program developed (1) to collect baseline inventories of basic biological and geophysical natural resources for all natural resource parks, and (2) to set up long-term monitoring programs that will be developed to efficiently and effectively monitor ecosystem status and trends over time at various spatial scales. The U.S. Geological Survey–National Park Service Vegetation Mapping Program collaborates closely with the I&M Program.

Influent: Water flowing into a reservoir, basin, or treatment plant.

Inorganic: Containing no carbon matter other than plant or animal.

Inorganic Chemical: A chemical substance of mineral origin not having carbon in its molecular structure.

Inorganic Soil: Soil with less than 20% organic matter in the upper 16 inches.

Ionic Bond: The attractive forces between oppositely charged ions—for example, the forces between the sodium and chloride ions in a sodium chloride crystal.

Instantaneous Discharge: The volume of water that passes a point at a particular instant of time.

Instream Use: Water use taking place within the stream channel for such purposes as hydroelectric power generation, navigation, water-quality improvement, fish propagation, and recreation. Sometimes called *nonwithdrawal use* or *in-channel use.*

Intermittent Stream: A stream that flows only when it receives water from rainfall runoff or springs, or from some surface source such as melting snow.

Internal Drainage: Surface drainage whereby the water does not reach the ocean, such as drainage toward the lowermost or central part of an interior basin or closed depression.

Intertidal: Alternately flooded and exposed by tides.

Intolerant Organisms: Organisms that are not adaptable to human alterations to the environment and thus decline in numbers where alterations occur.

Invertebrate: An animal having no backbone or spinal column.

Ions: Atoms that have lost or gained one or more negatively charged electrons.

Irrigation: Controlled application of water to arable land to supply requirements of crops not satisfied by rainfall.

Irrigation Return Flow: The part of irrigation applied to the surface that is not consumed by evapotranspiration or uptake by plants and that migrates to an aquifer or surface-water body.

Irrigation Withdrawals: Withdrawals of water for application on land to assist in the growing of crops and pastures or to maintain recreational lands.

Kill: Dutch term for stream or creek.

Lacustrine: Pertaining to, produced by, or formed in a lake.

Land Classification: The arrangement of land units into various categories based upon the properties of the land and its suitability for some particular purpose.

Land Cover Classification: A classification of the cultural, physical, and vegetation features that cover the Earth, commonly used with remote sensing technology. Vegetation classification is a subset of land cover classification.

Land Use Classification (LUC): A classification of the Earth's surface that defines the use that people are making of the land, commonly used with remote sensing technology, and commonly combined with land cover classification. Natural vegetation areas may be classified as *vacant, forest,* or *grazing.*

Land Use/Land Cover Classification (LU/LUC): A combination of a land use classification and land cover classification where the land use classification is used to classify areas that are under a definite land use, such as agriculture, residential, or mining. The land cover classification is used to classify lands that do not have definite land use, such as areas of bare rock, snow and ice, or open water. The Anderson Classification System is a land cover and land use classification.

Leachate: A liquid that has percolated through soil containing soluble substances and that contains certain amounts of these substances in solution.

Leaching: The removal of materials in solution from soil or rock; also refers to movement of pesticides or nutrients from land surface to groundwater.

Liebig's Law: The growth and reproduction of an organism are determined by the nutrient substance (oxygen, carbon dioxide, calcium, etc.) that is available in minimum quantity with respect to organic needs, the limiting factor.

Limnetic: The deepwater zone (greater than 2 meters deep).

Littoral: The shallow-water zone (less than 2 meters deep).

Load: Material that is moved or carried by streams, reported as weight of material transported during a specified period, such as tons per year.

Loam: The textural-class name for soil having moderate amounts of sand, silt, and clay.

Loess: An accumulation of wind-blown dust (silt), which may have undergone mild digenesis.

Look-Up Table (LUT): A computer file that relates the elements of one classification to another in a crosswalk. The values of a map classification are related to the associations of the National Vegetation Classification Standard in a park project.

Main Stem: The principal trunk of a river or a stream.

Map Accuracy: A measure of the maximum errors permitted in horizontal positions and elevations shown on maps. The National Map Accuracy Standard of the U.S. Geological Survey at 1:24,000 scale is the map accuracy standard for the program. This standard is that 90% of well-defined objects should appear within 40 feet (12.2 meters) of their true location.

Map Attribute: Collectively the map class (or map unit) code, the physiognomic modifier codes, and special modifiers if they are used: map unit code is that portion of the map attribute code defining the map unit (e.g., AB) the physiognomic modifier code—portion of map attribute code defining the vegetation community's structure (e.g., –1A3). The map attribute code is thus AB-1A3.

Map Class: The vegetation units that can be discerned on an aerial photograph. Often associations in an alliance cannot be distinguished on an aerial photograph because the differences are found in the understory, so map classes must be developed. For example, at Devils Tower National Monument there were five associations in the Ponderosa Pine Woodland Alliance, but it was necessary to create two ponderosa pine map classes because the associations could not be distinguished on the photography. Map classes may be complexes or mosaics of associations or map classes may also be the same as an association if that can be discerned on the photograph. Also known as *map unit*.

Map Scale: The relationship between a distance portrayed on a map and the same distance on the Earth (Dana, 1999). A map scale can be defined by a representative fraction (e.g., 1 unit on map/12,000 units on ground) or by a graphic scalebar.

Map Validation: The process of field checking and updating photo interpretation. This step is completed prior to accuracy assessment.

Marl: An earthy deposit consisting mainly of calcium carbonate, usually mixed with clay. Marl is used for liming acid soils. It is slower-acting than most lime products used for this purpose.

Marsh: A water-saturated, poorly drained area, intermittently or permanently water covered, having aquatic and grasslike vegetation.

Maturity (Stream): The stage in the development of a stream at which it has reached its maximum efficiency, when velocity is just sufficient to carry the sediment delivered to it by tributaries; characterized by a broad, open, flat-floored valley having a moderate gradient and gentle slope.

Maximum Contaminant Level (MCL): The maximum allowable concentration of a contaminant in drinking water, as established by state and federal regulations. Primary MCLs are health related and mandatory. Secondary

516 THE HANDBOOK OF GEOSCIENCE

MCLs are related to the aesthetics of the water and are highly recommended, but not required.

Mean Discharge: The arithmetic mean of individual daily mean discharges of a stream during a specific period, usually daily, monthly, or annually.

Membrane Filter Method: A laboratory method used for coliform testing. The procedure uses an ultra-thin filter with a uniform pore size smaller than bacteria (less than a micron). After water is forced through the filter, the filter is incubated in a special media that promotes the growth of coliform bacteria. Bacterial colonies with a green-gold sheen indicate the presence of coliform bacteria.

Metadata: Data about data. Metadata describes the content, quality, condition, and other characteristics of data. Its purpose is to help organize and maintain an organization's internal investment in spatial data, provide information about an organization's data holdings to data catalogues, clearinghouses, and brokerages, and provide information to process and interpret data received through a transfer from an external source (Federal Geographic Data Committee [FGDC], 1997). The FGDC sets standards for metadata content and structure.

Method Detection Limit: The minimum concentration of a substance that can be accurately identified and measured with current lab technologies.

Midge: A small fly in the family *Chironomidae*. The larval (juvenile) life stages are aquatic.

Mineralization: The conversion of an element from an organic form to an inorganic state as a result of microbial decomposition.

Minimum Mapping Unit (MMU): The smallest area that will be consistently delineated during photo interpretation. The MMU for the U.S. Geological Survey–National Park System Vegetation Mapping Program is 0.5 hectares.

Minimum Reporting Level (MRL): The smallest measured concentration of a constituent that may be reliably reported using a given analytical method. In many cases, the MRL is used when documentation for the method detection limit is not available.

Mitigation: Actions taken to avoid, reduce, or compensate for the effects of human-induced environmental damage.

Modes of Transmission of Disease: The ways in which diseases spread from one person to another.

Monitoring: Repeated observation, measurement, or sampling at a site, on a scheduled or event basis, for a particular purpose.

Monitoring Well: A well designed for measuring water levels and testing groundwater quality.

Mosaic: An intermixing of associations in an area that can be distinguished on the aerial photography, but is too intricate to delineate each association polygon. Compare with *complex.*

Multiple-Tube Fermentation Method: A laboratory method used for coliform testing, which uses a nutrient broth placed in a culture tubes. Gas production indicates the presence of coliform bacteria.

National Biological Information Infrastructure (NBII): A broad, collaborative program to provide increased access to data and information on the nation's biological resources. The NBII links diverse, high-quality biological databases, information products, and analytical tools maintained by NBII partners and other contributors in government agencies, academic institutions, nongovernment organizations, and private industry. Resource managers, scientists, educators, and the general public use the NBII to answer a wide range of questions related to the management, use, or conservation of this nation's biological resources. The US Geological Survey–National Park Service Vegetation Mapping Program is part of NBII.

National Biological Service (NBS): The agency that originated the U.S. Geological Survey–National Park Service Vegetation Mapping Program. Organized as the National Biological Survey in 1993, its name was changed to *National Biological Service* in 1995. It became the U.S. Geological Survey Biological Resources Division in 1996.

National Park Service (NPS): A U.S. Department of Interior agency, created in 1916, charged with preserving unimpaired the natural and cultural resources and values of the national park system for the enjoyment, education, and inspiration of this and future generations. NPS manages the National Parks and the Inventory and Monitoring Program, and works closely with the U.S. Geological Survey (USGS) to coordinate the USGS–NPS Vegetation Mapping Program

National Pollutant Discharge Elimination System (NPDES): A requirement of the Clean Water Act that discharges meet certain requirements prior to discharging waste to any water body. It sets the highest permissible effluent limits, by permit, prior to making any discharge.

National Primary Drinking Water Regulations (NPDWRs): Regulations developed under the Safe Drinking Water Act, which establish maximum contaminant levels, monitoring requirements, and reporting procedures for contaminants in drinking water that endanger human health.

Natural Resources Conservation Service (NRCS): (Formerly the Soil Conservation Service, formerly the Bureau of Soils.) The NRCS, U.S. Department of Agriculture, producer to the Nation's Soil Surveys, is the lead federal agency for conservation on private land and is a partner in conservation with many state, local and tribal organization.

Natural Vegetation: Plant life of an area that appears to be unmodified by human activities. Most existing vegetation has been subjected to some alteration by human activities, so a clear distinction between natural and cultural vegetation may be difficult (Grossman et al., 1998).

National Vegetation Classification Standard (NVCS): The Federal Geographic Data Committee's vegetation classification standard. It has been adopted to the formation level (as of June 2001); adoption of the floristic levels is pending. It is based on the Association for Biodiversity Information's United States National Vegetation Classification (USNVC) system.

Natural Heritage Programs: Natural Heritage Programs gather, manage, and distribute information about the biological diversity found within their jurisdiction. Most programs are part of government agencies—such as natural resources or fish and wildlife departments—although some are located within universities or nongovernmental organizations. The Association with Biodiversity Information works closely in partnership with Natural Heritage Programs to organize, store, and disseminate data collected by the programs. Natural Heritage Programs often collaborate on vegetation mapping projects in their states.

NatureServe: A website from the Association for Biodiversity Information that provides authoritative conservation information in a searchable database for more than 50,000 plants, animals, and ecological communities in the United States and Canada. Vegetation community data developed by the U.S. Geological Survey–National Park Service program is available on NatureServe (URL is http://www.natureserve.org/explorer/).

Near Coastal Water Initiative: This initiative was developed in 1985 to provide restoration and protection of the nation's coastal areas.

Nitrate: An ion consisting of nitrogen and oxygen (NO_3). Nitrate is a plant nutrient and is very mobile in soils.

Nitrogen Fixation: The biological conversion of elemental nitrogen (N_2) to organic combinations, or to forms readily utilized in biological processes.

Nonbiodegradable: Substances that do not break down easily in the environment.

Nonpoint Source: A source (of any water-carried material) from a broad area, rather than from discrete points.

Nonpoint-Source Contaminant: A substance that pollutes or degrades water that comes from lawn or cropland runoff, the atmosphere, roadways, and other diffuse sources.

Nonpoint-Source Water Pollution: Water contamination that originates from a broad area (such as leaching of agricultural chemicals from crop land) and enters the water resource diffusely over a large area.

Nonpolar Covalently Bonded: A molecule composed of atoms that share their electrons equally, resulting in a molecule that does not have polarity.

North American Datum (NAD): The datum for map projections and coordinates throughout North America (see also *datum*). Usually associated with a version, such as 1927 or 1983. This program uses the 1983 datum (NAD83), which is consistent with satellite location systems. The 1983 datum uses the GRS 80 spheroid, whereas the 1927 datum uses the Clarke 1866 spheroid (Portions from Environmental Systems Research Institute, 1994).

Nutrient: Any inorganic or organic compound needed to sustain plant life.

Observation Point: A field location point used to support map unit and vegetation classification development. These points are collected during reconnaissance and the mappers' subsequent fieldwork.

Organic: Containing carbon, but possibly also containing hydrogen, oxygen, chlorine, nitrogen, and other elements.

Organic Chemical: A chemical substance of animal or vegetable origin having carbon in its molecular structure.

Organic Detritus: Any loose organic material in streams—such as leaves, bark, or twigs—removed and transported by mechanical means, such as disintegration or abrasion.

Organic Soil: Soil that contains more than 20% organic matter in the upper 16 inches.

Organochlorine Compound: Synthetic organic compounds containing chlorine. As generally used, the term refers to compounds containing mostly or exclusively carbon, hydrogen, and chlorine.

Order: The second highest level in the National Vegetation Classification Standard hierarchy under division. The orders within the vegetated division are generally defined by dominant life-form (tree, shrub, dwarf shrub, herbaceous, or nonvascular) (Federal Geographic Data Committee, 1997).

Ortho Image: An aerial photograph that has had the distortions due to camera lens, topographic relief, tilt of the aircraft, and other factors common to aerial photography removed and has been registered to locations on the Earth. A digital ortho image can be placed in a geographic information system (GIS) and have other layers, such as vegetation, overlain on it. Aerial photo interpretation can also be registered to an ortho image in the process of registering and automating the data into a GIS. A digital orthophoto quadrangle is a digital ortho image covering 3.75 minutes by 3.75 minutes of the Earth's surface.

Outwash: Soil material washed down a hillside by rainwater and deposited on more gently sloping land.

Overland Flow: The flow of rainwater or snowbelt over the land surface toward stream channels.

Oxidation: When a substance either gains oxygen or loses hydrogen or electrons in a chemical reaction. One of the chemical treatment methods.

Oxidizer: A substance that oxidizes another substance.

Parent Material: The unconsolidated and more or less chemically weathered mineral or organic matter from which the solum of soils is developed by pedogenic processes.

Parts per Million (PPM): The number of weight or volume units of a constituent present with each 1 million units of the solution or mixture. Formerly used to express the results of most water and wastewater analyses, PPM is being replaced by milligrams per liter (M/L). For drinking water analyses, concentration in PPM and M/L are equivalent. A single PPM can be compared to a shot glass full of water inside a swimming pool.

Pathogens: Types of microorganisms that can cause disease.

Pattern: Configuration of vegetation features or across a landscape. One of the physiognomic modifiers classified in the U.S. Geological Survey–National Park Service Vegetation Mapping Program. Pattern in map units is classified as *evenly dispersed, clumped/bunched, gradational/transitional,* and *alternating.*

Ped: A unit of soil structure such as an aggregate, crumb, prism, block, or granule, formed by natural processes.

Pedogenic or Pedological Process: Any process associated with the formation and development of soil.

Perched Groundwater: Unconfined groundwater separated from an underlying main body of groundwater by an unsaturated zone.

Percolation: The movement, under hydrostatic pressure, of water through interstices of a rock or soil (except the movement through large openings such as caves).

Perennial Stream: A stream that normally has water in its channel at all times.

Periphyton: Microorganisms that coat rocks, plants, and other surfaces on lake bottoms.

Permeability: The capacity of a rock for transmitting a fluid; a measure of the relative ease with which a porous medium can transmit a liquid.

pH: The degree of acidity or alkalinity of the soil. Also referred to as *soil reaction,* this measurement is based on the pH scale where 7.0 is neutral—values from 0.0 to 7.0 are acid and values from 7.0 to 14.0 are alkaline. The pH of soil is determined by a simple chemical test where a sensitive indicator solution is added directly to a soil sample in a test tube.

Phosphorus: A nutrient essential for growth that can play a key role in stimulating aquatic growth in lakes and streams.

Photo Interpretation: The art and science of identifying and delineating objects on an aerial photograph. Photo interpreters in the U.S. Geological Survey–National Park Service Vegetation Mapping Program are knowledgeable about the vegetation in their project area and highly skilled in identifying vegetation map units accurately and consistently.

Photo Interpretation Key: A description of the distinguishing features that make up the signature of each map class. This description may include written clues, as well as graphic examples of the signatures.

Photo-Signature: Characteristics of an item on a photograph by which the item may be identified.

Photosynthesis: The process by which green leaves of plants, in the presence of sunlight, manufacture their own needed materials from carbon dioxide in the air and water, and minerals taken from the soil.

Physical Treatment: Any process that does not produce a new substance (e.g., screening, adsorption, aeration, sedimentation, and filtration).

Physiognomic Modifiers: Modifiers used for mapping to describe the physiognomic structure of the vegetation found within a mapped polygon (coverage density, coverage pattern, and height.

Physiognomy: The structure and life form of a plant community (Federal Geographic Data Committee, 1997).

PLANTS Database: A database maintained by the Natural Resource Conservation Service that is a single source of standardized information about plants. This database focuses on vascular plants, mosses, liverworts, hornworts, and lichens of the United States and its territories. The PLANTS Database includes names, checklists, automated tools, identification information, species abstracts, distributional data, crop information, plant symbols, plant growth data, plant materials information, plant links, references, and other plant information. This is the database that maintains current scientific names. The PLANTS database provides the Integrated Taxonomic Information System data for plants. The PLANTS database is the taxonomic authority for the U.S. Geological Survey–National Park Service Vegetation Mapping Program. For more information go to http://plants.usda.gov/.

Plot: A formal field location of a certain size where the data necessary to classify the vegetation communities is collected. The data generated from the plot data collection is subsequently entered into a database known as the PLOTS database. Plot size will vary depending on the vegetation physiognomy being sampled. Synonymous with vegetation plot. For more information see Grossman, et al., 1994, at http://biology.usgs.gov/npsveg/standards/fieldmethodsrpt.pdf.

PLOTS Database: A database in Microsoft Access format that contains the information collected from field plots with tables and fields for all the necessary data. The PLOTS database table structure and user's guide can be downloaded from http://biology.usgs.gov/npsveg/tools/plotsdatabase.html.

Point Source: Originating at any discrete source.

Polar Covalent Bond: The shared pair of electrons between two atoms are not equally held. Thus, one of the atoms becomes slightly positively charged and the other atom becomes slightly negatively charged.

Polar-Covalent Molecule: (water) One or more polar-covalent bonds result in a molecule that is polar covalent. Polar-covalent molecules exhibit partial positive and negative poles, causing them to behave like tiny magnets. Water is the most common polar-covalent substance.

Pollutant: Any substance introduced into the environment that adversely affects the usefulness of the resource.

Pollution: The presence of matter or energy whose nature, location, or quantity produces undesired environmental effects. Under the Clean Water Act, for example, the term is defined as artificial alteration of the physical, biological, and radiological integrity of water.

Polychlorinated Biphenyls (PCBs): A mixture of chlorinated derivatives of biphenyl, marketed under the trade name Aroclor with a number designating the chlorine content (such as Aroclor 1260). PCBs were used in transformers and capacitors for insulating purposes and in gas pipeline systems as a lubricant. Further sale or new use was banned by law in 1979.

Polycyclic Aromatic Hydrocarbon (PAH): A class of organic compounds with a fused-ring aromatic structure. PAHs result from incomplete combustion of organic carbon (including wood), municipal solid waste, and fossil fuels, as well as from natural or anthropogenic introduction of uncombusted coal and oil. PAHs included benzo(a)pyrene, fluoranthene, and pyrene.

Population: A collection of individuals of one species or mixed species making up the residents of a prescribed area.

Porosity: The ratio of the volume of voids in a rock or soil to the total volume.

Porosity, Soil: The volume percentage of the total bulk not occupied by solid particles.

Positional Accuracy: The nearness of a point in a spatial database to its actual location on the Earth's surface. The program standard for horizontal positional accuracy meets National Map Accuracy Standards at the 1:24,000 scale. This means that each well-defined object in the spatial database will be within 1/50 of an inch of its actual location or 40 feet (12.2 meters).

Potable Water: Water that is safe and palatable for human consumption.

Potential Vegetation: The vegetation structure that would become established if all successional sequences were completed without interference under the present climatic and edaphic conditions (Ecological Society of America, 1999).

Precipitation: Any or all forms of water particles that fall from the atmosphere, such as rain, snow, hail, and sleet. The act or process of producing a solid phase within a liquid medium.

Precision Lightweight GPS Receiver (PLGR): A small, handheld global positioning system receiver featuring selective availability–antispoofing and antijam capability.

Pretreatment: Any physical, chemical, or mechanical process used before the main water treatment processes. It can include screening, presedimentation, and chemical addition.

Primary Drinking Water Standards: Regulations on drinking water quality (under the Safe Water Drinking Act) considered essential for preservation of public health.

Primary Treatment: The first step of treatment at a municipal wastewater treatment plant. It typically involves screening and sedimentation to remove materials that float or settle.

Producers' Accuracy: The probability that a reference sample (the ground data) has been classified correctly, also known as *error of omission*. This quantity is computed by dividing the number of samples that have been classified correctly by the total number of reference samples in that class (Story & Congalton, 1986).

Profile, Soil: A vertical section of the soil through all its horizons and extending into the parent material.

Projection: A map or a geospatial database is a flat representation of data located on a curved surface. A projection is a device for producing all or part of a round body on a flat sheet. This projection cannot be done without distortion, so the cartographer must choose which characteristic (distance, direction, scale, area, or shape) that is to be emphasized at the expense of the other characteristics (Snyder, 1982). All spatial data in the program are represented in the universal transverse Mercator coordinate system that is based on the transverse Mercator projection applied between 84 degrees north and 80 degrees south latitude.

Public-Supply Withdrawals: Water withdrawn by public and private water suppliers for use within a general community. Water is used for a variety of purposes such as domestic, commercial, industrial, and public water use.

Public Water System: As defined by the Safe Drinking Water Act, any system, publicly or privately owned, that serves at least 15 service connections 60 days out of the year or serves an average of 25 people at least 60 days out of the year.

Publicly Owned Treatment Works (POTW): A waste treatment works owned by a state or local government unit or Indian tribe, usually designed to treat domestic wastewaters.

Quadrangle: A U.S. Geological Survey (USGS) paper map. Typically, a 7.5-minute USGS map. Informally known as a *quad.*

Quarter Quadrangle: A map or image that includes one-fourth the area of a 7.5-minute quadrangle and is organized in quadrants of the original quadrangle as follows: northeast, northwest, southeast, and southwest. The U.S.

Geological Survey digital orthophoto quadrangles cover one-fourth of a 7.5-minute quadrangle. Informally known as *quarter quad.*

Rain Shadow: A dry region on the lee side of a topographic obstacle, usually a mountain range, where rainfall is noticeably less than on the windward side.

Reach: A continuous part of a stream between two specified points.

Reaeration: The replenishment of oxygen in water from which oxygen has been removed.

Receiving Waters: A river, lake, ocean, stream, or other water source into which wastewater or treated effluent is discharged.

Recharge: The process by which water is added to a zone of saturation, usually by percolation from the soil surface.

Recharge Area (Groundwater): An area within which water infiltrates the ground and reaches the zone of saturation.

Rectify: Remove distortions common to aerial photographs in the process of automating the photo-interpreted information into a digital database. Distortions on aerial photographs are due to topographic relief on the ground, radial distortion in the geometry of the aerial photography, tip and tilt of the plane, and differences in elevation of the airplane from its nominal scale. This process may be separate or included in the registration process depending on the technology used.

Reduction: The gain of electrons, and therefore the loss of positive valence charge by a substance.

Reference Dose (RfD): An estimate of the amount of a chemical that a person can be exposed to on a daily basis that is not anticipated to cause adverse systemic health effects over the person's lifetime.

Register: The process of correlating objects on an aerial photograph with locations on the surface of the Earth using a defined coordinate system. This is necessary to be able to place the vegetation community data in a geographic information system with other appropriate data such as transportation, topography, soils, and so on. This process may be separate or included in the rectification process depending on the technology used.

Regolith: The unconsolidated mantle of weathered rock and soil material on the Earth's surface; loose Earth materials above solid rock.

Representative Sample: A sample containing all the constituents present in the water from which it was taken.

Return Flow: That part of irrigation water that is not consumed by evapotranspiration and that returns to its source or another body of water.

Reverse Osmosis (RO): Solutions of differing ion concentration are separated by a semipermeable membrane. Typically, water flows from the chamber with lesser ion concentration into the chamber with the greater ion concentration, resulting in hydrostatic or osmotic pressure. In RO, enough external pressure is applied to overcome this hydrostatic pressure, thus reversing the flow of water. This results in the water on the other side of the membrane becoming depleted in ions and demineralized.

Riffle: A shallow part of the stream where water flows swiftly over completely or partially submerged obstructions to produce surface agitation.

Riparian: Pertaining to or situated on the bank of a natural body of flowing water.

Riparian Rights: A concept of water law under which authorization to use water in a stream is based on ownership of the land adjacent to the stream.

Riparian Zone: Pertaining to or located on the bank of a body of water, especially a stream.

Rock: The material that forms the essential part of the Earth's solid crust, including loose, incoherent masses such as sand and gravel, as well as solid masses of granite and limestone.

Rock Cycle: The global geological cycling of lithospheric and crustal rocks from their igneous origins through all of any stages of alteration, deformation, resorption, and reformation.

Runoff: That part of precipitation or snowmelt that appears in streams or surface-water bodies.

Rural Withdrawals: Water used in suburban or farm areas for domestic and livestock needs. The water generally is self-supplied and includes domestic use, drinking water for livestock, and other uses such as dairy sanitation, evaporation from stock-watering ponds, and cleaning and waste disposal.

Safe Drinking Water Act (SDWA): A federal law passed in 1974 with the goal of establishing federal standards for drinking water quality, protecting underground sources of water, and setting up a system of state and federal cooperation to ensure compliance with the law.

Saline water: Water that is considered unsuitable for human consumption or for irrigation because of its high content of dissolved solids; generally expressed as milligrams per liter (mg/L) of dissolved solids; seawater is generally considered to contain more than 35,000 mg/L of dissolved solids. A general salinity scale is:

Concentration of dissolved solids in mg/L	
Slightly saline:	1,000–3,000
Moderately saline:	3,000–10,000
Very saline:	10,000–35,000
Brine:	More than 35,000

Salinization: The process of accumulation of salts in soil.

Sand: A soil particle between 0.05 and 2 mm in diameter; a soil textural class.

Saturated Zone: A subsurface zone in which all the interstices or voids are filled with water under pressure greater than that of the atmosphere.

Scale: The relationship between a distance portrayed on a map and the same distance on the Earth (Dana, 1999). A map scale can be defined by a representative fraction (e.g., 1 unit on map/12,000 units on ground) or by a graphic scale bar.

Screening: A pretreatment method that uses coarse screens to remove large debris from the water to prevent clogging of pipes or channels to the treatment plant.

Secondary Drinking Water Standards: Regulations developed under the Safe Drinking Water Act that established maximum levels of substances affecting the aesthetic characteristics (taste, color, or odor) of drinking water.

Secondary Maximum Contaminant Level (SMCL): The maximum level of a contaminant or undesirable constituent in public water systems that, in the judgment of the U.S. Environmental Protection Agency (EPA), is required to protect the public welfare. SMCLs are secondary (nonenforceable) drinking water regulations established by the EPA for contaminants that may adversely affect the odor or appearance of such water.

Secondary Treatment: The second step of treatment at a municipal wastewater treatment plant. This step uses growing numbers of microorganisms to digest organic matter and reduce the amount of organic waste. Water leaving this process is chlorinated to destroy any disease-causing microorganisms before its release.

Sedimentation: A physical treatment method that involves reducing the velocity of water in basins so that the suspended material can settle out by gravity.

Seep: A small area where water percolates slowly to the land surface.

Seiche: A sudden oscillation of the water in a moderate-size body of water, caused by wind.

Signature: The unique combination of color, texture, pattern, height, physiognomy, and position in the landscape used by any photo interpreters to identify map classes on an aerial photograph.

Silt: A soil separate consisting of particles between 0.05 and 0.002 mm in equivalent diameter. A soil textural class.

Sinuosity: The ratio of the channel length between two points on a channel to the straight-line distance between the same two points; a measure of meandering.

Slope: The degree of deviation of a surface from horizontal, measured in a numerical ratio, percent, or degrees.

Soil: An assemblage of loose and normally stratified granular minerogenic and biogenic debris at the land surface, it is the supporting medium for the growth of plants.

Soil Air: The soil atmosphere; the gaseous phase of the soil, being that volume not occupied by soil or liquid.

Soil Horizon: A layer of soil, approximately parallel to the soil surface, with distinct characteristics produced by soil-forming processes. These characteristics form the basis for systematic classification of soils.

Soil Moisture: Water occurring in the pore spaces between the soil particles in the unsaturated zone from which water is discharged by the transpiration of plants or by evaporation from the soil.

Soil Profile: A vertical section of the soil from the surface through all its horizons, including C horizons.

Soil Structure: The combination or arrangement of primary soil particles into secondary particles, units, or peds. These secondary units may be, but usually are not, arranged in the profile in such a manner as to give a distinctive characteristic pattern. The secondary units are characterized and classified on the basis of size, shape, and degree of distinctness into classes, types, and grades, respectively.

Soil Texture: The relative proportions of the various soil separates in a soil.

Soluble: Will dissolve easily in water.

Solum: (plural *sola*) The upper and most weathered part of the soil profile; the A, E, and B horizons.

Solution: Formed when a solid, gas, or another liquid in contact with a liquid becomes dispersed homogeneously throughout the liquid. The substance, called a *solute,* is said to dissolve. The liquid is called the *solvent.*

Solvated: When either a positive or negative ion becomes completely surrounded by polar solvent molecules.

Sorb: To take up and hold either by absorption or adsorption.

Sorption: General term for the interaction (binding or association) of a solute ion or molecule with a solid.

Spatial Data Transfer Standard (SDTS): A comprehensive transfer standard for Earth-referenced data endorsed by the Federal Geographic Data Committee. Spatial data in SDTS format consists of a group of files each with specific content and format.

Special Modifiers: Modifiers used as part of a map attribute to describe special features that are not part of the U.S. National Vegetation Classification. For example, a park may be interested in eagle nests, beaver dams, prairie dog towns, and forest blow-down. See *map attributes, map codes,* and *physiognomic modifiers.*

Specific Yield: The ratio of the volume of water that will drain under the influence of gravity to the volume of saturated rock.

Spring: Place where a concentrated discharge of groundwater flows at the ground surface.

Stratum: A horizontal layer of vegetation. A stratum may be defined by the life form of the vegetation (tree, shrub, herbaceous), or its actual height.

Structure (Vegetation): The spatial distribution pattern of life forms in a plant community, especially with regard to their height, abundance, or coverage within the individual layers (Ecological Society of America, 1999). Synonymous with *physiognomy.*

Subclass: The level in the National Vegetation Classification Standard hierarchy under class based on growth form characteristics (Grossman et al., 1998).

Subgroup: The level in the National Vegetation Classification Standard hierarchy below group, which divides each group into either a "natural/semi-natural" or "cultural" (planted/cultivated) subgroup (Grossman et al., 1998).

Subsoil: That part of the soil below the plow layer.

Surface Runoff: Runoff that travels over the land surface to the nearest stream channel.

Surface Tension: The attractive forces exerted by the molecules below the surface on those at the surface, resulting in them crowding together and forming a higher density.

Surface Water: All water naturally open to the atmosphere, and all springs, wells, or other collectors that are directly influenced by surface water.

Surface Water Treatment Rule (SWTR): A federal regulation established by the U.S. Environmental Protection Agency under the Safe Drinking Water Act that imposes specific monitoring and treatment requirements on all public drinking water systems that draw water from a surface water source.

Suspended Sediment: Sediment that is transported in suspension by a stream.

Suspended Solids: Different from suspended sediment only in the way that the sample is collected and analyzed.

Synthetic Organic Chemicals (SOCs): Generally applied to manufactured chemicals that are not as volatile as volatile organic chemicals. Included are herbicides, pesticides, and chemicals widely used in industries.

Thematic Accuracy: The correctness of the map classes in relation to the vegetation on the ground. This is determined through the accuracy assessment procedures and the program standard is 80% accurate for each map class at the 90% confidence interval.

The Nature Conservancy (TNC): A nonprofit conservation organization founded in 1951. Working with communities, businesses, and people, TNC protects millions of acres of valuable lands and waters worldwide. TNC was the original caretaker of the U.S. National Vegetation Classification, but those responsibilities have been spun off to the Association for Biodiversity Information. TNC no longer has an active role with the U.S. Geological Survey–National Park Service Vegetation Mapping Program.

Till: Unstratified glacial drift deposited directly by the ice and consisting of clay, sand, gravel, and boulders intermingled in any proportion.

Tilth: The physical condition of soil as related to its ease of tillage, fitness as a seedbed, and its impedance to seedling emergence and root penetration.

Topographic Quadrangles: U.S. Geological Survey paper maps showing the topography of an area along with roads, railroads, water bodies, building, urban developments, and wetlands. These come in a variety of scales, but commonly refer to 1:24,000 scale 7.5-minute quads. Informally referred to as *topo quads*.

Topsoil: The layer of soil moved in cultivation.

Total Head: The height above a datum plane of a column of water. In a groundwater system, it is composed of elevation head and pressure head.

Total Suspended Solids (TSS): Solids present in wastewater.

Transfer: The process of moving photo-interpreted data from an aerial photo overlay to an ortho image to register and rectify the data. This process varies depending on the type of technology used.

Transform(ation): The process of converting coordinates (map or image) from one coordinate system to another. This involves scaling, rotation, translation, and warping (images) (Environmental Systems Research Institute, 1994).

Transition Zone: An area where the vegetation composition and structure is intermediate between two associations. The transition zone may be small as the associations abruptly change due to a large shift in the landscape, such as a cliff, or it may be large as the physical environment changes gradually. Transition zones often are challenges to properly classify and map vegetation.

Transpiration: The process by which water passes through living organisms, primarily plants, into the atmosphere.

Trihalomethanes (THMs): A group of compounds formed when natural organic compounds from decaying vegetation and soil (such as humic and fulvic acids) react with chlorine.

Turbidity: A measure of the cloudiness of water caused by the presence of suspended matter, which shelters harmful microorganisms and reduces the effectiveness of disinfecting compounds.

Type: A generic term that can reference any vegetation level in the U.S. National Vegetation Classification, whether it is association, alliance, formation, and so on, and even a combination of levels, or no reference to a known "type" within the classification. It is a vague but useful term. It is correctly used when the focus is not on the vegetation "type," but rather when used loosely to explain some other point (e.g., "We do not have a good grasp of how vegetation types at Acadia link to the map units used for mapping"). *Types* refers to all levels of the classification, not specific association level. Plus the focus is on "not having a good grasp," not the vegetation type or map unit. Also known as *vegetation type*.

UKMET: United Kingdom forecast model.

Ultraviolet Radiation: The energy range just beyond the violent end of the visible spectrum. Although ultraviolet radiation constitutes only about 5% of the total energy emitted from the sun, it is the major energy source for the stratosphere and mesosphere, playing a dominant role in both energy balance and chemical composition.

Unconfined Aquifer: An aquifer whose upper surface is a water table free to fluctuate under atmospheric pressure.

United States Geological Survey (USGS): Established in 1879, the USGS is the natural science agency for the Department of the Interior. The USGS is one of the host agencies, along with the National Park Service, for the USGS–National Park Service Vegetation Mapping Program.

United States National Map Accuracy Standards (NMAS): U.S. Geological Survey accuracy standards for published maps, including horizontal and vertical accuracy, accuracy testing method, accuracy labeling on published maps, labeling when a map is an enlargement of another map, and basic information for map construction as to latitude and longitude boundaries. The following table shows the standard for some common map scales. To meet NMAS, maps must have less than 10% of the points tested (well-defined points) exceed the standard. Note that the conversion of paper maps into digital data usually creates additional error.

	ABI USNVC	FGDC NVCS	Example
Physiognomic Levels	System		Terrestrial
		Division	Vegetated
		Order	Tree
	Class	Class	Woodland
	Subclass	Subclass	Evergreen woodland
	Group	Group	Temperate or subpolar needle-leaved evergreen woodland
	Subgroup	Subgroup	Natural or seminatural vegetation
	Formation	Formation	Rounded-crowned temperate or subpolar needle-leaved evergreen woodland
Floristic Levels	Alliance	Alliance (not finally adopted)	Juniperus scopularum Woodland Alliance
	Association	Association (not finally adopted)	Juniperus scopularum/*Oryzopsis micrantha* Woodland

United States National Vegetation Classification (USNVC): The Association for Biodiversity's vegetation classification system. It is the basis for the Federal Geographic Data Committee National Vegetation Classification Standard. See Table 2 for comparison.

Unsaturated Zone: A subsurface zone above the water table in which the pore spaces may contain a combination of air and water.

Unstable Air: Air that rises easily and can form clouds and rain.

Unstable Atmosphere: Characterized by a high degree of turbulence.

Updraft: A small-scale current or rising air. This is often associated with cumulus and cumulonimbus clouds.

Upper Level System: A general term for any large-scale or mesoscale disturbance capable of producing upward motion (lift) in the middle or upper parts of the atmosphere.

Upper Midwest Environmental Sciences Center (UMESC): A U.S. Geological Survey Science Center concerned with large river issues and medicinal drugs for public aquaculture. UMESC has established a significant geospatial technologies capability and is managing several park projects for the U.S. Geological Survey-NPS Vegetation Mapping Program.

Horizontal Accuracy Examples

Scale	Engineering Scale	National Map Accuracy Standard
1:1,200	1"= 100'	+/– 3.33 feet
1:2,400	1" = 200'	+/– 6.67 feet
1:4,800	1" = 400'	+/– 13.33 feet
1:9,600	1" = 800'	+/– 26.67 feet
1:10,000		+/– 27.78 feet
1:12,000	1" = 1000'	+/– 33.33 feet
1:24,000	1" = 2000'	+/– 40.00 feet
1:63,360	1" = 1 mile	+/– 105.60 feet
1:100,000		+/– 166.67 feet

Universal Transfer Mercator (UTM): A map coordinate system (not a map projection) that is defined by the transverse Mercator projection, which has a set of zones defined by a central meridian as shown in the following figure for the United States (Portions from Environmental Systems Research Institute, 1994). All spatial data products developed by the program (vegetation spatial data, plot and accuracy assessment plot data locations) are in this coordinate system.

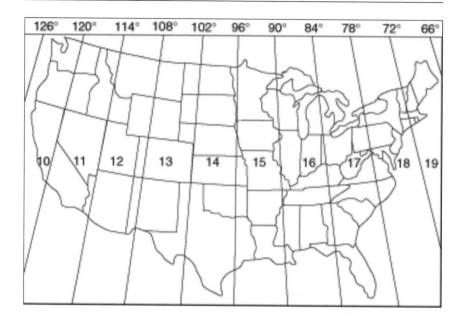

Upslope Flow: Air that flows toward higher terrain, and hence is forced to rise. The added lift often results in widespread low cloudiness and stratiform precipitation if the air is stable, or an increased chance of thunderstorm development if the air is unstable.

Upstream: Toward the source of the flow, or located in the area from which the flow is coming.

User's Accuracy: The probability that a sample from the mapped data actually represents that category on the ground, also known as *error of commission.* This quantity is computed by dividing the number of correctly classified samples by the total number of samples that were classified as belonging to that category (Story & Congalton, 1986).

UTC: Coordinated Universal Time. The time in the zero degree meridian time zone.

UVI: Ultraviolet index.

UVV: Upward vertical velocity.

Vector Data: Spatial (usually digital) data that consists of using coordinate pairs (x, y) to represent locations on the Earth. Features can take the form of single points, lines, arcs, or closed lines (polygons).

Vegetation: The collective plant cover over an area (Federal Geographic Data Committee, 1997).

Vegetation Characterization: The detailed portrayal of a vegetation association using diagnostic and dominant species, structure, and ecological processes. The program has a formal structure for association description based on the Association for Biodiversity Information model. Also known as *vegetation description.*

Vegetation Classification: The process of categorizing vegetation into repeatable and consistent elements. Also a document that lists and organizes the vegetation communities in an area. An example of a vegetation classification can be found at http://biology.usgs.gov/npsveg/agfo/agforpt.pdf.

Vegetation Mapping: The process of identifying, labeling, and placing in real-world coordinates vegetation communities.

Vehicle of Disease Transmission: Any nonliving object or substance contaminated with pathogens.

Vernal Pool: A small lake or pond that is filled with water for only a short time during the spring.

Wastewater: The spent or used water from individual homes, a community, a farm, or an industry that contains dissolved or suspended matter.

Water Budget: An accounting of the inflow to, outflow from, and storage changes of water in a hydrologic unit.

Water Column: An imaginary column extending through a water body from its floor to its surface.

Water Demand: Water requirements for a particular purpose, such as irrigation, power, municipal supply, plant transpiration, or storage.

Water Table: The top water surface of an unconfined aquifer at atmospheric pressure.

Waterborne Disease: Water is a potential vehicle of disease transmission, and waterborne disease is possibly one of the most preventable types of communicable illness. The application of basic sanitary principles and technology have virtually eliminated serious outbreaks of waterborne diseases in developed countries. The most prevalent waterborne diseases include *typhoid fever, dysentery, cholera, infectious hepatitis,* and *gastroenteritis.*

Water Softening: A chemical treatment method that uses either chemicals to precipitate or a zeolite to remove those metal ions (typically Ca^{2+}, Mg^{2+}, Fe^{3+}) responsible for hard water.

Watershed: The land area that drains into a river, river system, or other body of water.

Weathering: All physical and chemical changes produced in rocks, at or near the Earth's surface, by atmospheric agents.

Wellhead Protection: The protection of the surface and subsurface areas surrounding a water well or well field supplying a public water system from contamination by human activity.

Wetland: A location on the landscape that is characterized by either hydric soils or hydrophytic plants, or both. A wetland may be vegetated or nonvegetated. The vegetation description for each association includes it wetland status.

Worms: In stream ecology, the presence of certain species of worms in bottom sediment indicates stream pollution.

Xenobiotics: Any chemical present in a natural environment that does not normally occur in nature; for example, pesticides or industrial pollutants.

Yield: The mass of material or constituent transported by a river in a specified period of time divided by the drainage area of the river basin.

Young Pond: In the cycle of pond evolution, the initial or earliest phase.

Zigzag Lightning: Ordinary lightning of a cloud-to-ground discharge that appears to have a single lightning channel.

Zodiac: The position of the sun throughout a year as it appears to move through successive star groups or constellations.

Zonal Flow (Zonal Wind): Large-scale atmospheric flow in which the east–west component (i.e., latitudinal) is dominant.

Zone of Maximum Precipitation: The belt of elevation at which the annual precipitation is greatest in a mountain region.

Zone of Recent Pollution: In streams or rivers, the point of pollution discharge.

Zulu Time: Same as universal coordinated time (UTC). It is called *Zulu* because *Z* is often appended to the time to distinguish it from local time.

Glossary References

American Society of Photogrammetry (1984). *Multilingual dictionary of remote sensing and photogrammetry.* G. A. Rabchevsky (Ed.). Falls Church, VA: American Society of Photogrammetry.

Anderson, J. R., Hardy, E. E., & Roach, J. T. (1976). Land use and land cover classification system for use with remote sensing data. *Geological survey professional paper 964.* A revision of the land use classification system as presented in U.S. Geological Circular 671. U.S. Government Printing Office, Washington, D.C.

Cowardin, L. W., Carter, V., Golet. F. C, & LaRoe, E. T. (1979). Classification of wetlands and deepwater habitats of the United States. Biological Service Program, U.S. Fish and Wildlife Service, FWS/OBS 79/31. Washington, DC: Office of Biological Services, Fish and Wildlife Service, U.S. Department of Interior.

Dana, P. H. (1999). Map projection overview. http://www.colorado.edu/geography/gcraft/notes/mapproj/mapproj_f.html.

Daubenmire, R. F. (1978). *Plant geography, with special reference to North America.* New York: Academic Press.

Environmental Systems Research Institute, Inc. (1994). *Understanding GIS: The arc/info method.* Redlands, CA: Environmental Systems Research Institute.

Establishing standards for using existing vegetation data (1996). Prepared for the U.S. Geological Survey–National Park Service Vegetation Mapping Program by the Nature Conservancy, Arlington VA, and Environmental Science Research Institute, Redlands, CA.

Federal Geographic Data Committee (1997). Vegetation classification standard. http://
www.fgdc.gov/standards/projects/FGDC-standards-projects/vegetation/NVCS_V2_
FINAL_2008-02.pdf/.

Gabriel, H. W., & Talbot, S. S. (1984). Glossary of landscape and vegetation ecology for
Alaska. *Alaska Technical Report 10.* Washington, DC: Bureau of Land Management,
U.S. Department of the Interior.

Grossman, D. H., Faber-Langendoen, D., Weakley, A. S., Anderson, M., Bourgeron, P.,
Crawford, R., Goodin, K., Landal, S., Metzler, K., Patterson, K., Pyne, M., Reid, M.,
& Sneddon, L. (1998). *International classification of ecological communities: terrestrial
vegetation of the United States. Volume I.* Arlington, VA: The Nature Conservancy.

Grossman, D. H., Goodin, K. L., Li, X., Faber-Langendoen, D., & Anderson, M.
(1994). Developing and documenting a National Vegetation Classification Standard.
Prepared for the U.S. Geological Survey–National Park Service Vegetation Mapping
Program by The Nature Conservancy, Arlington VA, and Environmental Science
Research Institute, Redlands, CA.

Grossman, D. H., Goodin, K. L., Li, X., Faber-Langendoen, D., Anderson, M., &
Vaughan, R. (1994). Establishing standards for field methods and mapping proce-
dures. Prepared for the U.S. Geological Survey–National Park Service Vegetation
Mapping Program by The Nature Conservancy, Arlington VA, and Environmental
Science Research Institute, Redlands, CA.

Snyder, J. P. (1982). *Map projections used by the U.S. Geological Survey* (2nd ed.). Washington, DC: United States Government Printing Office.

Stadelmann, M., Curtis, A., Vaughan, R., & Goodchild, M. (1994). Producing rigorous
and consistent accuracy assessment procedures. Prepared for the U.S. Geological Survey–National Park Service Vegetation Mapping Program by The Nature Conservancy,
Arlington VA, and Environmental Science Research Institute, Redlands, CA.

Story, M., & Congalton, R. G. (1986). Accuracy assessment: A user's perspective. *Photogrammetric Engineering and Remote Sensing,* 52(3):397–399.

Index

water balance, 130
water cycle, 319–320
water quality, 322
water rights, 325
watershed, 322
wave-cut benches, 361
wave erosion, 361
waves, 360
weather, 144–147
weather generator, 144–146
weathering, 245–249; factors affecting, 245–246
wet shaking test, 396

wind deposition, 255–256
wind erosion, 251–265
wind forms, 140–141
wind transport, 254
warm front, 146
winds of the world, 134
work, 55, 81–85

xenon, 116–117; physical properties of, 117
xylene, 428

zone of saturation, 332

About the Authors

Frank R. Spellman is a retired U.S. naval officer with 26 years of active duty, a retired environmental safety and health manager for a large wastewater sanitation district in Virginia, and retired assistant professor of environmental health at Old Dominion University, Norfolk, Virginia. He is the author or co-author of 78 books and consults on environmental matters with the U.S. Department of Justice and various law firms and environmental entities across the world. He holds a BA in Public Administration, BS in Business Management, and an MBA and PhD in Environmental Engineering. In 2011, he traced and documented the ancient water distribution system at Machu Pichu, Peru, and surveyed several drinking water resources in Coco and Amazonia, Ecuador.

Melissa L. Stoudt is a radiation controls training instructor at an atomic power laboratory. Previously, she instructed students and officers of the U.S. Navy in proper radiation controls and the handling of radioactive materials.